# Convection an
# Conduction H
# Transfer

# Convection and Conduction Heat Transfer

*Editor*

Nicolaos Sabella

scitus
academics

**Convection and Conduction Heat Transfer**

Edited by **Nicolaos Sabella**

ISBN: 978-1-68117-235-4
Library of Congress Control Number: 2016934767


SCITUS Academics LLC,
www.scitusacademics.com
Box No. 4766, 616 Corporate Way,
Suite 2, Valley Cottage,
NY 10989



**Notice**



Printed in United States of America on Acid Free Paper

# Preface

All matter is made up of molecules and atoms. These atoms are always in different types of motion (translation, rotational, vibrational). The motion of atoms and molecules creates heat or thermal energy. All matter has this thermal energy. The more motion the atoms or molecules have the more heat or thermal energy they will have.Heat transfer is the exchange of thermal energy between physical systems. The rate of heat transfer is dependent on the temperatures of the systems and the properties of the intervening medium through which the heat is transferred. The three fundamental modes of heat transfer are conduction, convection and radiation. Heat transfer, the flow of energy in the form of heat, is a process by which a system changes its internal energy, hence is of vital use in applications of the First Law of Thermodynamics. Conduction is also known as diffusion, not to be confused with diffusion related to the mixing of constituents of a fluid. Heat energy transferred between a surface and a moving fluid at different temperatures is known as convection. In reality this is a combination of diffusion and bulk motion of molecules. Near the surface the fluid velocity is low, and diffusion dominates. Away from the surface, bulk motion increases the influence and dominates. Natural convection is caused by buoyancy forces due to density differences caused by temperature variations in the fluid. At heating the density change in the boundary layer will cause the fluid to rise and be replaced by cooler fluid that also will heat and rise. This continues phenomena is called free or natural convection. Conduction as heat transfer takes place if there is a temperature gradient in a solid or stationary fluid medium. With conduction energy transfers from more energetic to less energetic molecules when neighboring molecules collide. Heat flows in direction of decreasing temperatures since higher temperatures are associated with higher molecular energy. This book emphasizes on the principles of convection and conduction heat transfer.

# Table of Contents

# CHAPTER 1

# Heat Transfer Characteristics of Square Micro Pin Fins under Natural Convection

Naoko Matsumoto, Toshio Tomimura, Yasushi Koito
*Department of Advanced Mechanical Systems, Kumamoto University, Kumamoto, Japan*

## ABSTRACT

In order to comply with the recent demand for downsizing of the electric equipment, the minia- turization and the improvement in heat transfer performance of a heat sink under natural air-cooling are increasingly required. This paper describes the experimental and numerical investigations of heat sinks with miniature/micro pins and the effect of the pin size, pin height and the number of pins on heat transfer characteristics of heat sinks. Five types of basic heat sink models are investigated in this study. The whole heat transfer area of heat sinks having the different pin size, pin height and the number of pins respectively is kept constant. From a series of experiments and numerical analyses, it has been clarified that the heat sink temperature rises with increase in the number of pins. That is, the heat sink with miniaturized fine pins showed almost no effect on the heat transfer enhancement. This is because of the choking phenomenon occurred in the air space among the pin fins. Reflecting these results, it is confirmed that the heat transfer coefficient reduces with miniaturization of pins. Concerning the effects of the heat transfer area on the heat sink performance, almost the same tendency has been observed in other three series of large surface area, that is, higher pin height. Furthermore as a result of studying non-dimensional convection heat transfer performance, it was found that the relation between the Nusselt number (Nu) and the Rayleight number (Ra) is given by $Nu = 0.16\ Ra^{0.52}$.

## KEYWORDS

Natural Air-Cooling, Heat Sink, Micro Pin Fin, Heat Transfer Performance, Experiment, Numerical Analysis

## 1. INTRODUCTION

In electronic equipment applications with high heat fluxes, the heat transfer performance and the lifetime of components are often deteriorated due to temperature rise. To avoid those unfavorable problems, it is necessary to keep the component under a critical temperature by introducing a high performance heat sink. Furthermore, in order to comply with the recent demand for downsizing of the electronic equipment, the miniaturization and the improvement in heat transfer performance of the heat sink are required much more strongly than ever. But it is well known that the decrease in heat transfer area by downsizing leads to a decline of the heat sink performance.

Many studies of heat transfer characteristics using large scaled heat sinks have been carried out so far. For example, Aihara et al. [1] and Sara [2] revealed that the heat sink performance depends on the pin size, the population density of pin. Zografos and Sunderland [3] studied the effect of pin arrangement on the heat transfer performance, and reported that a choking phenomenon, which had a negative effect on heat transfer from a heat sink, was observed in the air space among pin fins. Huang et al. [4], Sparrow et al. [5] and Sertkaya et al. [6] investigated experimentally the dependence of the pin fin performance on a heat sink orientation.

In electronic equipment applications, heat sinks are placed within enclosed areas and the emphasis is on making the enclosures more compact than ever. Yu et al. [7] and Bocu et al. [8] reported natural convection heat transfer from heat sink attached to enclosures in their studies.

On the other hand, concerning small sized heat sinks or nano particles structure, the analytical and experimental studies were published by Narasimhan et al. [9] and Minakami et al. [10] , Kunugi et al. [11] . Although interesting findings on natural and forced convection, and radiation from heat sinks have been reported in the previous studies, the mechanism of the heat transfer has not been clarified yet.

Accordingly, in the present study, the effects of pin size, pin height and the number of pins on the heat transfer characteristics of heat sinks with miniature/micro pins have been investigated fundamentally.

## 2. HEAT SINK MODELS

Five types of basic heat sink models are shown in Figure 1. The first model has four square pins, and is named as Type 1. Each pin placed at equal space of the heat sink base (length: 25 mm, width: 25 mm, thickness: 2 mm) has the pin width $w_p$ = 6.25 mm and the pin height $h_p$ = 6.5, 14, 21.5, 29 mm at intervals of 7.5 mm. And in Type 2 model, the pin size is reduced to half the pin width and height of Type 1. That is, under the conditions of constant heat transfer area, the number of pins for Type 2 is quadrupled against Type 1. In the same manner, five types of heat sinks in total, which have the pin height from several hundred micro meters to a few millimetres and have the same heat transfer area, are prepared.

As listed in Table 1, the number of pins N is changed from 4 to 1024, and corresponding to this change in N, the pin width $w_p$ and the pin height $h_p$ vary from 0.39 mm to 6.25 mm and from 0.41 mm to 29 mm, respectively. The whole heat transfer area of the square pin series $A_{hs}$ is 1.48 ′ $10^{-3}$ $m^2$, 2.23 ′ $10^{-3}$ $m^2$, 2.98 ′ $10^{-3}$ $m^2$ and 3.73 ′ $10^{-3}$ $m^2$.

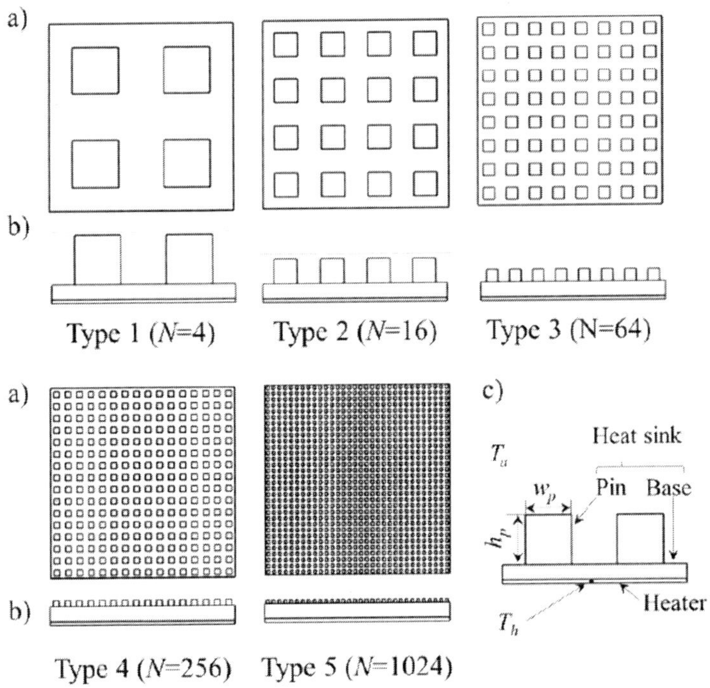


**Figure 1.** Heat sink models, (a) Top view, (b) Side view, (c) Parts and nomenclature.

**Table 1**. Dimensions of pin fin.

| Type | $N$ (-) | $w_p$ (mm) | $h_p$ (mm) | | | |
|---|---|---|---|---|---|---|
| | | | ● | □ | ◆ | △ |
| 1 | 4 | 6.25 | 6.5 | 14 | 21.5 | 29 |
| 2 | 16 | 3.13 | 3.25 | 7 | 10.75 | 14.5 |
| 3 | 64 | 1.56 | 1.63 | 3.5 | 5.38 | 7.25 |
| 4 | 256 | 0.78 | 0.81 | 1.75 | 2.69 | 3.63 |
| 5 | 1024 | 0.39 | 0.41 | 0.88 | 1.34 | 1.81 |

## 3. EXPERIMENTAL

Figure 2 shows a schematic diagram of experimental apparatus, which consists of a chamber, a vacuum pump, a vacuum gauge, and a heat sink with a heater used as the heat source. The chamber made of SUS is 400 mm in diameter and 170 mm in height. Here, its wall temperature which is regarded as the ambient temperature $T_a$ was kept at 25°C by circulating cooling water. The heat sink was hung in the air in the chamber by using fine wires to reduce the heat loss from the heat sink to the chamber wall by conduction. The heat sink was placed in the upward facings of the heated surface. The pressure in the chamber was controlled with the vacuum pump from the atmospheric pressure to the vacuum condition of approximately $3.5 \times 10^{-2}$ Pa, and the chamber pressure was measured using the vacuum gauge. Under the atmospheric pressure, heat is transferred from the heat sink to the surroundings by convection and thermal radiation. On the other hand, under the vacuum condition of approximately $3.5 \times 10^{-2}$ Pa, heat is transferred from the heat sink to the surrounding wall only by thermal radiation.

As shown in Figure 3, the heat sink, which is made of Aluminum alloy A6063 (the thermal conductivity l = 200 W/(m·K), the emissivity e = 0.89) with a base plate of 25 mm long, 25 mm wide and 2 mm thick, was placed on the same sized square copper block (thickness: 2 mm, l = 398 W/(m·K)) through a thermal interface material (thickness: 0.5 mm, l = 3.1 W/(m·K)). Furthermore, on the bottom part of the copper block, a ceramic heater is attached through the thermal interface material. The representative temperature $T_h$ was measured using a f 0.5 mm K-type thermocouple fixed in a small groove machined at the copper block center. In Figure 3, $T_a$ is the ambient temperature.

First, under the atmospheric pressure, the heat rate $Q_{in}$ = 1.25 W was applied to the heat sink by the ceramic heater. Second, the representative temperature $T_h$ was measured under the steady-state condition. Third, under the vacuum condition of approximately $3.5 \times 10^{-2}$ Pa, some amount of heat was applied to the heat sink by the heater until the heat sink temperature became the same temperature $T_h$ under the atmospheric pressure, and then the heat transferred by thermal radiation $Q_r$ from the heat sink under the steady-state conditions was measured based on the input power to the heater. Furthermore, in the same manner, $T_h$ and $Q_r$ for the heat sink consisting only of the base plate were also measured for reference.

# 4. NUMERICAL ANALYSIS

Figure 4 shows the boundary conditions for the large air space surrounding the heat sink placed in the upward facings of the heated surface. A series of numerical analyses have been performed for the physical models shown in Figure 1 and for the heat sink consisting only of the base plate (25 mm, 50 mm, 100 mm and 150 mm on a side) for reference. In this study, the fine wires for hanging the heat sink and the copper block, the thermal interface material were neglected for simplification, and only the heater (length: 25 mm, width: 25 mm, thickness: 0.5 mm), the heat sink, and the surrounding large air space were taken into consideration.

The present numerical calculations were carried out using the simulation software of Computational Fluid Dynamics (CFD). The following assumptions are applied in the analysis.

1) Fluid is incompressible.

2) Thermophysical properties are assumed to be constant except density (Boussinesq approximation).

3) Ambient temperature $T_a$ is constant.

Based on these assumptions, the following basic equations are obtained under the steady state conditions.

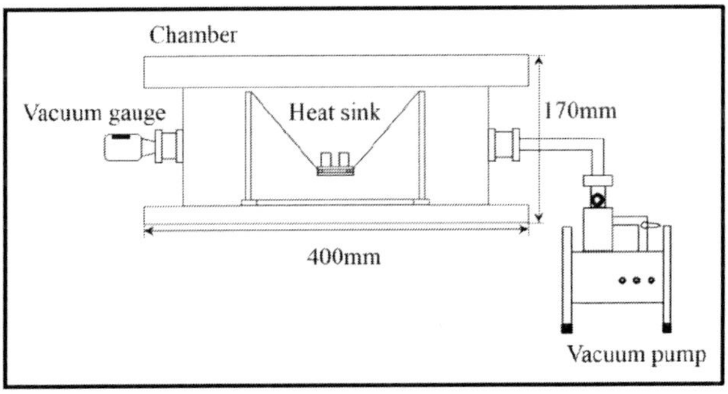


**Figure 2.** Schematic of experimental apparatus.

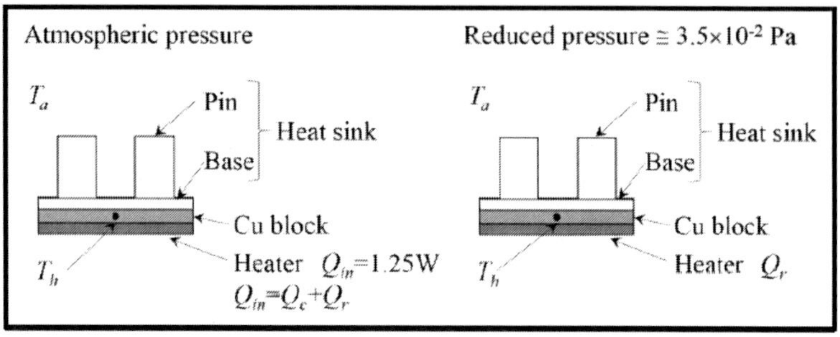


**Figure 3.** Configuration of tested heat sink.

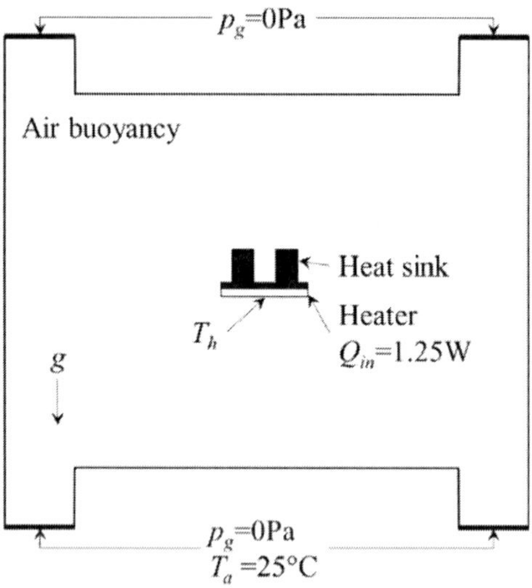


**Figure 4.** Boundary conditions of CFD (upward facing).

Continuity equation:

$$\frac{\partial u}{\partial x}+\frac{\partial v}{\partial y}+\frac{\partial w}{\partial z}=0 \tag{1}$$

Momentum equations:

$$u\frac{\partial u}{\partial x}+v\frac{\partial u}{\partial y}+w\frac{\partial u}{\partial z}=-\frac{1}{\rho}\frac{\partial p}{\partial x}+\upsilon\left(\frac{\partial^2 u}{\partial x^2}+\frac{\partial^2 u}{\partial y^2}+\frac{\partial^2 u}{\partial z^2}\right) \quad \text{(x-direction) (2)}$$

$$u\frac{\partial v}{\partial x}+v\frac{\partial v}{\partial y}+w\frac{\partial v}{\partial z}=-\frac{1}{\rho}\frac{\partial p}{\partial y}+\upsilon\left(\frac{\partial^2 v}{\partial x^2}+\frac{\partial^2 v}{\partial y^2}+\frac{\partial^2 v}{\partial z^2}\right)$$
(y-direction) (3)

$$u\frac{\partial w}{\partial x}+v\frac{\partial w}{\partial y}+w\frac{\partial w}{\partial z}=-\frac{1}{\rho}\frac{\partial p}{\partial z}+\upsilon\left(\frac{\partial^2 w}{\partial x^2}+\frac{\partial^2 w}{\partial y^2}+\frac{\partial^2 w}{\partial z^2}\right)+g\beta(T-T_a)$$
(z-direction) (4)

Energy equation:

$$u\frac{\partial T}{\partial x}+v\frac{\partial T}{\partial y}+w\frac{\partial T}{\partial z}=\alpha\left(\frac{\partial^2 T}{\partial x^2}+\frac{\partial^2 T}{\partial y^2}+\frac{\partial^2 T}{\partial z^2}\right)$$
(5)

Boundary conditions are given as follows. The same heat rate $Q_{in}$ = 1.25 W as the experiment was applied to the heat sink. The velocity is zero on all the solid surfaces, that is u = v = w = 0. The gauge pressure $p_g$ of the inlet and the outlet of the large surrounding space is 0 Pa, the inlet air temperature is 25°C, and other remaining surfaces are adiabatic. The amount of heat transfer rate by convection and thermal radiation is equal to that by conduction through the solid.

The equation of conduction within the solid is given by the following equation.

$$\frac{\partial^2 T}{\partial x^2}+\frac{\partial^2 T}{\partial y^2}+\frac{\partial^2 T}{\partial z^2}=0$$
(6)

In the numerical analysis, the same heat rate $Q_{in}$ = 1.25 W as the experiment was applied to the heater. In the case of large sized heat sink consisting only of the base plate, the heat rate added to the base plate was kept constant.

Then the representative temperature $T_h$ of the heat sink under the steady-state conditions and the heat rate released from the heat sink by thermal radiation $Q_r$ were calculated using the simulation software of CFD. In addition, the velocity vectors and the temperature fields at the central lateral plane of the heat sink were obtained.

# 5. RESULTS AND DISCUSSION

## 5.1. Heat Transfer Performance of Heat Sinks

Figure 5 and Figure 6 show the measured and simulated results on the effects of the number of pins N, the pin height $h_p$ on the temperature rise DT of the heat sink. Here, DT is defined as the difference between the representative temperature $T_h$ and the ambient temperature $T_a$. Accordingly, the lower DT means a higher heat transfer performance. The symbols show the results for the heat sink with different pin height, that is, heat transfer area. The broken line shows the DT for the heat sink consisting only of the base plate. Further, the number of pins N corresponds to the heat sink type from Type 1 to Type 5.

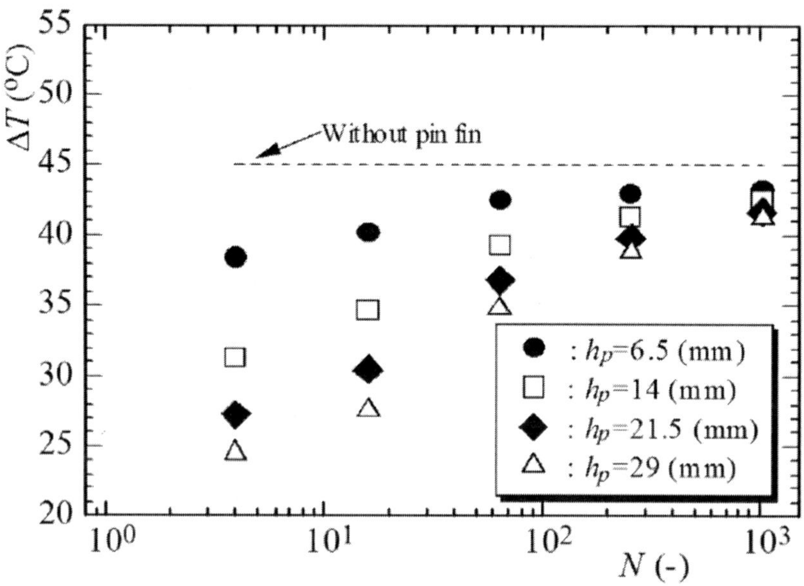


**Figure 5**. Effects of N and $h_p$ on DT (measured results).

Concerning Type 1 model having the smallest number of pins, it is found that the heat sink performance of Type 1 model is the highest and as the pin height $h_p$ is higher, the heat sink performance is better. In a series of heat sinks, the heat transfer rate increases with the pin height of Type 1. Each representative temperature $T_h$ of the heat sink rises with increase in the number of pins N. Especially, in the case of large N, the temperature rise DT of the heat sink comes close to that of the base plate level. In other words, the heat sink with miniaturized pins has almost no effect on the heat transfer enhancement. As seen from the figure, the simulated results agree qualitatively well with the measured results. That is, the dependence of the temperature rise DT on the number of pins N agrees fairly well with each other. Unfortunately, however, a quantitative agreement is not obtained between them. This is because the present physical model cannot reproduce the experimental heat sink system exactly.

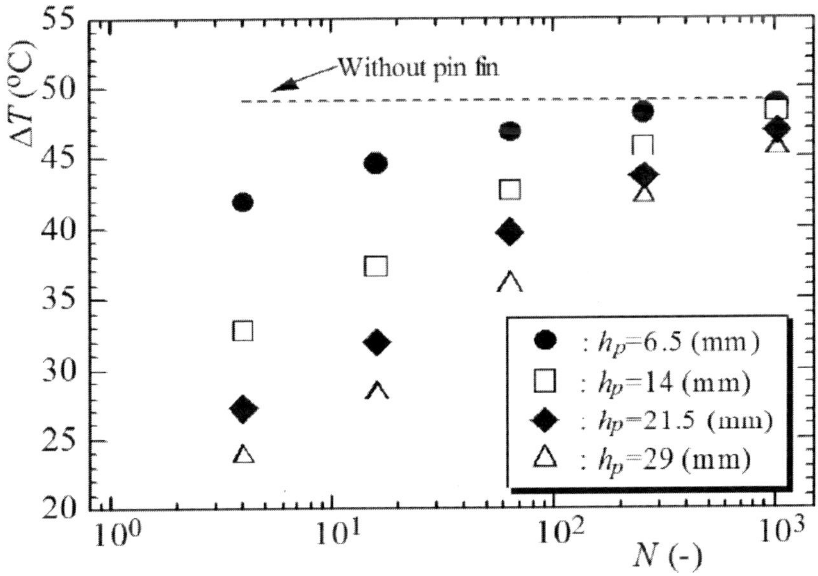


**Figure 6**. Effects of N and $h_p$ on DT (calculated results).

### 5.2. Velocity Vector and Temperature Field

The velocity vector in air space surrounding the heat sink model at central longitudinal section, which are obtained from numerical simulation, are shown in Figure 7. From this figure, it can be seen that the velocity vector among pins becomes small with increase in the number of pins. In particular, the air velocity of Type 5 models becomes almost zero.

The temperature fields of the heat sink and its surrounding air space are shown in Figure 8. The representative temperature $T_h$ at the heater surface is the highest. From this figure, it can be seen that the heater tem- perature rises from Type 1 to Type 5. Namely, this means that the increase in the number of pins causes the deterioration of the heat sink performance.

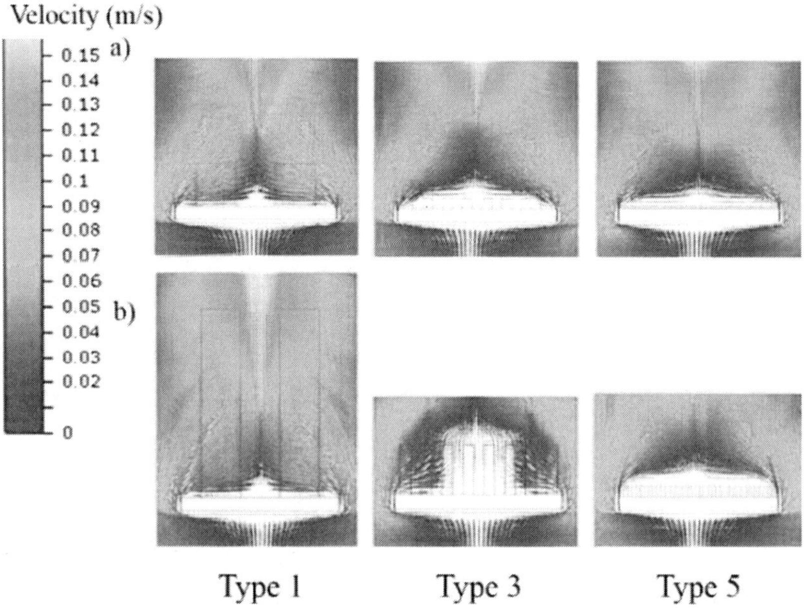


**Figure 7**. Flow field along center line of heat sink; (a) $h_p = 6.5$ mm, (b) $h_p = 29$ mm series.

### 5.3. Air Velocity between Pins

Figure 9 shows the air velocity between pins which was numerically obtained for Type 1 to Type 5 models. The present air velocity means the average value obtained at central lateral plane of heat sink. As shown in the figure, the air velocity decreases with increase in the number of pins. This result shows the same tendency as the velocity vectors. In Type 3 to Type 5 models, regardless of the pin height, the air velocity becomes almost zero. This behavior must be the choking phenomena observed in the previous study using the large sized heat sinks done by Zografos et al. [3], and as an inevitable consequence, it leads to low heat transfer performance of the heat sink.

### 5.4. The Ratio of Convective and Radiative Heat Transfer Rates

In order to estimate the heat sink performance from the view point of the heat transfer, the ratio of convective and radiative heat transfer rates in whole heat sink system has been also investigated experimentally and analytically, and the measured and calculated results are shown in Figure 10 and Figure 11 respectively. The total amount of heat transfer rate into and out has to keep the energy balance expressed by Equation (7). Under the conditions of the experiments and numerical calculations, $Q_{in}$ is 1.25 W. In addition, the output energy from heat sink is the sum of convective and radiative heat transfer rates $Q_c$ and $Q_r$, so the amount of heat transferred by convection $Q_c$ is given by the following Equation (8).

$$Q_{in} = Q_{out} \tag{7}$$

$$Q_c = Q_{in} - Q_r \tag{8}$$

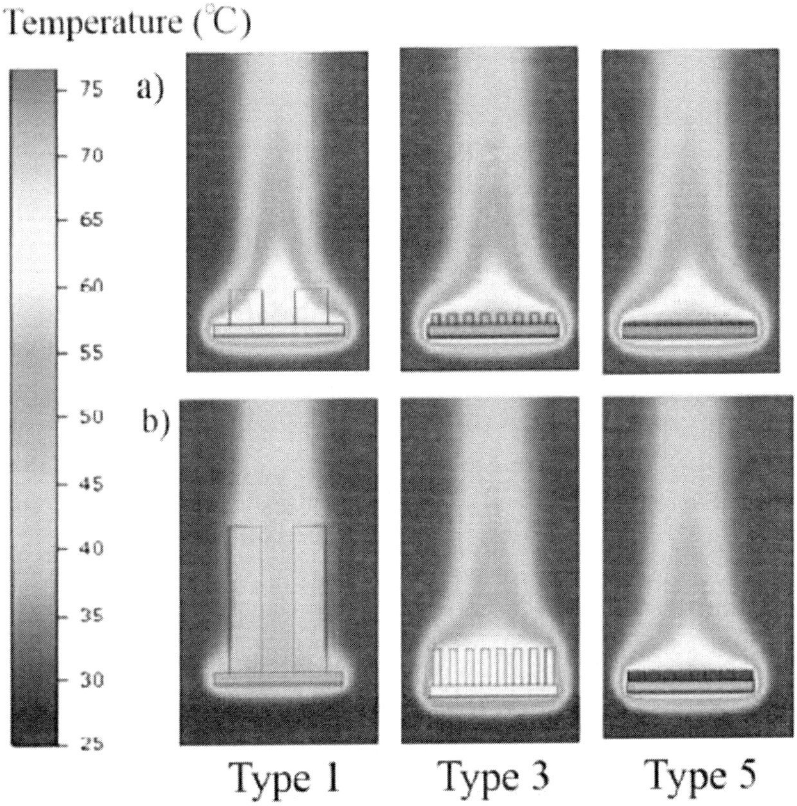


**Figure 8.** Temperature field along center line of heat sink; (a) $h_p$ = 6.5 mm, (b) $h_p$ = 29 mm series.

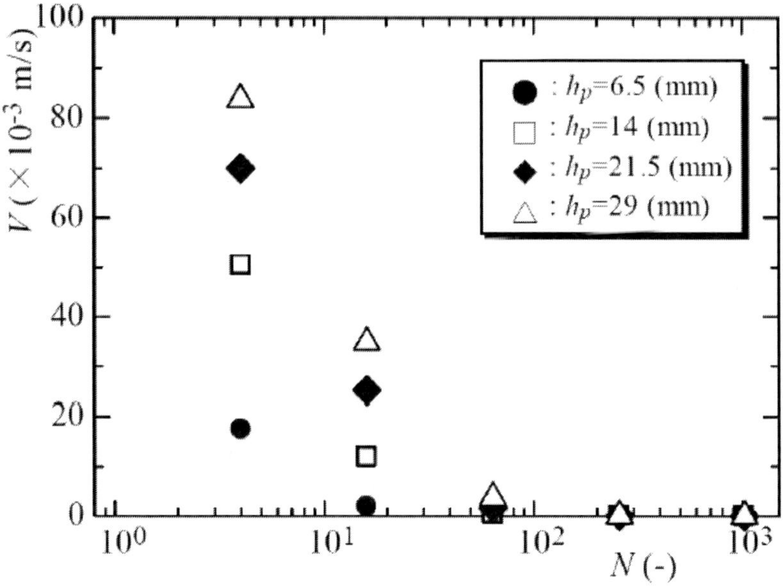


**Figure 9**. Flow velocity between pins on heat sink.

As seen from Figure 10, it is clear that the convective component $Q_c$ which is the dominant parameter in this case attains its maximum by Type 2, and reduces to Type 5 and then becomes saturated. It can be observed that the convection component $Q_c$ of Type 2 is exactly proportional to the pin height $h_p$, in other words the heat rate transferred by thermal radiation $Q_r$ vary inversely proportional to the pin height $h_p$. And in other heat sink models there is no correlation between the convection component $Q_c$ and the pin height $h_p$. Furthermore, the heat rate transferred by convection $Q_c$ occupies approximately 60% of the total input or output energy $Q_{in}$ or $Q_{out}$, the heat rate transferred by thermal radiation $Q_r$ is about 40%, and the result agrees very well with the results reported by Sparrow et al. [5] . And from Figure 11, it is found that the calculated results agree well with the measured results qualitatively. However a quantitative agreement is not obtained between them, as previously noted, this is because the present physical model cannot reproduce the experimental heat sink system exactly.

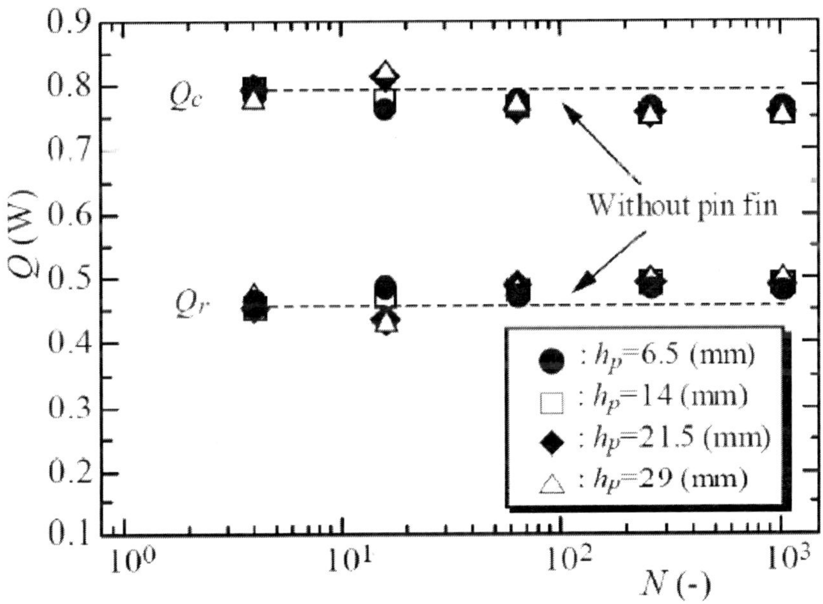


**Figure 10**. Effects of N and $h_p$ on Q (measured results).

### 5.5. Heat Transfer of Heat Sink under Natural Convection

In numerical simulations, the total amount of heat transfer rates of the heater and the heat sink was investigated. The convective component of heat sink only is represented by Equation (9) using computable parameters of the total amount of convective heat transfer rate $Q_c$ and the heat transfer from heater by convection $Q_{c\_ht}$.

$$Q_{c\_hs} = Q_c - Q_{c\_ht} \tag{9}$$

The convective heat transfer rate of the heater is calculated by Equation (10),

$$Q_c = \frac{Nu\lambda_a}{L} A\left(T_h - T_a\right) \tag{10}$$

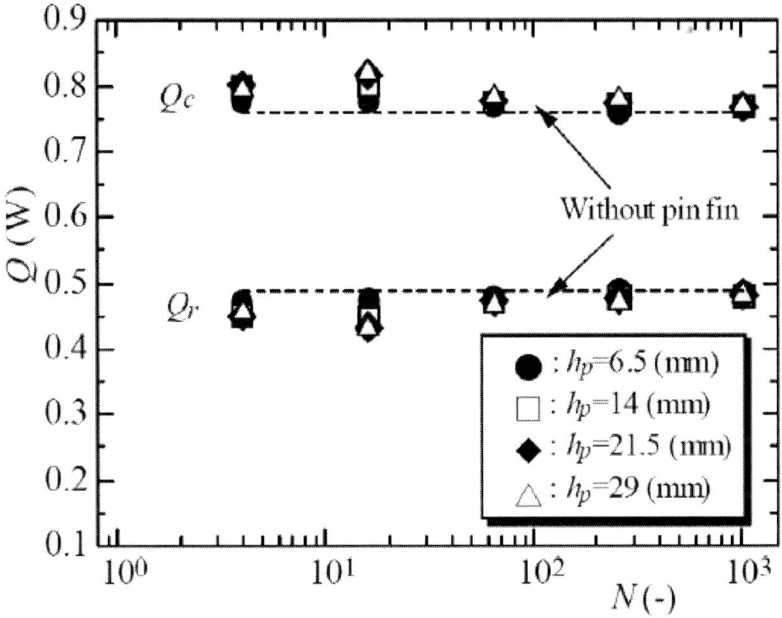


**Figure 11**. Effects of N and $h_p$ on Q (calculated results).

$$h = \frac{Nu\lambda_a}{L} \tag{11}$$

where Nu is the Nusselt number, $l_a$ is thermal conductivity of air, L is characteristic dimension and A is heat transfer surface of the heater. Equation (11) gives the heat transfer coefficient h.

It is found that the heater has isothermal heating surface according to the simulated results, the Nusselt numbers at the lower and side surfaces of the heater are calculated by previously-proposed Equation (12) [12] , Equation (13) [13] and Equation (14), where $Nu_1$ is the Nusselt number of lower surface of heated plate, $Nu_2$ is that of vertical surface of heated plate. And Ra is the Rayleigh number which is represented as the correlation parameter between the Grashof number and the Prandtl number.

$$Nu_1 = 0.27Ra^{0.25} \tag{12}$$

$$Nu_2 = 0.59Ra^{0.25} \tag{13}$$

where

$$Ra = GrPr = \frac{g\beta(T_h - T_a)L^3}{\upsilon}\frac{\upsilon\rho c_p}{\lambda_a} \tag{14}$$

Two types of characteristic lengths L will be employed here. In the case of the lower surface of cooled plate, L is taken as L = A/P where A is the plate surface area and P is the perimeter which surrounds the area [14] . In the vertical flat plate, L is taken as 0.5 mm which is the height of heater. Using these numbers, $Q_{c\_ht}$ of the each side is calculated by substituting $Nu_1$ and $Nu_2$ into Equation (10). And then $Q_{c\_hs}$ is determined by Equation (9) using known $Q_c$.

Figure 12 shows the effects of the number of pins and the pin height on the convective heat transfer coefficient. Here, in order to evaluate the improvement in the convective heat transfer performance for heat sink base surface, the surface area of the top of the base is adopted as A ($6.25 \times 10^{-4}$ $m^2$ constant). In this figure, it is clear that each heat transfer coefficient h of the heat sink decreases with increase in the number of pins N and in the case of Type 5 model it comes close to that of the base plate level. Since Figure 12 agrees well with the results of the temperature rise, it is considered that the factor of decreasing heat transfer performance due to the miniaturization is the convective heat transfer coefficient.

Finally, the calculated data of heat sink, expressed in terms of Nu and Ra, are plotted in Figure 13. In this figure, the lower solid line represents Equation (15) [14] in which $Nu_3$ is the Nusselt number of upper surface of heated plate and the symbols (′) show the calculated results of the base plate.

$$Nu_3 = 0.54Ra^{0.25} \quad \left(10^4 < Ra < 10^7\right) \tag{15}$$

From these results, it is confirmed that Equation (15) applied to a range of the low-Rayleigh number. Using Equation (11), the Nusselt number of the heat sinks was calculated. Here, L was estimated to be the total value of the pin height $h_p$ and the base height 2 mm. As seen this figure, regardless of the pin size, the pin height, the number of pins, it is revealed that Nu and Ra of the heat sink are expressed by an approximate equation, such as Equation (16).

$$Nu = 0.16Ra^{0.52} \tag{16}$$

Since the calculated results were shown between ±20% of the approximate expression and with miniaturization of pins the Nusselt number Nu of the heat sink comes close to that of the base plate level, it is considered that the proposed correlation implies high accuracy.

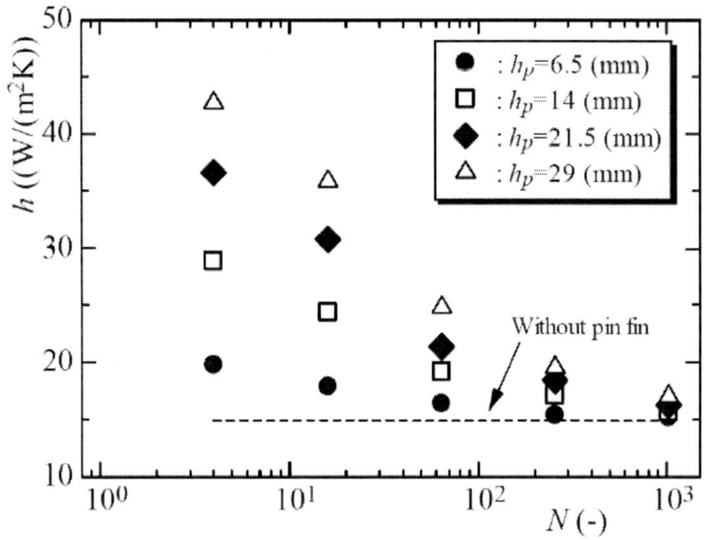


**Figure 12**. Effects of N and $h_p$ on h (calculated results).

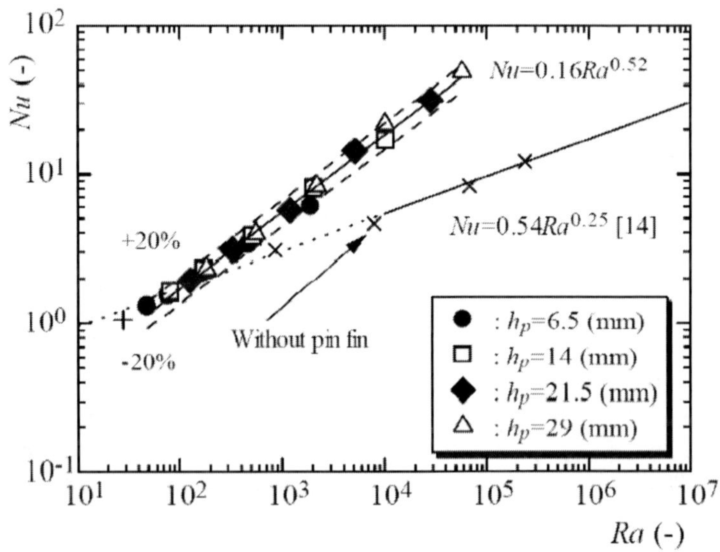


**Figure 13**. Correlation of Ra and Nu of the heat sink based on the base plate.

# 6. CONCLUSIONS

In this study, experimental and numerical studies have been done on the effects of the pin size, pin height, the number of pins on the heat transfer performance of the heat sink.

Concerning the heat sinks with the same heat transfer area, it has been confirmed that the heat sink performance changes depending on the population density of pin and the pin size. From the measured and calculated results, the heat sink temperature has been shown to rise with increase in the number of pins. Especially, the heat sink with miniaturized pins has almost no effect on the heat transfer enhancement. It is considered that these characteristics of the heat sink cause the decrease in the convective component, which occupies approximately 60% of total heat transfer rate, and the heat transfer coefficient due to the decrease in the flow velocity when the pin to pin clearances are reduced.

Regardless of the pin size, the pin height, and the number of pins, it has been revealed that the Nusselt number and the Reyleigh number of the heat sink are represented by an approximate equation.

# REFERENCES

1. Aihara, T., Maruyama, S. and Kobayakawa, S. (1990) Free Convective/Radiative Heat Transfer from Pin Fin Arrays with a Vertical Base Plate (General Representation of Heat Transfer Performance). International Journal of Heat Mass Transfer, 33, 1223-1232.

2. Sara, O.N. (2003) Performance Analysis of Rectangular Ducts with Staggered Square Pin Fins. Energy Conversion and Management, 44, 1787-1803.

3. Zografos, A.I. and Sunderland, J.E. (1990) Natural Convection from Pin Fin Arrays. Experimental Thermal and Fluid Science, 3, 440-449.

4. Huang, R.T., Sheu, W.J. and Wang, C.C. (2008) Orientation Effect on Natural Convective Performance of Square Pin Fin Heat Sinks. International Journal of Heat and Mass Transfer, 51, 2368-2376.

5. Sparrow, E.M. and Vemuri, S.B. (1986) Orientation Effects on Natural Convection/Radiation Heat Transfer from Pin- Fin Arrays. International Journal of Heat and Mass Transfer, 29, 359-368.

6. Sertkaya, A.A., Bilir, S. and Kargici, S. (2011) Experimental Investigation of the Effects of Orientation Angle on Heat Transfer Performance of Pin-Finned Surfaces in Natural Convection. Energy, 36, 1513-1517.

7. Yu, E. and Joshi, Y. (2002) Heat Transfer Enhancement from Enclosed Discrete Components Using Pin-Fin Heat Sinks. International Journal of Heat and Mass Transfer, 45, 4957-4966.

8. Bocu, Z. and Altac, Z. (2011) Laminar Natural Convection Heat Transfer and Air Flow in Three-Dimensional Rectangular Enclosures with Pin Arrays Attached to Hot Wall. Applied Thermal Engineering, 31, 3189-3195.

9. Narasimhan, S. and Majdalani, J. (2002) Characterization of Compact Heat Sink Models in Natural Convection. IEEE Transactions on Components and Packaging Technologies, 25, 78-87.

10. Minakami, K., Mochisuki, S., Murata, A., Yagi, Y. and Iwasaki, H. (1993) Heat Transfer Characteristics of Pin-Fins with In-Line Arrangement (1st Report, Effect of the Pin Pitch). Japan Society of Mechanical Engineers, 59, 300-307.

11. Kunugi, T., Muko, K. and Shibahara, M. (2004) Ultrahigh Heat Transfer Enhancement Using Nano-Porous Layer. Superlattices and Microstructures, 35, 531-542.

12. Kreith, F. and Bohn, M.S. (2000) Principles of Heat Transfer. 6th Edition, Brooks/Cole.

13. Nakayama, A. (2008) Problems in Heat Transfer. Japan Society of Mechanical Engineers. Maruzen Publishing Co., Ltd.

14. Goldstein, R.J. and Sparrow, E.M. (1973) Natural Convection Mass Transfer Adjacent to Horizontal Plates. International Journal of Heat and Mass Transfer, 16, 1025-1035. f

## NOMENCLATURES

A: heat transfer area ($m^2$);

$c_p$: specific heat (J/(kg×K));

Gr: Grashof number (–);

g: gravitational acceleration ($m/s^2$);

h: heat transfer coefficient (W/($m^2$×K));

$h_p$: height of pin (mm);

L: characteristic length (m);

N: number of pins (–);

Nu: Nusselt number (–);

P: perimeter (m);

Pr: Prandle number (–);

p: pressure (Pa);

Q: heat transfer rate (W);

Ra: Rayleigh number (–);

T: temperature (°C or K);

u: velocity component in x-direction (m/s);

V: air velocity (m/s);

v: velocity component in y-direction (m/s);

w: velocity component in z-direction (m/s);

$w_p$: width of pin (mm).

## GREEK SYMBOLS

α: thermal diffusivity ($m^2/s$);

b: coefficient of volume expansion ($K^{-1}$);

e: emissivity (–);

ΔT: temperature difference (°C or K);

λ: thermal conductivity (W/(m×K));

υ: kinematic viscosity ($m^2/s$);

ρ: density ($kg/m^3$).

### Subscripts

a: ambient or air;

c: convection;

g: gauge;

h: representative;

ht: heater;

hs: heat sink;

in: input;

out: output;

r: radiation;

1: lower surface of heated plate;

2: vertical surface of heated plate;

3: upper surface of heated plate.

# CHAPTER 2

# Modelling Turbulent Heat Transfer in a Natural Convection Flow

Claudia Zimmermann, Rodion Groll

*Center of Applied Space Technology and Microgravity, University of Bremen, Bremen, Germany*

## ABSTRACT

In this paper a numerical study of a turbulent, natural convection problem is performed with a compressible Large-Eddy simulation. In a natural convection the fluid is accelerated by local density differences and a resulting pressure gradient. Directly at the heated walls the temperature distribution is determinate by increasing temperature gradients. In the centre region convective mass exchange is dominant. Density changes due to temperature differences are considered in the numerical model by a compressible coupled model. The obtained numerical results of this study are compared to an analogue experimental setup. The fluid properties profiles, e.g. temperature and velocity, show an asymmetry which is caused by the non-Boussinesq effects of the fluid. The investigated Rayleigh number of this study lies at $Ra = 1.58 \times 10^9$.

## KEYWORDS

Turbulent Natural Convection, Compressible Flow, Large-Eddy Simulation, Heat Transfer

# 1. INTRODUCTION

Convection flows are one of the fundamental problems in fluid dynamics. They play a decisive role in meteorology where they appear as wind caused by solar radiation in the atmosphere. Moreover, they are important in industrial applications, where they are used, to mention just one example, in cooling systems to reduce possible noise exposures and technical failures. Although many theoretical, experimental and numerical investigations have been performed in the past, these flows are not quite understood and they are still of substantial interest. The main focus of this paper lies on a Large-Eddy simulation (LES) of the effective lift due to the occurred convection and the following thermal mixing of the flow. Therefore, a rectangular air-filled container with plane walls is chosen as computational geometry. The vertical side walls of the container are heated isothermally with a constant temperature difference $\Delta T$. The computational test case is chosen analogously to a comparable experimental test case of Tian and Karayiannis in [1].

# 2. TEST CASE CONFIGURATION

To investigate the natural convection in the simulation, the computational test case is built of a rectangular, air-filled, enclosed container, with a length (L) of 0.75 m, a height (H) of 0.75 m and a depth (D) of 1.5 m. A scheme of the computational test case is displayed in Figure 1, left picture.

The vertical sidewalls of the setup are heated differently, but homogeneously with a constant temperature difference of $\Delta T = T_{\text{hot}} - T_{\text{cold}} = 323.15\text{ K} - 283.15\text{ K} = 40\text{ K}.$ The vertical orientation of the heated walls implies a quasi-steady state flow in the setup. Due to a non-slip condition, the velocity field at all walls is zero $(\boldsymbol{u} \equiv 0)$. The experimental test case consists of highly conducting sidewalls [1]. This boundary condition is modelled in the simulation by an ideal linear temperature distribution between the hot and cold wall. Also an adiabatic temperature condition is analysed. The convection cell can be described by the parameters of the Rayleigh number Ra, the Prandtl number Pr, the Nusselt number Nu and its aspect ratios $\Gamma_x, \Gamma_z$. These parameters are defined as

$$\text{Ra} = \frac{g\Delta T\beta H^3}{\alpha\upsilon}, \text{Pr} = \frac{\upsilon}{\alpha}, \text{Nu} = \left.\frac{\partial T}{\partial y}\right|_{\text{w}} \frac{H}{\Delta T}, \Gamma_x = \frac{L}{H} = 1, \Gamma_z = \frac{D}{L} = 2 \tag{1}$$

where g is the gravitational acceleration, $\Delta T$ the temperature difference between the horizontal walls, $\beta$ the coefficient of thermal expansion, H the height of the container, $\alpha$ the thermal diffusivity, $\upsilon$ the kinematic viscosity and $\partial T/\partial y$ the temperature gradient directly at the heated walls (index w) The Nusselt number characterises the heat flux in the container. All parameters in Equation (1) correspond to the mean temperature field $T_{\text{mean}} = T_{\text{cold}} + \Delta T/2 = 303.15\text{ K}$ between both heated walls of the setup. The internal temperature field in the container is chosen as the mean temperature field, $T = T_{\text{IF}} = T_{\text{mean}}$, analogously to the experiment in [1] . If the fluid exceeds a critical value of the Rayleigh-number, it starts moving, controlled by the temperature difference of the vertical, heated walls. The Prandtl number lies in this study at Pr = 0.71. The simulation assumes a nonBoussinesq fluid, consequently, temperature dependent fluid properties are calculated by the Sutherland model as stated in [2] . The computational geometry consists of a Cartesian block-structured mesh with 27 blocks. A scheme of the mesh can be seen in Figure 1, right picture. This mesh partition enables an exterior zone where a finer resolution can easily be chosen independently of the other blocks. Directly at the walls the mesh is clustered and the cell ratios decrease to the walls. In this way, all relevant turbulent scales can be resolved and no wall functions have to be considered. In the middle of the geometry, large scales of the flow are dominant. Thus, the mesh can have a coarser resolution in this region than close to the walls. The final mesh consists of $(180 \times 180 \times 270) = 8{,}748{,}000$ cells in (x,y,z)-direction. The first grid point is located at $y_{w1} = 5 \times 10^{-4}\,\text{m}$ in vertical distance to the hot, respectively cold, wall. To analyse possible grid dependencies, additionally to the 3D simulation also a 2D simulation is performed with a mesh resolution of $(750 \times 750) = 562{,}500$ cells in (x,y)-direction. The cells are equally spaced over the plane. The first grid point is located at $y_{w1} = 6.9 \times 10^{-4}\text{ m}$ in vertical distance to the hot, respectively cold, wall. The non-dimensional distance from the wall $y^+$ is given by

$$y^+ = \frac{u_\tau}{\upsilon} y_{w1} \quad \text{with } u_\tau = \sqrt{\frac{\tau_\omega}{\rho}} \text{ and } \tau_\omega = \mu \left|\frac{\partial u}{\partial y}\right|_{y=0}, \tag{2}$$

where $u_\tau$ is the wall shear rate, $\upsilon$ the kinematic viscosity and $\mu$ dynamic viscosity. $\partial u / \partial y|_{y=0}$ marks the velocity gradient directly at the wall. Table 1 lists values of $y^+$ estimated in the first cell midpoint $y_{w1}$ in the full turbulent flow for both simulation types and both temperature boundary conditions.

The different values between the simulation types are caused by different velocities close to the heated walls.

## 3. GOVERNING EQUATIONS

Thermally driven flow in the setup is described as a turbulent compressible problem. "Compressible" referred in this case to density changes in the fluid due to temperature differences and not to a definition in terms of the Mach number. The turbulent, 3-dimensional flow field is calculated using a transient LES with the open-source simulation tool OpenFOAM®. In a LES the flow is separated in large and small scales, the so-called subgrid scales (marked by index "sgs" in the following). The large scales are solved directly on the discretisation grid, while the subgrid scales are modelled with an applicable turbulence model, which is in this study the so-called model of Fureby [3] . This model is based on the compressible Smagorinsky model. It assumes a thermodynamic ideal gas

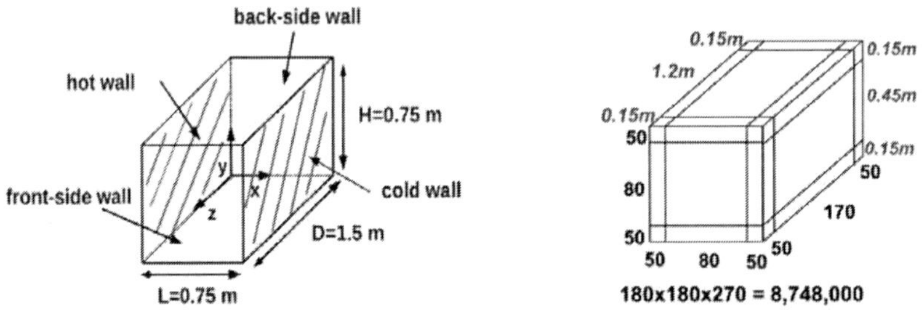


**Figure 1.** Left: Configuration of the computational test case. Right: Computational mesh resolution with $(180 \times 180 \times 270) = 8,748,000$ cells.

**Table 1.** Temperature boundary conditions and values of the non-dimensional wall distance $y^+$.

| Boundary condition | 3D | 2D |
|---|---|---|
| Adiabtic | $y^+ = 0.112$ | $y^+ = 0.0482$ |
| Linear temperature | $y^+ = 0.332$ | $y^+ = 0.190$ |

$$\bar{\rho} = \left(\bar{p}M\right) / \left(R_0 \bar{T}\right) \quad \text{and} \quad \bar{T} = \frac{1}{c_p}\left(\tilde{h} - \tilde{h}_0\right). \tag{3}$$

For the LES a space-time variable is separated and subsequent filtered as follows

$$\psi = \bar{\psi} + \psi' \quad \text{with} \quad \overline{\psi(\boldsymbol{x},t)} = \int_{-\infty}^{\infty}\int_{-\infty}^{\infty} \psi(\boldsymbol{r},t') G(\boldsymbol{x} - \boldsymbol{r}, t - t') \mathrm{d}t' \mathrm{d}\boldsymbol{r}. \tag{4}$$

Additionally, the dense weighted Favre-filtering of Vreman [4] is used

$$\tilde{\psi} = \frac{\overline{\rho \psi}}{\bar{\rho}}. \tag{5}$$

Hence, the governing equations are

Ÿ Compressible conservation of mass

$$\frac{\partial \bar{\rho}}{\partial t} + \nabla \cdot \left(\bar{\rho} \tilde{\boldsymbol{u}}\right) = 0, \tag{6}$$

Ÿ Compressible conservation of momentum

$$\frac{\partial \bar{\rho}\tilde{\boldsymbol{u}}}{\partial t}+\nabla\cdot\left(\bar{\rho}\tilde{\boldsymbol{u}}\tilde{\boldsymbol{u}}\right)-\nabla\cdot(\mu+\mu_{\text{sgs}})\left(\nabla\tilde{\boldsymbol{u}}+\left(\nabla\tilde{\boldsymbol{u}}\right)^{\mathrm{T}}-\frac{2}{3}\nabla\cdot\tilde{\boldsymbol{u}}\right)=-\nabla\bar{p}+\bar{\rho}\boldsymbol{g}, \tag{7}$$

Ÿ Compressible conservation of enthalpy

$$\frac{\partial \bar{\rho}\tilde{h}}{\partial t}+\nabla\cdot\left(\bar{\rho}\tilde{\boldsymbol{u}}\tilde{h}\right)-\nabla\cdot\left(\left(\alpha+\alpha_{\text{sgs}}\right)\nabla\tilde{h}\right)=\frac{\partial \bar{p}}{\partial t}+\tilde{\boldsymbol{u}}\cdot\nabla\bar{p}. \tag{8}$$

Additional force terms, which model the natural convection in the turbulent flow as in [5] [6] or a Boussinesq-approximation as in [7] [8] are not included in equation (7). The subgrid-scale dynamic viscosity $\mu_{\text{sgs}}$ and the subgrid-scale thermal diffusivity $\alpha_{\text{sgs}}$ are defined as

$$\mu_{\text{sgs}}=c_k\bar{\rho}\Delta\sqrt{k_{\text{sgs}}} \quad \text{and} \quad \alpha_{\text{sgs}}=\frac{\mu_{\text{sgs}}}{\mathrm{Pr}_{\text{sgs}}}. \tag{9}$$

The turbulent Prandtl number $\mathrm{Pr}_{\text{sgs}}$ is taken with $\mathrm{Pr}_{\text{sgs}}=0.4$ according to [6] [9] [10] . $k_{\text{sgs}}$ is the subgridscale turbulent kinematic energy which is defined as

$$k_{\text{sgs}}=\left(\frac{-\frac{2}{3}\frac{\partial\tilde{u}_i}{\partial x_i}+\sqrt{\left(\frac{2}{3}\frac{\partial\tilde{u}_i}{\partial x_i}\right)^2+8c_e c_k\left(\tilde{S}_{ij}\tilde{S}^*_{ij}\right)}}{2\cdot\frac{c_e}{\Delta}}\right)^2=0,\quad \tilde{S}^*_{ij}=\tilde{S}_{ij}-\frac{1}{3}\tilde{S}_{kk}\delta_{ij}, \tag{10}$$

with the deviatoric part of the strain rate tensor $\tilde{S}^*_{ij}$, the filtered strain rate tensor

$$\tilde{S}_{ij} = \frac{1}{2}\left(\frac{\partial \tilde{u}_i}{\partial x_j} + \frac{\partial \tilde{u}_j}{\partial x_i}\right) \tag{11}$$

and two model coefficients, $c_e = 1.046, c_k = 0.02.$ The temperature-dependence of the fluid properties is described by the Sutherland model [2] . For ideal gases it gives a relation between the dynamic viscosity $\mu$ and the absolute temperature $T$

$$\mu = \mu_{\mathrm{ref}}\left(\frac{T}{T_{\mathrm{ref}}}\right)^{\frac{3}{2}}\left(\frac{T_{\mathrm{ref}} + T_S}{T + T_S}\right). \tag{12}$$

$\mu_{\mathrm{ref}}$ is a reference dynamic viscosity at a reference temperature $T_{\mathrm{ref}}$ and $T_S$ is the so-called Sutherland temperature. In this study these coefficients are chosen as $\mu_{\mathrm{ref}} = 1.827 \times 10^{-5}$ kg/ms , $T_{\mathrm{ref}} = 291.15$ K and $T_S = 120$ K.

# 4. RESULTS

## 4.1. Temperature Profile between the Heated Walls

The flow movement of the convection is initialised by the temperature difference between the vertical walls. Thus, the temperature profile is one important aspect to understand the dynamic of the flow. Figure 2 displays the typical temperature and vertical velocity profile at the xy-midplane for a natural convection in the investigated setup. Near to the vertical walls, the temperature and vertical velocity can be described by a linear law. This region is called the inner layer of the boundary layer. In the following, the time-averaged temperature profile is estimated between the heated walls at the xy-midplane at $x = 0.75$ m , $z = 0.75$ m for three different heights, $y_1 = 0.1\mathrm{H},\ y_2 = 0.5\mathrm{H}$ and $y_3 = 0.9\mathrm{H}$ . The time-averaged results of the 3D and 2D simulation are plotted against the experimental data of Tian and Karayiannis [1] in Figure 3.

The thermal boundary layer near the heated walls and a variation of its thickness along the heated walls can clearly be seen. Further, a dependence on the particular boundary condition is visible. The boundary layer increases in the clockwise flow direction. Steep temperature gradients appear, before they reach their minimum at ca. 0.03 m afar from the hot wall (respectively 0.73 m from the cold wall), where the bulk region begins. Here, the temperature is almost stationary and almost equal to the mean temperature $T_{mean}$. The simulation profiles are not formed anti-symmetrically as in the experiment, but even asymmetrically. According to the study of Gifford in [11] , the anti-symmetrical profile vanishes for a non-Boussinesq-approximation. At midheight, all simulations approximate well the bulk temperature of the measured data. In vicinity of both horizontal walls the adiabatic results deviate significantly from the experiment, due to the adiabatic condition. Adiabatic walls entail no heat flux which implies lower temperatures. According to Tian and Karayiannis in [1] , the vertical thermal boundary layer of an experiment with an adiabatic boundary condition is thicker than the boundary layer of one with a conducting condition. This is fulfilled in the numerical results for the adiabatic case compared to the data of [1] . One exception is position $y_3 = 0.9H$ at the hot wall. This aspect could be a possible hint at shifted circulation zones compared to the experiment. The 2D plot shows only slight deviations to the 3D plot. The linear results approximate the experimental data closer than the adiabatic results, due to the similar boundary condition. The results of the 2D-simulation lie even closer to the experimental data than the results of the 3D simulation.

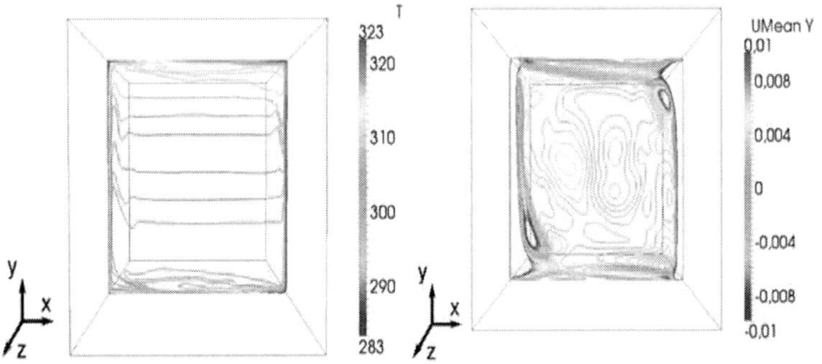


**Figure 2.** Instantaneous temperature (left) and vertical velocity (right) profile between the heated walls, linear temperature condition.

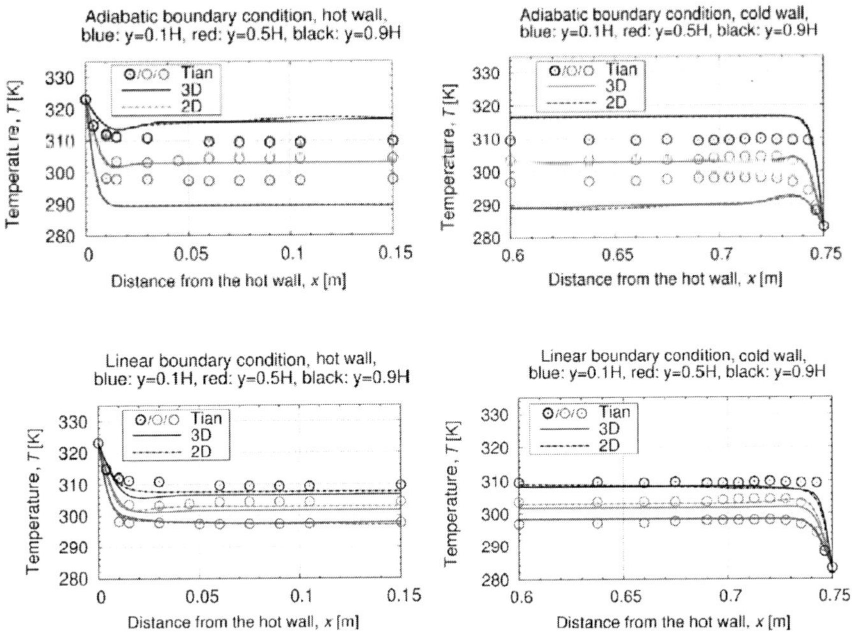


**Figure 3**. Time-averaged temperature profile at the vertical xy-midplane along the horizontal axis, at $x$ = 0.75 m, $z$ = 0.75 m and different heights. Top: Adiabatic *bc*. Bottom: Linear *bc*. Detailed plot of the hot/cold wall. In all pictures: $y_1$ = 0.1H (blue), $y_2$ = 0.5H (red) and $y_3$ = 0.9H (black), – (solid line): 3D, – – – (dashed line): 2D. study Tian and Karayiannis [1] , Ra = 1.58 × $10^9$.

## 4.2. Nusselt Number Profile along the Heated Walls

Besides the temperature distribution, also values of the Nusselt number are evaluated along the isothermal heated walls and plotted in Figure 4. With Equation (1) the heat flux is defined positive in direction from the hot wall into the container and in direction to the cold wall (respectively from the bottom to the top wall). The adiabatic 3D simulation matches at midheight of the hot wall exactly with the values of [1] . The cold wall value is slightly lower than in [1] . The time-averaged values at midheight deviate at most about 5% from the experiment, which can be neglected. From midheight on, the values underrun the experimental results due to lower temperature gradients as seen before in Figure 3. The 2D results display an equal tendency as the 3D ones and show only differences in vicinity to the heated walls. The time-averaged results of both simulation types of the linear case reveal an analogue profile to [1] , as it was expected due to the similar boundary conditions. The 3D case presents higher results than the 2D case, which is founded in steeper temperature gradients. The data of the 3D case are slightly lower at the hot wall than at the cold one. But the values at midheight deviate in both simulations significantly from the data in [1] , of about at most 20%. This might possibly be caused by lower temperature gradients. The numerical data approximate well the numerical study of Mergui, Penot and Tuhault [12] .

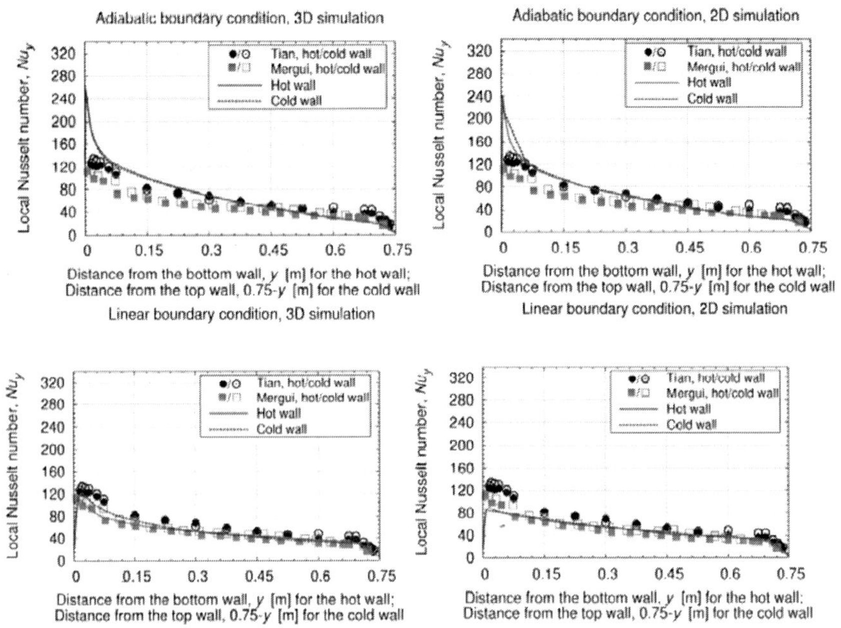


**Figure 4.** Time-averaged profile of the local Nusselt number along the heated walls. Top: Adiabatic bc. Bottom: Linear bc. Left: 3D simulation. Right: 2D simulation. In each picture: – (red solid line): hot wall, – – – (blue dashed line): cold wall. Study Tian and Karayiannis[1] : ●: hot wall, cold wall, Ra = 1.58 × $10^9$. Study Mergui, Penot and Tuhault [12] : ■ : hot wall, □ : cold wall, Ra = 1.34 × $10^9$.

### 4.3. Vertical Velocity Profile between the Heated Walls

Figure 2 shows a snapshot of the instantaneous mean velocity distribution at the xy-midplane in the container for the linear boundary condition. The plot reveals an exterior circulation zone as well as an interior one. Both circulations are reversed to each other. Furthermore, smaller circulations appear additionally in the left bottom and top right corner as well as top left and bottom right corner. In this section, the time-averaged vertical velocity profile at the xy-midplane between the temperatured walls is investigated and plotted in Figure 5. The profiles are estimated at the same positions as the temperature profiles before in Figure 3. For a better demonstration a constant was added to the results of each height, which are: $y_1: +0,\ y_2: +1,\ y_3: +2.$

As in the temperature profile, a clockwise circulation direction and an increase of the boundary layer along the heated walls in flow direction are visible. The boundary layer reaches its maximum width at the hot wall at height $y_3 = 0.9\text{H}$ (respectively at the cold wall at height $y_1 = 0.1\text{H}$). The thermal boundary layer is smaller than the velocity boundary layer. As in [1] , negative values appear between the inner and outer layer of the profile in the simulation results which indicate possibly a reverse flow. The vertical velocity component reveals at every height high peak values close to the vertical, heated walls. Minor decreases with negative values can be noticed outside the boundary layer at the hot wall (vice versa at the cold wall). According to [1] , this decrease/ increase might indicate two further possible clockwise circulation zones. The adiabatic profile shows a slightly smaller boundary layer than in the experiment, as it was expected. The peak values are located closer to the heated walls and lie beneath the experimental ones. No significant differences arise between the 2D and 3D simulation results. The profiles of the linear boundary case approximate well the results of [1] . Due to the different temperature conditions at the horizontal walls between the simulation and experiment of [1] , the peak values of the 3D and 2D simulation are higher than in the experimental study.

## 4.4. Wall Shear Stress along the Heated Walls

According to Tian and Karayiannis in [1] , a cubic law can be assumed for the velocity profile in the outer region of the inner layer. Therefore, the shear stress profile is described as

$$\tau_w = \mu \left. \frac{\partial u_y}{\partial x} \right|_w ,[\tau_w] = \text{N/m}^2, \tag{13}$$

with the gradient of the vertical velocity component directly at the heated walls (index w) and a depending dynamic viscosity $\mu$ . Besides the experimental data of [1] also another experimental study of King [13] is compared to the numerical data of this study in Figure 6.

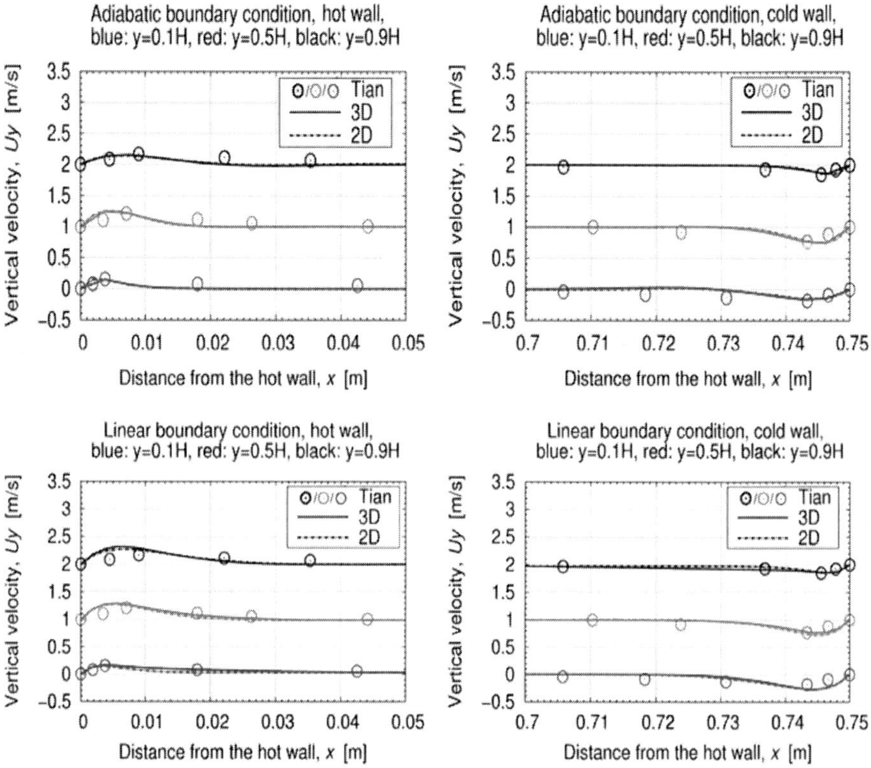


**Figure 5.** Time-averaged vertical velocity profile at the xy-midplane, at x = 0.75 m, z = 0.75 m and different heights. Top: Adiabatic bc. Bottom: Linear bc. Detailed plot of the hot/cold wall. In all pictures: $y_1$ = 0.1H (blue), $y_2$ = 0.5H (red) and $y_3$ = 0.9H (black), – (solid line): 3D, – – – (dashed line): 2D. ⊙ : study Tian and Karayiannis [1] , Ra = $1.58 \times 10^9$.

As Figure 6 shows, the time-averaged shear stress rises, analogously to the velocity, along the vertical, heated walls to its peak value and back to zero in the corners. The results of [1] reveal an asymmetry in the corner regions. The data of King [13] reveals higher values than in the study of Tian and Karayiannis [1] , which are possibly caused by a higher investigated Rayleigh number of King [13] according to [1] . The simulation profiles show an asymmetrical form, which is founded in the asymmetrical velocity profile. All simulations reveal negative values in the top hot and bottom cold corner which indicate anti-clockwise vortex regions as mentioned in [1] . The values along the heated walls in this study are higher than the data of [1] , due to higher velocity gradients and higher dynamic viscosity values. The high values of this study approximate rather the data of King[13] than of Tian and Karayiannis [1] .

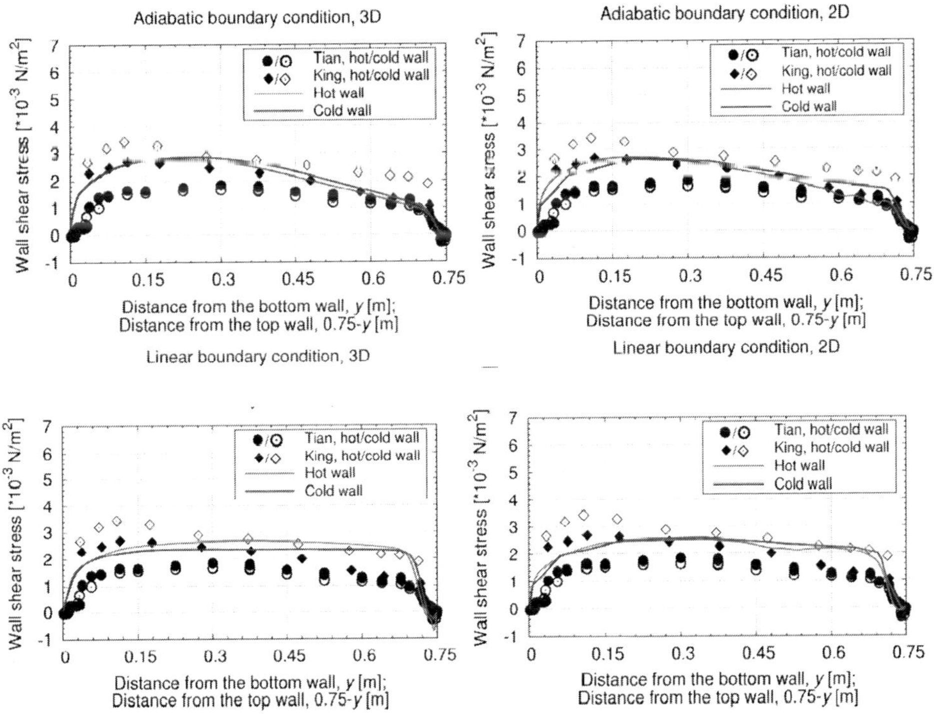


**Figure 6**. Time-averaged wall shear stress profile along the heated walls. Top: Adiabatic bc. Bottom: Linear bc. Left: 3D simulation. Right: 2D simulation. In all pictures: – (red solid line): hot wall, – (blue solid line): cold wall. Study Tian and Karayiannis [1] : ●: hot wall, ⊙: cold wall, Ra = $1.58 \times 10^9$. Study King [13] : ♦: hot wall, ◊: cold wall, Ra = $4.5 \times 10^9$.

## 5. CONCLUSION

A non-Boussinesq, compressible LES was performed for a turbulent, natural convection in an air-filled enclosed container with vertical heated walls. The results were compared to an analogue experimental setup of Tian and Karayiannis [1] . In the experiment, the setup consisted of conducting later walls, while in the numerical simulation two different related boundary conditions were tested. Additionally to the 3-dimensional simulations, 2-dimensional simulations were executed to investigate possible grid dependencies. Taking the different boundary conditions and the non-Boussinesq fluid assumption of the computational study into consideration, the results of the investigated fluid properties in all 3-dimensional simulations approximated well the experimental results of [1] . The simulation data showed the same profile tendencies as in the experiment [1] . The results of the 2- and 3-

dimensional simulations revealed in most cases similar results and only small deviations to each other which appeared mainly in vicinity to the heated walls. These deviations have to be further investigated in future studies to find an additional explanation as the one of grid dependencies. Summarising, the presented LES is very qualified to model the flow structures and fluid properties profiles of a turbulent, natural convection flow in the investigated configuration of a rectangular container where the vertical walls are heated.

## ACKNOWLEDGEMENTS

The authors wish to thank the North-German Supercomputing Alliance (HLRN) for providing their hardware during the numerical simulations of this study.

## REFERENCES

1. Tian, Y.S. and Karayiannis, T.G. (2000) Low Turbulence Natural Convection in an Air Filled Square Cavity, Part I: The Thermal and Fluid Flow Fields. International Journal of Heat and Mass Transfer, 43, 849-866.
2. Sutherland, W. (1893) The Viscosity of Gases and Molecular Force. Philosophical Magazine, Series 5, 36, 507-531.
3. Fureby, C. (1996) On Subgrid Scale Modeling in Large Eddy Simulations of Compressible Fluid Flow. Physics of Fluids, 8, 1301-1311.
4. Vreman, B. (1995) Direct and Large-Eddy Simulation of the Compressible Turbulent Mixing Layer. Ph.D. Thesis, University of Twente, Enschede.
5. Kenjeres, S. and Hanjalic, K. (1999) Transient Analysis of Rayleigh-Bénard Convection with a RANS Model, International Journal of Heat and Fluid Flow, 20, 329-340.
6. Sergent, A., Joubert, P. and Quéré, P. Le (2003) Development of a Local Subgrid Diffusivity Model for Large-Eddy Simulation of Buoyancy Driven Flows: Application to a Square Differentially Heated Cavity. Numerical Heat Transfer, Part A: Applications, 44, 789-810.
7. Shishkina, O. and Wagner, C. (2008) Analysis of Sheet-Like Thermal Plumes in Turbulent Rayleigh-Bénard Convection. Journal of Fluid Mechanics, 599, 383-404.

8. Shishkina, O. and Thess, A. (2009) Mean Temperature Profiles in Turbulent Rayleigh-Bénard Convection of Water. Journal of Fluid Mechanics, 633, 449-460.

9. Kosovic, B., Pullin, D.I. and Samtaney, R. (2002) Subgrid-Scale Modeling for Large-Eddy Simulations of Compressible Turbulence, Physics of Fluids, 14, 1511-1522.

10. Erlebacher, G., Hussaini, M.Y., Speziale, C.G. and Zang, T.A. (1992) Toward the Large-Eddy Simulation of Compressible Turbulent Flows. Journal of Fluid Mechanics, 238, 155-185.

11. Gifford, W.A. (1991) Natural Convection in a Square Cavity without the Boussinesq-Approximation. 49th Annual Technical Conference-ANTEC'91, Montreal, 5-9 May 1991, 2448-2454.

12. Mergui, S., Penot, F. and Tuhault, J.H. (1993) Experimental Natural Convection in an Air-Filled Square Cavity at . In: Henkes, R.A.W.M. and Hoogendoorn, C.J., Eds., Turbulent Natural Convection in Enclosures—A Computational and Experimental Benchmark Study, Proceedings of the Eurotherm Seminar no 22, Editions Européennes Thermique et Industrie, Paris, 9-108.

13. King, K.J. (1989) Turbulent Natural Convection in Rectangular Air Cavities. Ph.D. Thesis, Queen Mary and Westfield College, University of London, London, UK.

# CHAPTER 3

# Numerical Simulation of Laminar Forced Convection of Pin-Fin Heat-Sink Array in a Channel by Using Porous Approach

Tzer-Ming Jeng and Sheng-Chung Tzeng *

*Department of Mechanical Engineering, Chienkuo Technology University, Changhua 500, Taiwan*

## ABSTRACT

This work used a porous approach model to numerically investigate the fluid flow and heat transfer characteristics of the pin-fin heat-sink array in a rectangular channel with in-line arrangement. The air flow through the channel was laminar. The pin-fin heat sinks with various porosities and pin-fin numbers were employed. The relative center-to-center longitudinal and transverse distances between adjacent heat sinks were changed. The results indicate that the Nusselt number of various heat-sink arrays increased with decreasing the relative center-to-center transverse distance, but not varied with the relative center-to-center longitudinal distance. For the typical pin-fin heat-sink arrays, the Nusselt number changed slightly for the heat sinks with 0.358–0.556 porosity, but increased by 11.7%–24.8% when the porosity increased from 0.556 to 0.750, and then dropped obviously when the porosity exceeded 0.750. Increasing the number of pin fins continuously could increase Nusselt number. However, when the number of pin fins was large, the Nusselt number increased with the number of pin fins slowly. The present numerical simulation has been validated by the typical experiment. Finally, a semi-empirical correlation of Nusselt number for each heat sink in the heat-sink array was proposed.

## KEYWORDS

porous approach; pin-fin heat sink; heat-sink array; heat transfer

## 1. INTRODUCTION

Since the invention of integrated circuit (IC) in 1964, the speed of computers has rapidly increased. As one IC can hold tens of electronic modules, the function of an electronic module is equivalent to a transistor or vacuum tube, so that the computer becomes smaller and more powerful. At present, the IC has evolved from early SSI (Small Scale Integrated Circuit, 10–100 modules/per unit chip) into LSI (Large Scale Integrated Circuit, 1000–10,000 modules/per unit chip), then VLSI (Very Large Scale Integrated Circuit, 100,000–1,000,000 modules/per unit chip), and finally even ULSI (Ultra Large Scale Integrated Circuit, more than a million modules/per unit chip). Such a high density package of semiconductor chip increases the heat density greatly, and the reliability or lifetime of semiconductor chip decays as the temperature rises, usually in the range of below junction temperature 100 °C. A lower temperature leads to a lower failure rate of the chip. Therefore, a good thermal control cooling design is increasingly important for high-speed computer. Multilayer PCB substrate is mounted inside general superhuge computer. Each PCB substrate is about tens of square centimeter, and carries a hundred LSIs. The cooling technique is that the finned heat sink is mounted on the LSI chip for forced air cooling. In such a configuration, there are gaps on both sides and top of the finned heat sink. A part of forced cold flow enters the internal flow channels of heat sink for cooling, and a part of cold flow leaves passing by the heat sink through the gaps. The bigger by-pass gap results in the smaller amount of the effective cold flow through the heat sink, which is disadvantageous to the overall cooling of the heat sink. Therefore, the by-pass gap and heat sink have optimal relative dimensions, so that there is maximum cooling air entering into the heat sink at fixed pumping power. This is an important topic considering actual forced air cooling.

The issue about the heat transfer characteristics of finned heat sinks without [1,2,3,4,5,6] or with bypass effect [7,8,9,10,11,12] attracts wide attention continuously. This work is to discuss putting the pin-fin heat-sink array into a rectangular channel in in-line arrangement with laminar side-bypass effect. This configuration is similar to the PCB substrate of LSI array with heat sinks inside the general supercomputer. Figure 1 shows the typical flow field of such cooling system with seven rows of heat sinks. Considering the periodical and symmetric configuration of the present system, it only shows the upper half of the single-column flow field. According to Figure 1, when the cold flow enters the heat sink array, a part of fluid exchanges heat with the heat sink, and the other part of fluid passes by the heat sink and passes through the gaps between adjacent heat-sink columns. The cold flow in the heat sink leaks out of the heat sink from side continuously under the flow-resistance effect of pin-fins, so that the downstream heat sink has less incoming cold flow. This phenomenon is especially obvious when the permeability of heat sink is low, even forming recirculating flow between downstream heat sinks. There is also a large recirculation zone behind the heat sink array. Such a change of flow field is mainly related to the flow resistance of pin-fin heat sink and the relative spacing between heat sinks, and affects the heat transfer characteristics of each heat sink significantly.

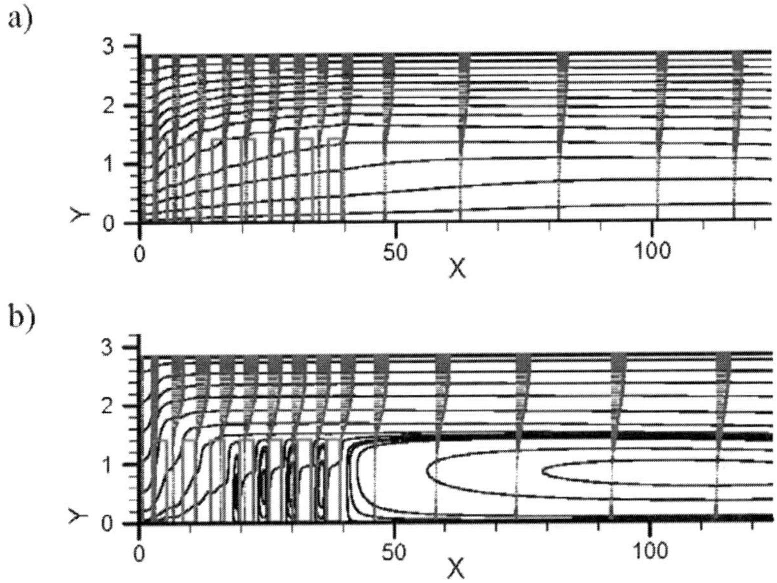


**Figure 1.** The upper half of the single-column flow field in seven-rows heat-sink array (the direction of pin fins is pointing the *Z*-axis vertical to paper surface): (**a**) high-permeability heat-sink array and (**b**) low-permeability heat-sink array.

In the numerical simulation of heat transfer and fluid flow characteristics of finned heat sink without bypass effect, some studies regard the finned heat sink as a porous medium, using the assumption of the volume-averaging technique for computations of the porous approach model [13,14,15,16,17]. Besides, Jeng [18] used porous approach model to simulate the fluid flow and heat transfer characteristics of single square pin-fin heat sink with laminar side-bypass effect. The Brinkman-Forchheimer model was used for fluid flow and two-equation model for heat transfer. Narasimhan *et al.* [19,20] numerically investigated the thermal behaviors of distributed porous blocks in a channel by using a *bi-disperse* porous medium approach. They indicated that the Nusselt number increasing with Reynolds number changed from non-linear to linear as number of blocks increased from one to 81, with corresponding insignificant pumping power increase. Feng *et al.* [21] present a porous medium model for forced air convection in pin/plate-fin heat sinks subjected to non-uniform heating of an impinging jet. The forced convection is considered by employing the non-Darcy model for fluid flow and the thermal non-equilibrium model for heat transfer. They showed that the inline square pin-fin heat sink has topological advantage over the plate-fin heat sink, although the heat spreading through the plate-fins on reducing the peak temperature on the substrate is pronounced.

The current study extends the porous approach simulation method discussed in the previous study [18] to simulate the hydrodynamic and thermal behaviors of the pin-fin heat-sink array situated in a rectangular channel in an in-line arrangement with various longitudinal and transverse distances between heat sinks. The in-line pin-fin arrays of the heat sinks with various porosities and numbers of pin fins, confined uniformly within a square spreader whose side length is 67 mm, were employed. In the current study, the effects of various parameters, such as the porosity of heat sink, the number of pin fins and the relative longitudinal and transverse distances between heat sinks, on the fluid flow and heat transfer were analyzed.

## 2. NUMERICAL METHOD

As the above-mentioned statement, this study uses the porous approach method proposed by Jeng [18] to simulate the fluid flow and heat transfer characteristics of multiple pin-fin heat sinks situated in a rectangular channel in an in-line arrangement similar to the PCB substrate of LSI array with heat sinks. In general, side-bypass and top-bypass flows exist in such issue at the same time. However, this work discusses the extreme configuration without any top-pass space above the heat-sink array. Figure 2a describes the corresponding physical configurations for the present and previous [18] studies, respectively. The previous study [18] only investigated the single pin-fin heat sink situated in a rectangular channel with laminar side-bypass flow. The originality and contribution of this work are to show the difference between single and array of pin-fin heat sink and the relevant hydrodynamic and thermal behaviors of the downstream heat sinks. Because the present study only considers the side-pass effect, it can be simplified to be two-dimensional problem with the assumption of ignoring the 3D vortex flow around the roots of pin fins. Furthermore, for the present in-line heat-sink array, each heat sink is responsible for cooling its corresponding bottom heat source. The heat flux is firstly transmitted upwards through the pin fins by heat conduction, and then transmitted to the air passing through the heat sink from the extended surface of the pin fins by heat convection. This work sets each pin-fin heat sink as a porous-medium block. Since that this work has been treated as a 2D issue, the heat dissipated from pin fins is regarded as the heat generated in the solid matrix of the porous block. Figure 2b shows the 2D configuration ($x$–$y$ plane) cut from the middle of pin fins. The top view of multiple pin-fin heat sinks with in-line arrangement is presented. The center-to-center longitudinal and transverse distances ($S_L^*$ and $S_T^*$) between the adjacent heat sinks are variable. Figure 2a also displays the 3D configuration of a typical pin-fin heat sink. The center-to-center longitudinal distance between the adjacent pin-fins is $S_L$, the center-to-center transverse distance is $S_T$, the number of pin-fins is $n$, the thickness of each square pin-fin is $d$, the size of pin-fin heat sink is $L \times L \times H$, and the height of the channel is $H$. Other assumptions are as follows: (1) the porous medium is homogenous and isotropic; (2) the fluid flow is steady state, laminar and incompressible; and (3) the thermophysical properties of the fluid and porous media do not depend on temperature. The dimensionless volume-averaged governing equations, including the Brinkman-Forchheimer model for momentum equation and two-equation model for energy equations, can be expressed as follows [18].

$$-\omega = \frac{\partial^2 \Psi}{\partial X^2} + \frac{\partial^2 \Psi}{\partial Y^2} \tag{1}$$

$$U\frac{\partial \omega}{\partial X} + V\frac{\partial \omega}{\partial Y} = -\frac{\varepsilon^2}{\mathrm{Re}\cdot Da}\omega - \frac{\varepsilon^2 C_F U_M}{\sqrt{Da}}\omega + \frac{\varepsilon^2 C_F}{\sqrt{Da}}\left(U\frac{\partial U_M}{\partial Y} - V\frac{\partial U_M}{\partial X}\right) + \frac{\varepsilon}{\mathrm{Re}}\left(\frac{\partial^2 \omega}{\partial X^2} + \frac{\partial^2 \omega}{\partial Y^2}\right) \tag{2}$$

$$U\frac{\partial \theta_f}{\partial X} + V\frac{\partial \theta_f}{\partial Y} = \frac{Nu_{fs}}{\mathrm{Re}\cdot \mathrm{Pr}}(\theta_s - \theta_f) + \frac{1}{\mathrm{Re}\cdot \mathrm{Pr}}\frac{k_f^*}{k_f}\left(\frac{\partial^2 \theta_f}{\partial X^2} + \frac{\partial^2 \theta_f}{\partial Y^2}\right) \tag{3}$$

$$0 = Nu_{fs}(\theta_f - \theta_s) + \frac{k_s^*}{k_f}\left(\frac{\partial^2 \theta_s}{\partial X^2} + \frac{\partial^2 \theta_s}{\partial Y^2}\right) + \delta \tag{4}$$

and the relevant dimensionless parameters are as follows:

$$X = \frac{x}{H}\ , \quad Y = \frac{y}{H}\ , \quad U = \frac{u}{u_i}\ , \quad V = \frac{v}{u_i}\ , \quad \Pr = \frac{\mu / \rho_f}{k_f / (\rho C_p)_f}\ , \quad Da = \frac{K}{H^2}\ ,$$

$$\mathrm{Re} = \frac{\rho_f u_i H}{\mu}\ , \ U_M = \frac{\sqrt{u^2 + v^2}}{u_i}\ , \ \theta = \frac{T - T_i}{q_w H / k_f}\ , \ Nu_{fs} = \frac{h_{fs} a_{fs} H^2}{k_f} \tag{5}$$

where $H$ is the channel (or pin-fin) height, $K$ is the permeability; $a_{fs}$ is the surface area of the fluid-solid interface per unit bulk volume of the pin-fin heat sink; $h_{fs}$ is the heat transfer coefficient between the fluid stream and the solid matrix; $q_w$ is the heat flux on the bottom of the pin-fin heat sink; ω is the vorticity, and ψ is the stream function.

$$\omega = -\frac{\partial U}{\partial Y} + \frac{\partial V}{\partial X}\ , \ U = \frac{\partial \Psi}{\partial Y}\ , \ V = -\frac{\partial \Psi}{\partial X} \tag{6}$$

In Equations (1)–(4), $C_F$ is the inertial coefficient; ε is the porosity of the pin-fin heat sink (*i.e.*, the total void volume divided by the total volume occupied by the pin fins and the void volume); $k_s^*$ is the effective conductivity of the solid matrix; and $k_f^*$ is the effective conductivity of the fluid. Notably, in the clear fluid region, $Da$ is set to infinity and ε to unity. In addition, $Nu_{fs} = k_s^* = 0$ and $k_f^* = k_f$. The δ is a function set equal to one to account for heat generation in the porous media of pin-fins and to zero elsewhere.

The relevant empirical coefficients in the present porous medium of pin-fins, such as $K$, $C_F$, $k_s^*$, $k_f^*$ and $h_{fs}$, generally do not have universal values. This work lists the porous properties of pin-fin heat sinks studied herein in Table 1 by using the relevant empirical formula reported in the open literatures [18,22,23]. Some heat sinks in the present study are the same as those used in previous study [18]. This is for verification purposes since the single heat sink in the previous study [18] might be like the first row of heat sinks in the present study. The $h_{fs}$ of the square pin-fin heat sink with uniform in-line arrangement is predicted by employing the empirical equations suggested by Kim *et al.* [24]:

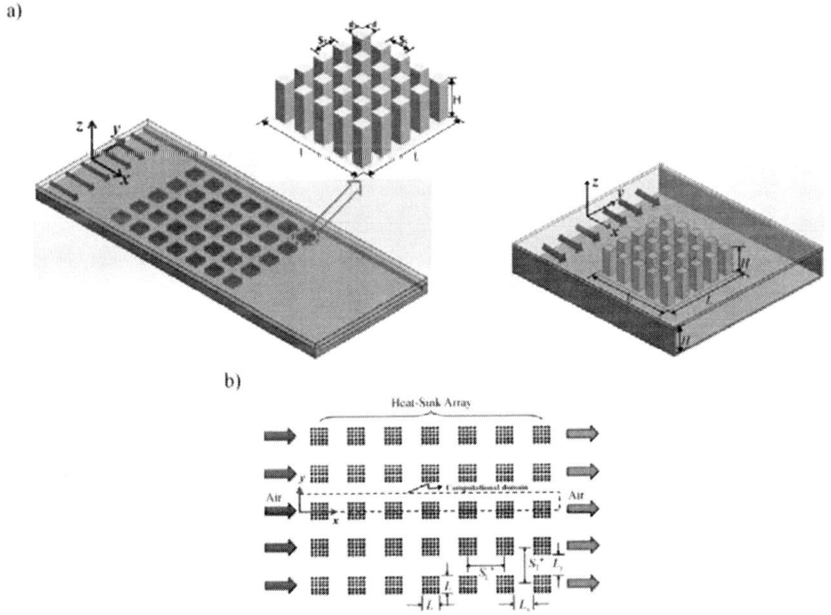


**Figure 2.** Physical configuration: (**a**) 3D configurations (left: present study; right: previous study [18]); and (**b**) 2D configuration ($x$–$y$ plane) cut from the middle of fins.

$$h_{fs} = \frac{\varepsilon k_f}{d}\left(0.36283\,\mathrm{Re}_d^{\;0.54219}\right) \text{ for } Re_d < 1000 \tag{7}$$

$$h_{fs} = \frac{\varepsilon k_f}{d}\left(0.04433\,\mathrm{Re}_d^{\;0.82934}\right) \text{ for } Re_d \geq 1000 \tag{8}$$

where $Re_d$ is defined as $\rho_f u_{max} d/\mu$; $u_{max}$ is the average maximum velocity between pin-fins, and $d$ is the pin-fin thickness.

Figure 3a schematically depicts the numerical domain and the boundary conditions. The relevant assumptions of boundary conditions for the vorticity ($\omega$) and the stream function ($\Psi$) refer Roache [25], who brilliantly discussed various boundary conditions for $\omega$ and $\Psi$. The fluid is assumed to enter the inlet with a uniform temperature ($T_i$) and a uniform velocity ($u_i$). The top and bottom faces are symmetric to the adjacent zones (not included in the present computational domain). The fluid phase at the outlet face is assumed to meet the zero diffusion condition since the downstream length was chosen to be sufficiently long.

**Table 1.** Porous properties of pin-fin heat sinks used herein.

| In-Line Square Pin-Fin Heat Sink with Uniform Distributions ($S_L = S_T$), $L$ = 0.067 m, $H$ = 0.0237 m | | | | | | | | |
|---|---|---|---|---|---|---|---|---|
| **Test Specimens** | **Sample 1** | **Sample 2** | **Sample 3** | **Sample 4** | **Sample 5** | **Sample 6** | **Sample 7** | **Sample 8** |
| $d$ (m) | 0.00596 | 0.00496 | 0.00372 | 0.00479 | 0.0067 | 0.00298 | 0.00248 | 0.00305 |
| $n$ | 9 × 9 | 9 × 9 | 9 × 9 | 7 × 7 | 5 × 5 | 9 × 9 | 9 × 9 | 11 × 11 |
| $\varepsilon$ | 0.358 | 0.556 | 0.750 | 0.750 | 0.750 | 0.84 | 0.889 | 0.75 |
| $K$ (m²) | $4.37 \times 10^{-8}$ | $2.40 \times 10^{-7}$ | $1.03 \times 10^{-6}$ | $1.63 \times 10^{-6}$ | $2.95 \times 10^{-6}$ | $2.13 \times 10^{-6}$ | $3.36 \times 10^{-6}$ | $7.03 \times 10^{-7}$ |
| $C_F$ | 0.0808 | 0.0785 | 0.0451 | 0.0486 | 0.0543 | 0.0721 | 0.0862 | 0.0426 |
| $k_f^*$ (W·m⁻¹·K⁻¹) | 0.0064 | 0.0116 | 0.018 | 0.018 | 0.018 | 0.0216 | 0.0231 | 0.018 |
| $k_s^*$ (W·m⁻¹·K⁻¹) | 0.129 | 0.0694 | 0.036 | 0.036 | 0.036 | 0.0237 | 0.0173 | 0.036 |
| $a_{fs}$ (m²/m³) | 445 | 381 | 300 | 241 | 181 | 251 | 217 | 360 |

This work applies the power-law developed by Patankar [26] to disperse the equations and employs the SIS (Strong Implicit Solver) algorithm proposed by Lee [27] to solve the related dispersal equations. All of the resolutions undergo the grid independence test and convergence test. The grid systems are separated into various classes to fit the physical model. The grid points, as shown in Figure 3b, are 1224 × 49–1440 × 77. The grid size falls as $X_T^*$ and $X_L^*$ values decrease. The $X_T^*(=S_T^*/L)$ and $X_L^*(= S_L^*/L)$ are the relative center-to-center longitudinal and transverse distances between the adjacent heat sinks, respectively. The iteration ends when the variables meet the criterion,

$$\sum_{i,j} \left| \frac{F_{i,j}^{r+1} - F_{i,j}^{r}}{F_{i,j}^{r+1}} \right| \leq 1 \times 10^{-5} \tag{9}$$

where $F$ represents Ψ, ω or θ. The subscripts $i$ and $j$ refer to the $i$-th and the $j$-th grid-nodes in the $X$ and $Y$ directions, respectively. The superscript $r$ indicates the $r$-th iteration.

The major parameters of heat transfer performance observed herein are the Reynolds number ($Re$), the dimensionless pressure drop ($\Delta P$) and the average Nusselt number ($Nu$), which are defined as follows.

$$\mathrm{Re} = \frac{\rho u_i H}{\mu} \tag{10}$$

$$\Delta P=\frac{p_i-p_e}{\rho_f u_i^2}=\int_{L_{in}/H}^{(L_{in}+L^*)/H}\left(\frac{1}{\mathrm{Re}\,Da}+\frac{C_F}{\sqrt{Da}}U_M U-\frac{1}{\varepsilon\,\mathrm{Re}}\frac{\partial^2 U}{\partial X^2}+\frac{U}{\varepsilon^2}\frac{\partial U}{\partial X}\right)_{Y=0}dX \quad (11)$$

$$Nu=\int_0^L\int_0^{L/2}Nu_{x,y}dydx/(L^2/2) \quad (12)$$

$$\overline{Nu}=\sum_{i=1}^{7}Nu_i/7 \quad (13)$$

Where

$$Nu_{x,y}=\frac{q_w H}{(T_w-T_i)k_f}=\frac{q_w H}{(T_s-T_i)k_f}\eta=\frac{\eta}{\theta_s},\ \eta=\frac{\tanh(mH)}{mH},\ m=\sqrt{\frac{h_{fs}a_{fs}}{(1-\varepsilon)k_s}} \quad (14)$$

The $L_{in}$ is the length of channel upstream from the pin-fin heat sink.

# 3. EXPERIMENTAL METHOD

In order to validate the present numerical simulation, an experimental setup to investigate the smoke flow field and heat transfer characteristics of 3 × 3 pin-fin heat-sink array in a rectangular channel is built. The experimental setup displayed inFigure 4. The relevant experimental setup, procedure and heat-loss estimation are similar with those reported in the authors' previous work [28]. The test section is made of a straightener and a rectangular channel with the 20 mm-thick Plexiglas cover and 20 mm-thick Bakelite bottom base. The configuration and dimensions of test section and positions of measured points are shown in Figure 5. In total, 9 pin-fin heat sinks are situated in the rectangular channel with $X_T^*(=S_T^*/L)=1.5$ and $X_L^*(=S_L^*/L)=2.0$. The Sample 5 heat sink shown in Table 1 is selected for the experiment herein. These pin-fin heat sinks are made of aluminum 6061. Each heat sink has a corresponding film heater fixed on its bottom surface. The wall heat flux supplied to each film heater from the electronic power is 0.49 W/cm² for 143 liter/min flow rate. Fifth thermocouple (Model: T-TT-30, OMEGA Engineering Inc., Stamford, CT, USA) for measuring the wall temperatures are embedded on the bottom surfaces of the center-column heat sinks. The ambient temperature and the air temperature at the channel inlet are also monitored by other thermocouples. Besides, local air velocity is measured by a hot-wire anemometer (Model: IEL VA-20, I Denshi Giken Co., Ltd., Ebina, Japan). Uncertainties in parameters are estimated by using the root-sum-square method of Moffat [29]. The experimental data obtained herein revealed that the uncertainties in the Reynolds number and the Nusselt number (as shown in Equations (10), (12) and (14)) were ±5.91% and ±6.83%, respectively.

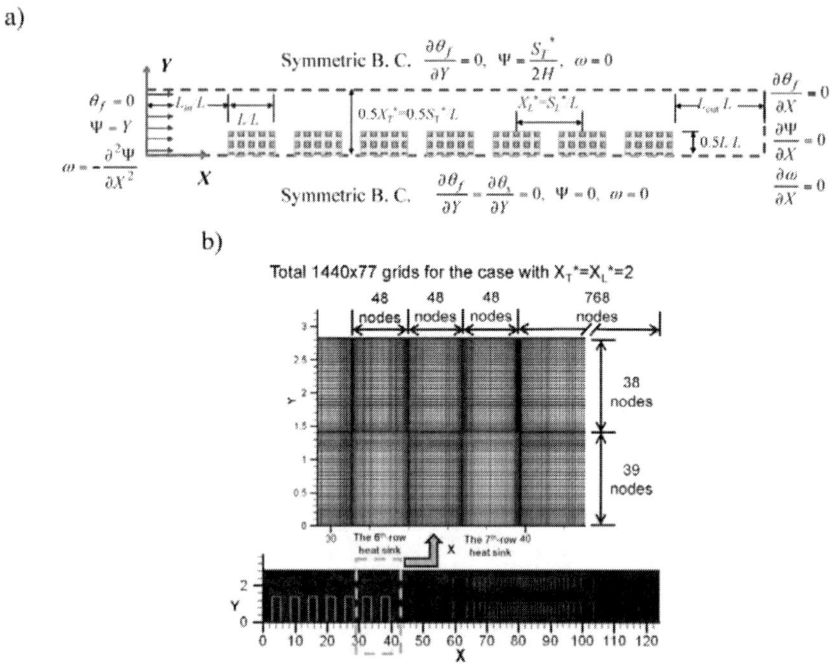


**Figure 3.** Computational configuration (the direction of pin fins is pointing the $Z$-axis vertical to paper surface): (**a**) boundary conditions and (**b**) mesh (the scales of $X$-axis and $Y$-axis are different).

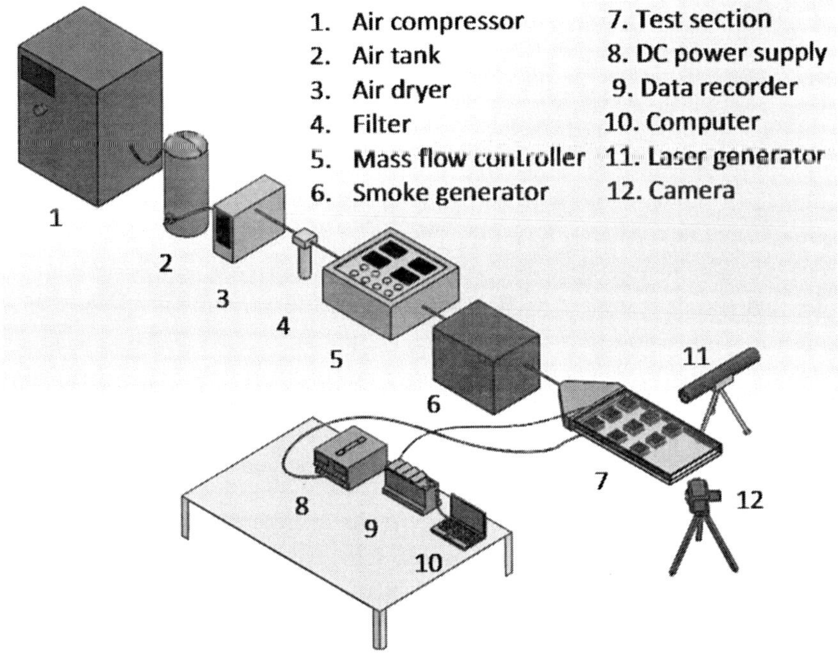


**Figure 4.** Experimental setup.

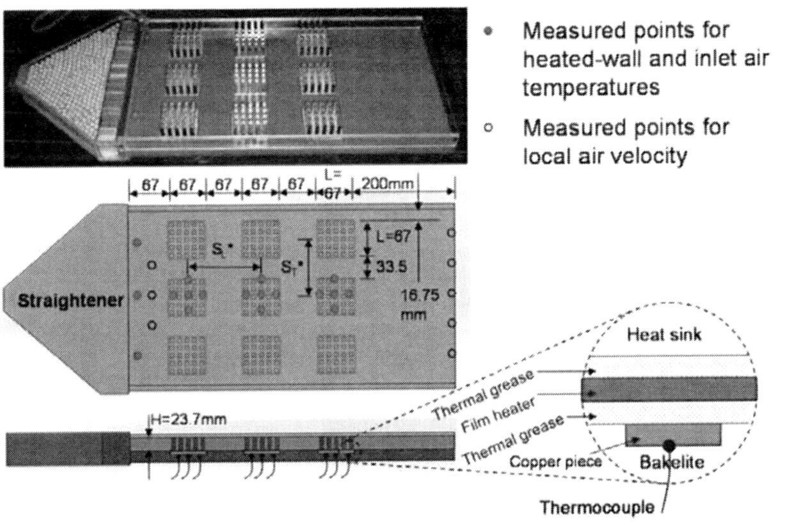


**Figure 5.** Configuration and dimensions of test section and positions of measured points.

# 4. RESULTS AND DISCUSSION

## 4.1. Flow Behaviors

Figure 6 shows the flow visualizations of 3 × 3 pin-fin heat-sink array in a rectangular channel in the A-A and B-B sectional views of laser sheet (top-view surface tangential to inside of heat-sink array). From the visualizations of initial, interim and later phases, it can be found that the smoke flows mainly through each longitudinal passage between in-line heat sinks; minority plume goes through the heat sinks. The flow visualizations of various top views give the similar flow behaviors. Of course, the real flow pattern in the present rectangular channel with 3 × 3 pin-fin heat-sink array cannot be two-dimensional completely due to the end-wall effect. However, by comparing with the variations of fluid flow in the $x$ and $y$ directions, the change of fluid flow in the $z$ direction is small and can be ignored. In other words, the present 2D simulations actually express the mean results of different cross sections along the $z$-axis. Figure 7 depicts the measured values of air velocity at the inlet and outlet of the center column of 3 × 3 heat-sink array. It generally agrees to the numerical simulation.

Figure 8 shows the simulation result of streamline and velocity vector of heat-sink array in different models with Reynolds number $Re = 500$ and relative distances between heat sinks $X_T^*(= S_T^*/L) = X_L^*(= S_L^*/L) = 1.5$. It is observed that when the uniform air flows into the heat sink array, a part of air flows through the side-bypass gap between transverse adjacent heat sinks, the other part of air flows into the heat sink. In addition, due to the flow resistance, the air flow through the heat sink leaks from the side of heat sink to the side-bypass gap continuously. Therefore, the by-pass gap in more downstream location contains faster air flow; this phenomenon is more obvious in the heat sink array with bigger flow resistance. For example, the heat sink with smaller porosity or larger number of pin fins has this feature. For the heat sink array with bigger flow resistance, there are vortices formed earlier behind the last row of heat-sink array. Generally speaking, the heat-sink array with bigger flow resistance has more bypass air flow, which is adverse to heat transfer. The heat transfer of the downstream heat sink is thus worsened.

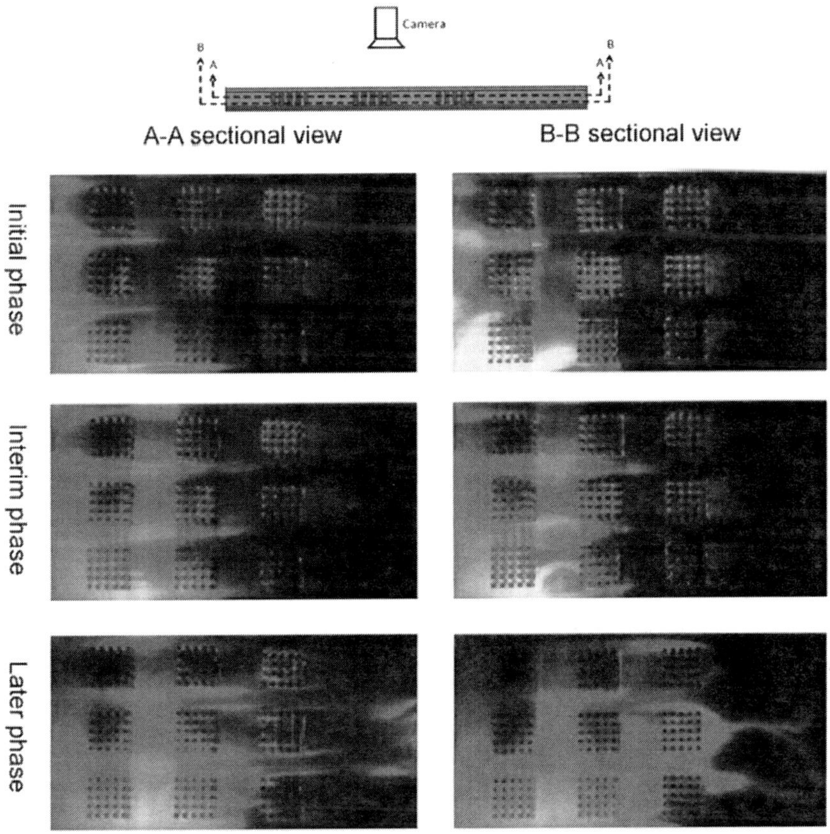


**Figure 6.** Flow visualization images. ($Re$ = 52; 15 liter/min).

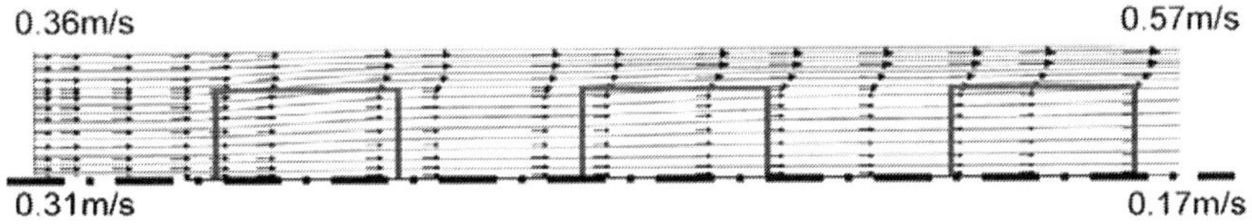


**Figure 7.** Measured values of air velocity at the center column of 3 × 3 heat-sink array. ($Re$ = 500).

Figure 9 shows the influence of variation in the $X_T^*$ and $X_L^*$ values on the streamline and velocity vector taking Sample 1 heat sink array as an example. According to Figure 9a–c, when the $X_L^*$ value is fixed at 2, a larger $X_T^*$ represents larger relative transverse distance between heat sinks, *i.e.*, larger lateral by-pass gap. In other words, a majority of air flow will pass through this gap, so that the air flow entering the heat sink decreases greatly (see Figure 9c). A recirculating flow is even formed between adjacent front and back rows of heat sinks in the downstream region, which is very adverse to heat transfer. According to Figure 9c–e, in the arrangement of $X_T^*$ value fixed at 2 but different $X_L^*$ values, the air velocity in the lateral by-pass gap is almost the same. This suggests that although the relative longitudinal distance between adjacent front and back rows of heat sinks is changed, the bypass air flow passing by the heat sink is not influenced (*i.e.*, total air flow amount through heat sinks has not changed).

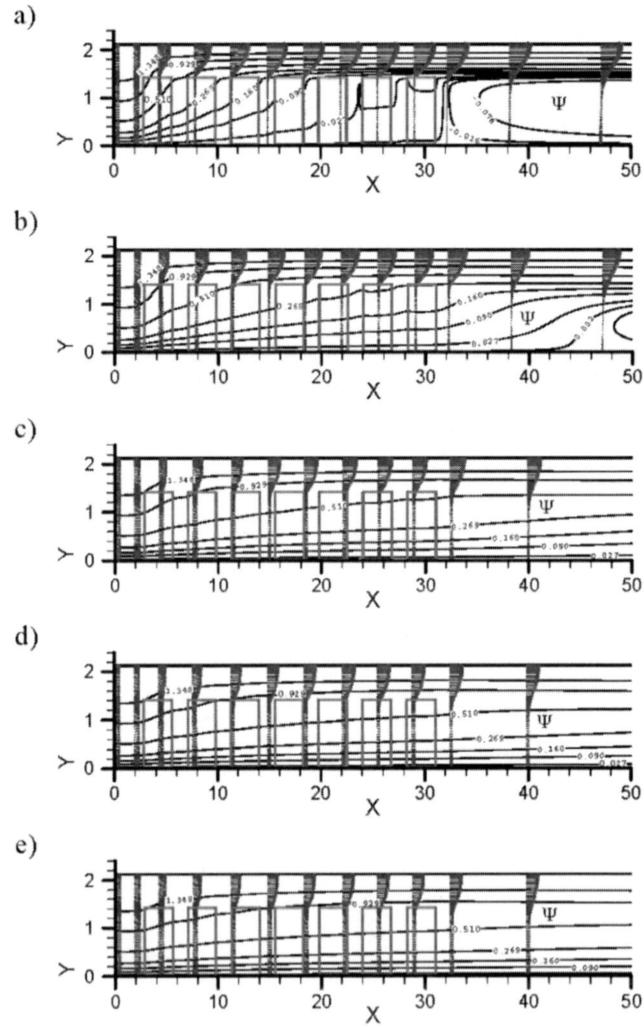


**Figure 8.** Streamlines and velocity vectors for various test samples with $X_T^* = X_L^* = 1.5$ (the scales of $X$-axis and $Y$-axis are different): (**a**) Sample 1; (**b**) Sample 2; (**c**) Sample 3; (**d**) Sample 4; and (**e**) Sample 5.

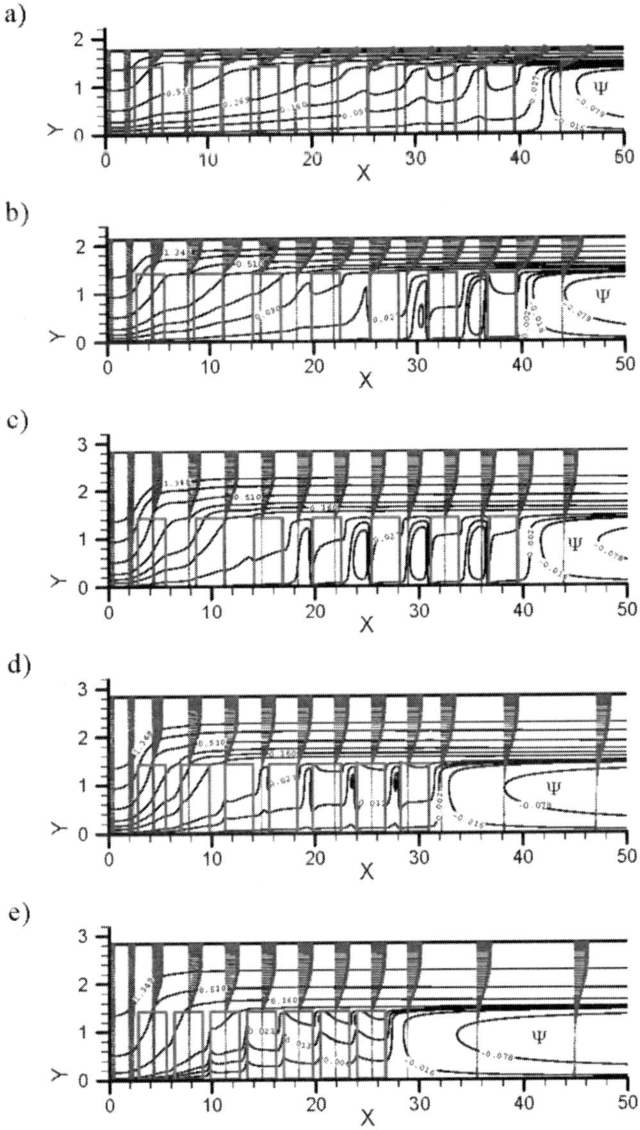


**Figure 9.** Streamlines and velocity vectors for test Sample 1 with various values of $X_T^*$ and $X_L^*$ (the scales of $X$-axis and $Y$-axis are different): (**a**) $X_T^* = 1.25$; $X_L^* = 2$; (**b**) $X_T^* = 1.5$; $X_L^* = 2$; (**c**) $X_T^* = 2$; $X_L^* = 2$; (**d**) $X_T^* = 2$; $X_L^* = 1.5$; and (**e**) $X_T^* = 2$; $X_L^* = 1.25$.

Figure 10 shows the average dimensionless velocity of air flow into windward side ($U_w$) of each row of heat sink array in the in-line arrangement formed of different heat sinks when $Re = 500$. As the flow resistance of passages between pin fins of heat sink is higher than the lateral by-pass gap between heat sinks, the air flow leaks into the by-pass gap continuously after it enters the first row of heat sink array. The leakage flow of upstream heat sink is higher than downstream heat sink. The heat sink with higher flow resistance (*i.e.*, heat sink with lower porosity or larger number of pin fins) has less air inflow and more leakage flow. The simulation results also indicate that the influence of current relative longitudinal distance ($X_L^*$) between heat sinks on the air flow entering the heat sink and the leakage flow can be neglected. However, when the relative transverse distance ($X_T^*$) between heat sinks is large, the air flow entering the heat sink is low, and the leakage flow is high, thus reducing the forced convection heat transfer capability of heat sink array.

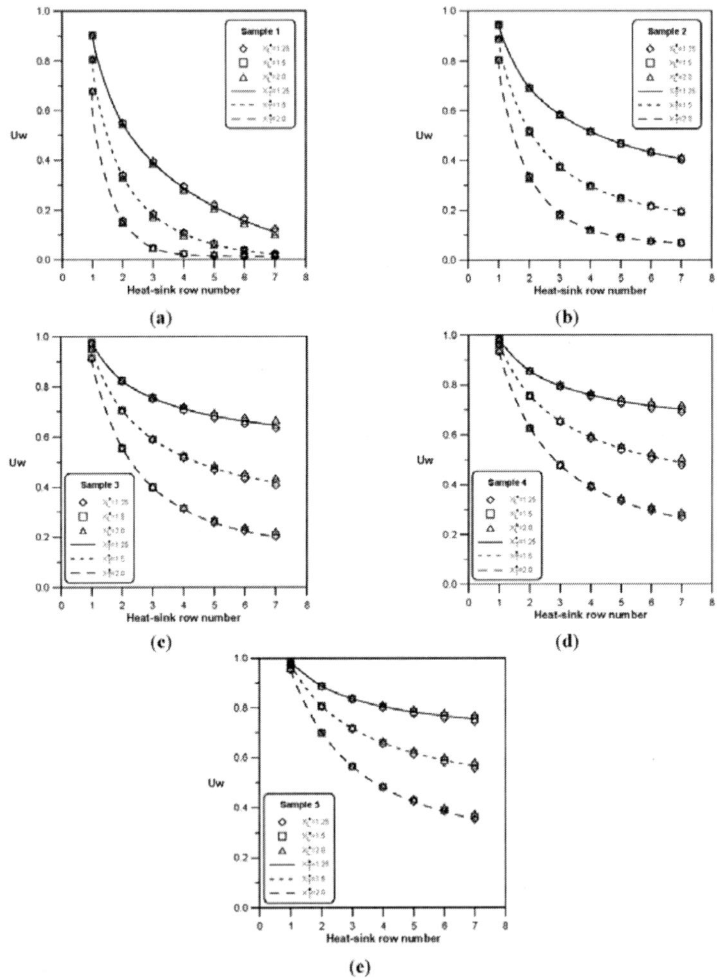


**Figure 10.** Dimensionless velocities of fluid flow through windward sides of various heat sinks: (**a**) Sample 1; (**b**) Sample 2; (**c**) Sample 3; (**d**) Sample 4; and (**e**) Sample 5.

Figure 11 shows the dimensionless pressure drop when the air flow at $Re = 500$ passes through different heat sink arrays with different $X_T^*$ and $X_L^*$ values. The result indicates that changing $X_L^*$ did not affect the dimensionless pressure drop, but increasing $X_T^*$ would reduce dimensionless pressure drop significantly. Because an increasing $X_T^*$ increases the air flow passing by the heat sink and reduces the air flow through the heat sink, the overall dimensionless pressure drop decreases. In addition, the array composed of heat sinks with higher flow resistance (*i.e.*, heat sink with lower porosity or larger number of pin fins) has a larger dimensionless pressure drop.

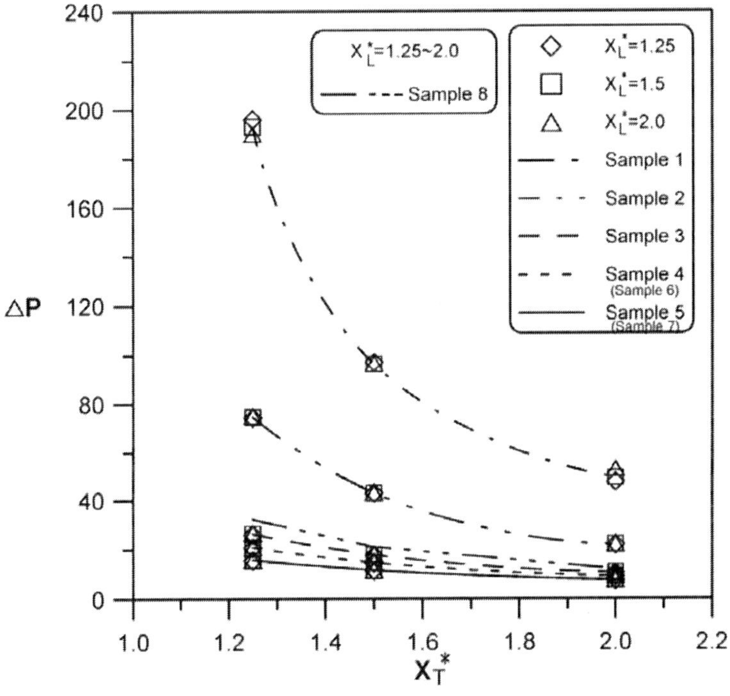


**Figure 11.** Dimensionless pressure drops for different heat-sink arrays with various $X_T^*$ and $X_L^*$.

### 4.2. Thermal Characteristics

Figure 12 shows the dimensionless temperature contours of different heat sink arrays when Reynolds number $Re = 500$ and $X_T^* = X_L^* = 1.5$. First, the dimensionless fluid temperature and dimensionless solid temperature in the heat sink area are compared, there is significant difference, the dimensionless solid temperature is higher than the dimensionless fluid temperature, meaning in the present numerical simulation, the solid and fluid have not yet reached local thermal equilibrium. Therefore, the two-equation model meets the thermophysics in the current test system better than one-equation model. Secondly, the dimensionless solid temperature distributions of different heat sink arrays are compared (as shown inFigure 12a–c). It is observed that among the heat sink arrays with the same number of pin fins (Sample 1, Sample 2 and Sample 3), the heat sink array with higher porosity has lower overall dimensionless solid temperature, suggesting that the heat is transferred to the fluid effectively. This is because in the present cooling system, the heat considering fin efficiency radiates in each unit volume of heat sink, and then it is transmitted to the fluid by convection heat transfer. Therefore, it is basically conjugate heat transfer of thermal conduction and convection. When the porosity of heat sink is low, the upward thermal conduction capability of pin fins is better and the fin efficiency is good, thus contributing to overall heat transfer (Equation (13)). However, as shown in Figure 8, the heat sink array with lower porosity generates higher flow resistance, so that less fluid flows into the heat sink, thereby reducing the convection heat transfer of heat sink. Therefore, the heat sink Sample 1 with the smallest porosity accumulates heat in the pin-fins, so that it has the highest overall dimensionless solid temperature. The heat sink Sample 3 with the largest porosity has more cold inflow, and the dimensionless solid temperature is lower. In addition, the heat sink arrays with the same porosity (Sample 3, Sample 4 and Sample 5) are compared, as shown in Figure 12c–e. The overall dimensionless solid temperature of the Sample 3 heat sink array with more pin fins is lower than that of the Sample 4 and Sample 5 heat sink arrays with fewer pin fins, especially in the first row and the second row of heat sink array. In other words, Sample 3 with more pin fins has more heat carried away, so it is unlikely to accumulate heat. However, according to the conjugate heat transfer principle stated in Figure 12a–c, as Sample 3, Sample 4 and Sample 5 have the same porosity, they have the same upward thermal conduction capability of pin fins. However, according to Figure 10, the Sample 5 heat sink array with fewer pin fins has lower flow resistance, and it is supposed to have higher air inflow. The reason that the dimensionless solid temperature of heat sink is slightly higher than that of Sample 3 is explained below. First, the heat exchange area in unit volume of Sample 3 with more pin fins is far larger than that of Sample 4 and Sample 5 (see Table 1). This makes up the air inflow during convection heat transfer. Secondly, as shown in Equations (7) and (8), Kim *et al.* [24] indicated that the heat sink with smaller pin-fin diameter has larger convection heat transfer coefficient. Therefore, the dimensionless solid temperature of Sample 3 heat sink array with smaller pin-fin diameter is slightly lower than that of Sample 5.

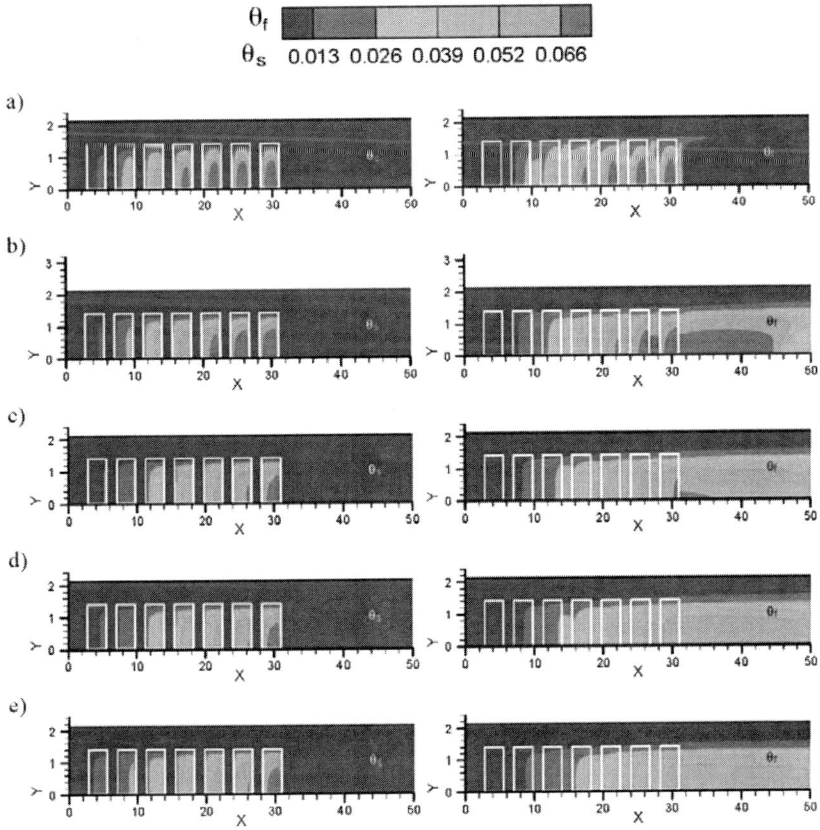


**Figure 12.** Dimensionless temperature contours for various test samples with $X_T^* = X_L^* = 1.5$ (the scales of $X$-axis and $Y$-axis are different): (**a**) Sample 1; (**b**) Sample 2; (**c**) Sample 3; (**d**) Sample 4; and (**e**) Sample 5.

Figure 13 shows the influence of $X_T^*$ and $X_L^*$ on the dimensionless temperature contours taking Sample 1 heat sink array as an example when Reynolds number $Re = 500$. At fixed $X_L^* = 2$ and variable $X_T^* = 1.25$, 1.5 and 2, the dimensionless solid temperature of heat sink increases with $X_T^*$. As the increase of $X_T^*$ amplifies the bypass effect of air flow, the air flow entering the heat sink decreases, and the heat removed decreases, so the solid part of heat sink accumulates heat and has higher temperature. Relatively, when the $X_T^*$ is larger, as the air flowing into the heat sink is lower, the fluid has larger temperature rise though the pin fins transfer less heat to the fluid. This result is reflected in the dimensionless fluid temperature contours. At fixed $X_T^* = 2$ and variable $X_L^* = 1.25$, 1.5 and 2, the dimensionless solid temperature and dimensionless fluid temperature in the same position of various heat sinks have not varied with $X_L^*$, meaning $X_L^*$ is not a sensitive parameter to the overall temperature distribution.

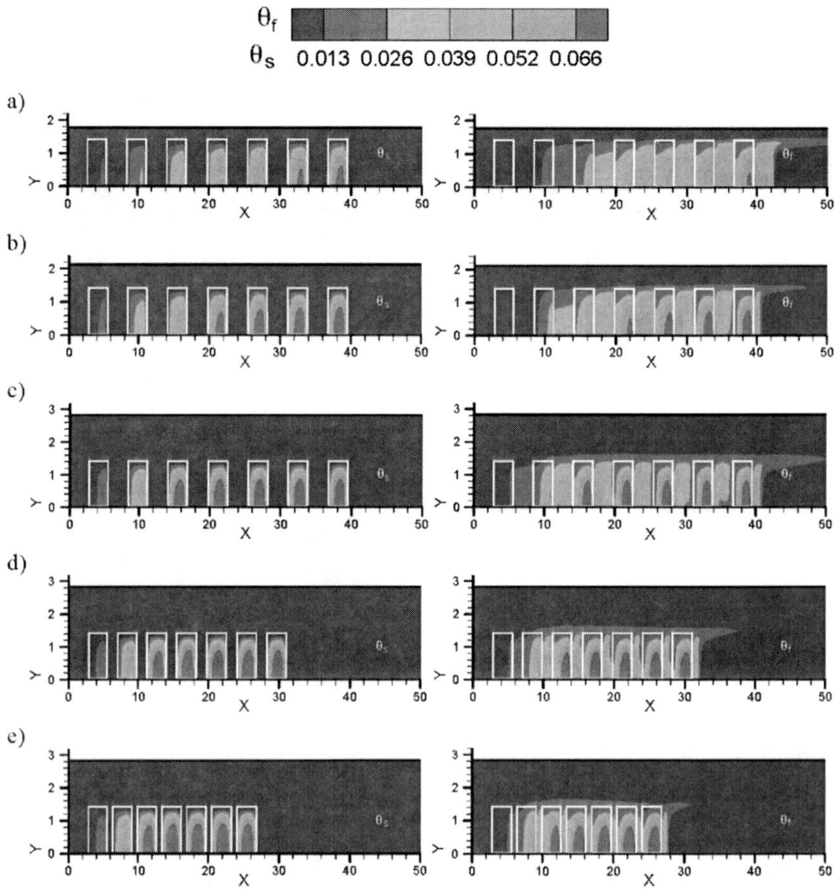


**Figure 13.** Dimensionless temperature contours for test Sample 1 with various values of $X_T^*$ and $X_L^*$ (the scales of X-axis and Y-axis are different): (**a**) $X_T^* = 1.25$; $X_L^* = 2$; (**b**)$X_T^* = 1.5$; $X_L^* = 2$; (**c**) $X_T^* = 2$; $X_L^* = 2$; (**d**) $X_T^* = 2$; $X_L^* = 1.5$; and (**e**) $X_T^* = 2$; $X_L^* = 1.25$.

Figure 14 shows the Nusselt numbers of various rows of heat sink array at different $X_T^*$ and $X_L^*$ values of Samples 1–5 when Reynolds number $Re = 500$. The number of pin fins of Sample 3 heat sink is 9 × 9, the porosity is 0.75. This heat sink array has a higher Nusselt number among all the heat sink arrays, and the Nusselt number of various heat sink arrays increases with decreasing $X_T^*$, but not varies with $X_L^*$. This result conforms to the dimensionless temperature contours inFigure 12 and Figure 13.

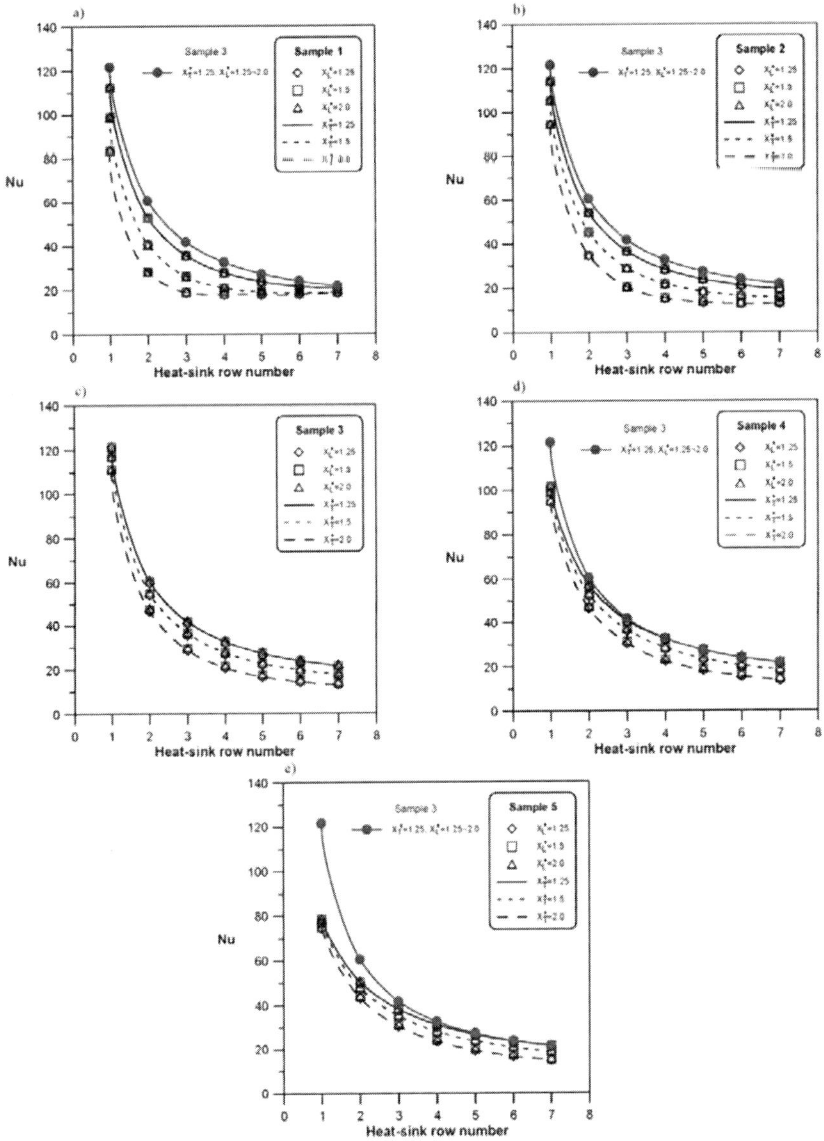


**Figure 14.** Nusselt number distributions along the main stream direction for various heat-sink arrays: (**a**) Sample 1; (**b**) Sample 2; (**c**) Sample 3; (**d**) Sample 4; and (**e**) Sample 5.

Figure 15 shows the relationship of average Nusselt number to the porosity and number of pin fins of heat sink array when Reynolds number $Re = 500$. In order to clarify the influences of porosity and number of pin fins on the average Nusselt number, the heat transfer characteristics of heat sink arrays with higher porosity (Sample 6 and Sample 7, see Table 1) and heat sink array with more pin fins (Sample 8, see Table 1) were also simulated. Figure 15a indicates that when the porosity of heat sink with 9 × 9 pin-fins increases from 0.358 to 0.556, the average Nusselt number changes slightly. When the porosity increases from 0.556 to 0.750, the average Nusselt number is increased significantly by 11.7% (for cases with $X_T^* = 1.25$)–24.8% (for cases with $X_T^* = 2$). However, when the porosity increases continuously from 0.750, the average Nusselt number decreases obviously. This is because the overall heat exchange area decreases with the pin diameter of pin-fin heat sink, which is adverse to convection heat transfer, and the upward thermal conduction capability of pin-fins declines, thus reducing the fin efficiency. As a result, the average Nusselt number decreases on the contrary. Figure 11b indicates that when the porosity is fixed at 0.75, increasing the number of pin fins continuously can increase the average Nusselt number. In order to keep the fixed porosity, the pin diameter must be reduced while the number of pin fins is increased; at this point, the overall heat transfer area still increases with the number of pin fins, which is advantageous to the convection heat transfer. According to empirical Equations (7) and (8) suggested by Kim *et al.* [24], the decrease in pin diameter increases the convection heat transfer coefficient. Therefore, the average Nusselt number increases with the number of pin fins. However, at the Reynolds number $Re = 500$, when the number of pin fins is large, the average Nusselt number increases with the number of pin fins slowly. The pressure drop of the heat sink array increases with the number of pin fins, thus preventing the air flow from entering the heat sink effectively for cooling.

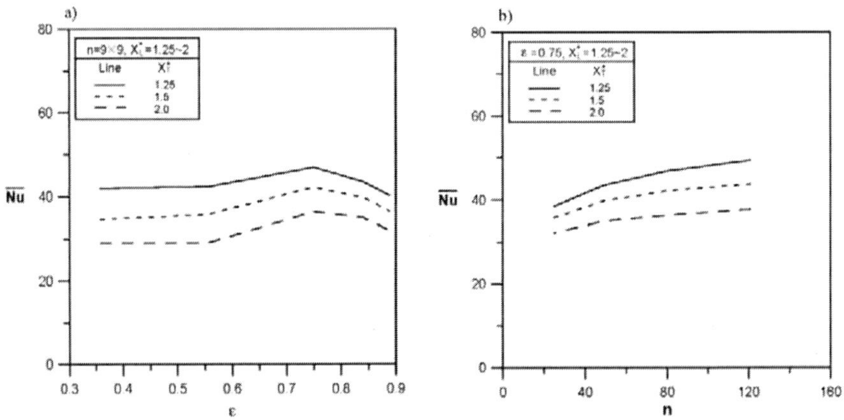


**Figure 15.** Average Nusselt numbers of pin-fin heat-sink array: (**a**) effect of heat-sink porosity and (**b**) effect of pin-fin numbers.

### 4.3. Semi-Empirical Correlation

Figure 16 illustrates the energy-balance schematic diagram for the air flow heated via the heat sink in this issue. The inlet air flow has the uniform temperature ($T_i$) and velocity ($u_i$). According to Figure 10, the average velocity of air flow into windward side ($u_{w,k}$, *i.e.*, the average velocity of flow-in air flow) of each heat sink decreases along the downward stream. In other words, a part of flow-in air continuously results in leakage air at each downward heat sink to join the bypass flow. Therefore, the energy balance for the flow-in air heated via each heat sink can be expressed as follows:

$$\rho \cdot u_i \cdot \left( \frac{U_{w,k} + U_{w,k+1}}{2} \right) \cdot H \cdot L \cdot C_p \cdot \left( T_{o,k} - T_{i,k} \right) = Q_{w,k}^{*} \qquad (15)$$

$$T_{b,k} = \frac{T_{o,k} + T_{i,k}}{2}, \quad U_{w,k} = \frac{u_{w,k}}{u_i} \qquad (16)$$

where $T_{i,k}$ and $T_{o,k}$ are the average air temperatures at the inlet and outlet faces of the *k*th heat sink, respectively; $T_{b,k}$ is the bulk mean temperature of the air through the *k*th heat sink; and $Q_{w,k}^{*}$ is the heat transferred to the air flow through the *k*th heat sink. Because that the leakage-flow effect, the mean velocity of air flow through the *k*th heat sink is expressed as $u_i(U_{w,k}+ U_{w,k+1})/2$. Besides, the $Q_{w,k}^{*}$ is less than the total heat ($Q_{w,k}$) dissipated from the *k*th heat sink since that the heat carried by the leakage air does not join to result in the $T_{o,k}$.

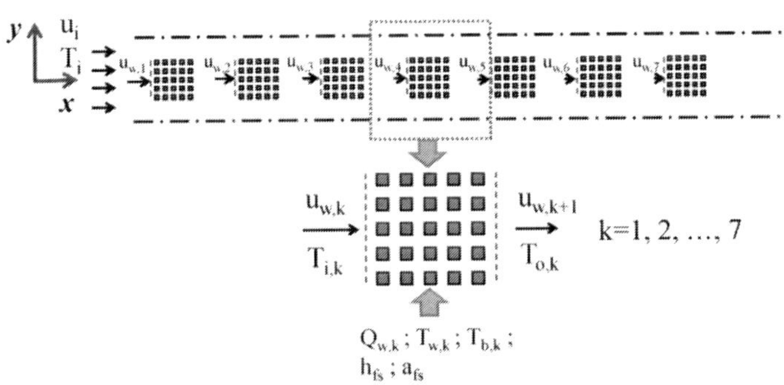


**Figure 16.** Energy-balance schematic diagram for the air flow heated via the heat sink.

Equations (17) and (18) are the definitions of the average Nusselt number ($Nu_{i,k}$ and $Nu_{b,k}$) of the $k$th heat sink separately based on $T_{i,1}$ (*i.e.*, $T_i$) and $T_{b,k}$. Substitute Equations (17) and (18) into Equation (15), the relationship between $Nu_{i,k}$ and $Nu_{b,k}$can be written as Equation (19).

$$Nu_{i,k} = \frac{Q_{w,k} \cdot H}{(T_{w,k} - T_{i,1}) \cdot k_f \cdot L^2} \tag{17}$$

$$Nu_{b,k} = \frac{Q_{w,k} \cdot H}{(T_{w,k} - T_{b,k}) \cdot k_f \cdot L^2} \tag{18}$$

$$Nu_{i,k}^{-1} = Nu_{b,k}^{-1} + \sum_{1}^{k} \frac{Q_{w,k}^{*}}{Q_{w,k}} \cdot \frac{L}{2\,\mathrm{Re}(\frac{U_{w,k} - U_{w,k+1}}{2}) \cdot \mathrm{Pr} \cdot H} \tag{19}$$

Finally, a semi-empirical correlation (see Equation (20)) can be completed to be a closed form by combined with Equations (7) and (8) suggested by Kim *et al.* [24], the present numerical data of $U_{w,k}$ plotted in Figure 10, and the assumption of$Q_{w,k}^{*}/Q_{w,k} = U_{w,k+1}/U_{w,k}$. For the present numerical simulation, $C_1$ and $C_2$ are 0.363 and 0.542, respectively (see Equation (7)).

$$\begin{aligned} Nu_{b,k} &= \left(\frac{h_{fs} \cdot d}{k_f}\right) \cdot \left(\frac{H}{d}\right) \cdot \left(\frac{a_{fs} \cdot L^2 \cdot H}{L^2}\right) \\ &= \left[\varepsilon \cdot C_1 \left(\mathrm{Re} \cdot \frac{U_{w,k} + U_{w,k+1}}{2} \cdot \frac{d}{H} \cdot \frac{S_T}{S_T - d}\right)^{C2}\right] \cdot \left(\frac{H}{d}\right) \cdot (a_{fs} \cdot H) \end{aligned} \tag{20}$$

$$\frac{Q_{w,k}^{*}}{Q_{w,k}} = \frac{U_{w,k+1}}{U_{w,k}} \tag{21}$$

The comparison results of the present numerical Nusselt number with the semi-empirical predictions and the experimental data for Sample 3 and Sample 5 heat-sink arrays with $Re = 500$ are plotted in Figure 17. The predicted $Nu$values generally agree with those numerical data. In addition, the numerical data and the predictions of Equation (20) also agree with the typical experimental results, demonstrating the correctness of the present study.

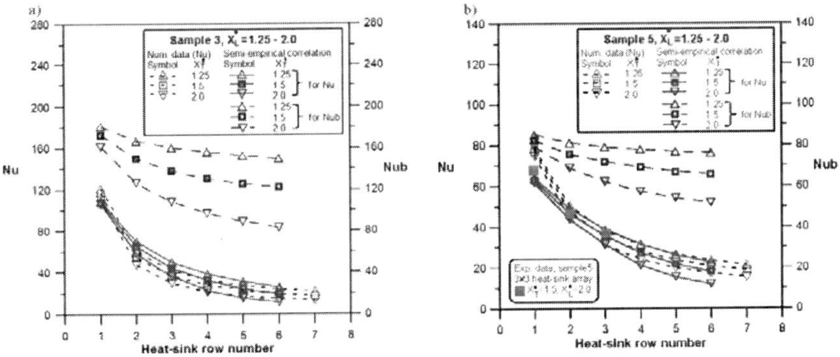


**Figure 17.** Comparison results of Nusselt number with semi-empirical predictions and experimental data: (**a**) Sample 3 and (**b**) Sample 5.

## 5. CONCLUSIONS

This work successfully proposed a porous model to numerically investigate the fluid flow and heat transfer behaviors of the pin-fin heat-sink array situated in a rectangular channel in an in-line arrangement with various relative center-to-center longitudinal and transverse distances between heat sinks. The present porous approach used the Brinkman-Forchheimer model for fluid flow and two-equation model for heat transfer. The pin-fin heat sinks with various porosities and pin-fin numbers were employed. Some major conclusions are summarized as follows:

(1) By comparing with downstream heat sink, more air inflow leaks from the upstream heat sink. The heat sink with lower porosity or larger number of pin fins has less air inflow and more leakage flow. The air flow entering the heat sink is low and the leakage flow is high as the relative transverse distance ($X_T^*$) between heat sinks is large. However, the effect of the relative longitudinal distance ($X_L^*$) between heat sinks on the air flow entering the heat sink and the

leakage flow is insignificant. Changing $X_L^*$ did not affect the dimensionless pressure drop, but decreasing $X_T^*$ would increase dimensionless pressure drop significantly. In addition, the array composed of heat sinks with lower porosity or larger number of pin fins has a larger dimensionless pressure drop.

(2) The dimensionless solid temperature is higher than the dimensionless fluid temperature, meaning the solid and fluid have not yet reached local thermal equilibrium in the present numerical simulation. Therefore, the two-equation model used for this issue is necessary. The Nusselt number of various heat-sink arrays increases with decreasing $X_T^*$, but does not vary with $X_L^*$. When the porosity of the typical heat sink increases from 0.358 to 0.556, the average Nusselt number changes slightly. However, the average Nusselt number is increased significantly by 11.7%–24.8% when the porosity increases from 0.556 to 0.750, and then drops obviously when the porosity exceeds 0.750. Besides, increasing the number of pin fins continuously can increase the average Nusselt number. However, when the number of pin fins is large, the average Nusselt number increases with the number of pin fins slowly.

(3) The present numerical simulations agree with the smoke flow visualization and heat transfer measurement of the typical experiment, proving the correctness of the present study. Finally, a semi-empirical correlation of Nusselt number for each heat sink in the present heat-sink array is proposed.

## ACKNOWLEDGMENTS

The authors would like to thank the Ministry of Science and Technology of the Republic of China for financially supporting this research under Contract Nos. MOST 103-2632-E-270-001-MY3, MOST 104-2221-E-270-002, MOST 104-2622-E-270-005-CC3 and MOST 104-2221-E-270-005.

## AUTHOR CONTRIBUTIONS

Tzer-Ming Jeng led this work and conducted the numerical simulation, experimental measurement and theoretical analysis; and Sheng-Chung Tzeng participated discussions and was responsible for correspondence.

## REFERENCES

1. Tahat, M.; Kodah, Z.H.; Jarrah, B.A.; Probert, S.D. Heat transfers from pin-fin arrays experiencing forced convection.*Appl. Energy* **2000**, *67*, 419–442.

2. Chang, S.W.; Yang, T.L.; Huang, C.C.; Chiang, K.F. Endwall heat transfer and pressure drop in rectangular channels with attached and detached circular pin-fin array. *Int. J. Heat Mass Transfer* **2008**, *51*, 5247–5259.

3. Zhuan, R.; Wang, W. Boiling heat transfer characteristics in a microchannel array heat sink with low mass flow rate.*Appl. Therm. Eng.* **2013**, *51*, 65–74.

4. Jeng, T.M.; Tzeng, S.C.; Liu, C.H.; Lin, D.J.; Yang, B.J. Design and Manufacture of an Energy-saving LED Lantern with Paper-cut Figure Projection Function. *Smart Sci.* **2014**, *2*, 13–19.

5. Ritchey, S.N.; Weibel, J.A.; Garimella, S.V. Local measurement of flow boiling heat transfer in an array of non-uniformly heated microchannels. *Int. J. Heat Mass Transfer* **2014**, *71*, 206–216.

6. Taji, S.G.; Parishwad, G.V.; Sane, N.K. Enhanced performance of horizontal rectangular fin array heat sink using assisting mode of mixed convection. *Int. J. Heat Mass Transfer* **2014**, *72*, 250–259.

7. El-Sayed, S.A.; Mohamed, S.M.; Abdel-latif, A.M.; Abouda, A.E. Investigation of turbulent heat transfer and fluid flow in longitudinal rectangular-fin arrays of different geometries and shrouded fin array. *Exp. Therm. Fluid Sci.* **2002**, *26*, 879–900.

8. Moores, K.A.; Kim, J.; Joshi, Y.K. Heat transfer and fluid flow in shrouded pin fin arrays with and without tip clearance. *Int. J. Heat Mass Transfer* **2009**, *52*, 5978–5989.

9. Dogan, M.; Sivrioglu, M. Experimental and numerical investigation of clearance gap effects on laminar mixed convection heat transfer from fin array in a horizontal channel—A conjugate analysis. *Appl. Therm. Eng.* **2012**, *40*, 102–113.

10. Zhang, Y.; Liu, J.; Chong, D.; Yan, J. Experimental investigation on the heat transfer and flow performances of the fin array with shield in bypass. *Int. J. Heat Mass Transfer* **2013**, *56*, 674–682.

11. Roth, R.; Lenk, G.; Cobry, K.; Woias, P. Heat transfer in freestanding microchannels with in-line and staggered pin fin structures with clearance. *Int. J. Heat Mass Transfer* **2013**, *67*, 1–15.

12. Mei, D.; Lou, X.; Qian, M.; Yao, Z.; Liang, L.; Chen, Z. Effect of tip clearance on the heat transfer and pressure drop performance in the micro-reactor with micro-pin-fin arrays at low Reynolds number. *Int. J. Heat Mass Transfer* **2014**,*70*, 709–718.

13. Morega, A.M.; Bejan, A.; Lee, S.W. Free stream cooling of a stack of parallel plates. *Int. J. Heat Mass Transfer* **1995**, *38*, 519–531.

14. Narasimhan, S.; Majdalani, J. Compact heat sink simulations in both forced and natural convection flows. In Proceedings of the 8th AIAA/ASME Joint Thermophysics and Heat Transfer Conference, St. Louis, MI, USA, 24–26 June 2002; pp. 24–26.

15. Kim, S.J.; Kim, D. Forced convection in microstructures for electronic equipment cooling. *J. Heat Transfer* **1999**, *121*, 639–645.

16. Kim, D.; Kim, S.J. Compact modeling of fluid flow and heat transfer in straight fin heat sinks. *J. Electron. Packag.* **2004**,*126*, 247–255.

17. Yu, E.; Joshi, Y. Heat transfer enhancement from enclosed discrete components using pin-fin heat sinks. *Int. J. Heat Mass Transfer* **2002**, *45*, 4957–4966.

18. Jeng, T.M. A porous model for the square pin-fin heat sink situated in a rectangular channel with laminar side-bypass flow. *Int. J. Heat Mass Transfer* **2008**, *51*, 2214–2226.

19. Narasimhan, A.; Reddy, B.V.K. Laminar forced convection in a heat generating *bi*-disperse porous medium channel.*Int. J. Heat Mass Transfer* **2011**, *54*, 636–644.

20. Narasimhan, A.; Reddy, B.V.K.; Dutta, P. Thermal management using the *bi*-disperse porous medium approach. *Int. J. Heat Mass Transfer* **2012**, *55*, 538–546.

21. Feng, S.S.; Kim, T.; Lu, T.J. Numerical investigation of forced convection in pin/plate-fin heat sinks heated by impinging jet using porous medium approach. *Int. J. Numer. Methods Heat Fluid Flow* **2013**, *23*, 88–107.

22. Holman, J.P. *Heat Transfer*; Mcgraw-Hill: New York, NY, USA, 1990; pp. 32–33.

23. Jeng, T.M.; Tzeng, S.C. A semi-empirical model for estimating permeability and inertial coefficient of pin-fin heat sinks. *Int. J. Heat Mass Transfer* **2005**, *48*, 3140–3150.

24. Kim, D.; Kim, S.J.; Ortega, A. Compact modeling of fluid flow and heat transfer in pin fin heat sinks. *J. Electron. Packag.* **2004**, *126*, 342–350.

25. Roache, P.J. *Computational Fluid Dynamics*; Hermosa: Albuquerque, NM, USA, 1972; pp. 139–173.

26. Patankar, S.V. *Numerical Heat Transfer and Fluid Flow*; McGraw Hill-Hemisphere: New York, NY, USA, 1980; pp. 113–137.

27. Lee, S.L. A strong implicit solver for two-dimensional elliptic differential equations. *Numer. Heat Transfer Part B* **1989**,*16*, 161–178.

28. Jeng, T.M.; Tzeng, S.C.; Tang, F.Z. Fluid flow and heat transfer characteristics of the porous metallic heat sink with a conductive cylinder partially filled in a rectangular channel. *Int. J. Heat Mass Transfer* **2010**, *53*, 4216–4227.

29. Moffat, R.J. Contributions to the theory of single-sample uncertainty analysis. *J. Fluids Eng.* **1982**, *104*, 250–258.

# CHAPTER 4

# Analysis of Fuel Cell Driven Ground Source Heat Pump Systems in Community Buildings

Jae-Ki Byun [1], Dong-Hwa Jeong [2], Young-Don Choi [1,*] and Jong-Keun Shin [3]

[1] *Department of Mechanical Engineering, Korea University, Anam-dong, Seongbuk-gu, Seoul 136-713, Korea*
[2] *NSSS Division, KEPCO Engineering & Construction Company Inc., 1045 Daedeok-daero, Yuseong-gu, Daejeon 305-353, Korea*
[3] *Department of Automotive Engineering, Hanzhong University, San 119, Jiheung-dong, Donghae, Gangwon-do 240-713, Korea*

## ABSTRACT

In the present study, a fuel cell driven ground source heat pump (GSHP) system is applied in a community building and heat pump system performance is analyzed by computational methods. Conduction heat transfer between the brine pipe and ground is analyzed by TEACH code in order to predict the performance of the heat pump system. The predicted coefficient of performance (COP) of the heat pump system and the energy cost were compared with the variation of the location of the objective building, the water saturation rate of the soil, and the driven powers of the heat pump system. Compared to the late-night electricity driven system, a significant reduction of energy cost can be accomplished by employing the fuel cell driven heat pump system. This is due to the low cost of electricity production of the fuel cell system and to the application of the recovered waste heat generated during the electricity production process to the heating of the community building.

## KEYWORDS

ground source heat pump (GSHP); community building; fuel cell

## 1. INTRODUCTION

The development of renewable energy is becoming an urgent issue as global warming is accelerated due to resource depletion and environmental pollution. Geothermal energy, which is one of the renewable energy sources recently receiving attention, can be used to run an energy saving cooling and heating system. The system is called the geothermal energy heat pump system. The geothermal heat pump system has high capability of utilization in both the summer and winter. Its annual cost of maintenance is low compared to present cooling and heating apparatus. It is also environmentally friendly as it emits less household gases such as $CO_2$. However, despite these benefits, various limits are involved in the ground heat exchanger installation site and the initial installation cost.

On the other hand, the air source heat pump, which can perform heating and cooling alternatively with a simple directional diverter valve, is acknowledged as a typical energy saving device and is widely used in leading countries. Ground heat, which is used as the heat source for the heat pump, is superior to other heat sources in its ease of applicability and lower maintenance cost. However, the weakness of the air source heat pump is that the heat pump performance is greatly influenced by the outdoor air temperature condition [1].

Thus, in order to supplement the weaknesses of the ground source and air source heat pumps and to improve system performance, a hybrid ground source heat pump was developed. The ground heat exchanger and a hybrid heat pump with an assistant heat source are widely used in America and Europe. Generally, a hybrid ground source heat pump is a system that uses a ground heat exchanger along with an assistant heat source apparatus such as a cooling tower and boiler. The ground heat exchanger is utilized at a certain load, and the cooling tower is utilized simultaneously when the load is higher than that of the selected load. During heating, the boiler or solar panels used for general heating facilities are additionally installed in the geothermal heat pump and utilized as the assistant heat source. A system using a ground heat exchanger and an air source as the assistant heat source can also be considered as a hybrid geothermal heat pump system. These systems are developed to assist condenser heat radiation during cooling and evaporator heat absorption during heating, thus increasing system efficiency, resulting in a saving in electricity.

Previous studies include geothermal heat pump system design [2], heat pump system efficiency and economic feasibility analysis [3], hybrid geothermal system performance analysis [4] and design construction and construction criteria [5].

Healy *et al.* [3] analyzed a geothermal heat pump operation and performance characteristic through a numerical study of various parameters that effect geothermal heat pump performance. Stoecker [6] proposed a data processing model for a vapor compression system simulation analysis. The model is used to predict the total system performance based on the performance characteristics of each component (evaporator, condenser, compressor, and expansion devices).

Mei *et al.* [7] conducted a study on a high-efficiency ground coil heat exchanger, analyzing the water phase change and the ground temperature variation of the horizontal ground coil due to the seasonal change in the soil.

In America, research carried out on geothermal energy increased after the 1980s. In 1984, Bose [8,9] proposed that for a vertical ground heat exchanger it is valid to use an equivalent diameter and to consider the two tubes in the U-tube as one tube while deriving heat characteristics. Later, in 1988, Bose was also the first to describe soil temperature variation around a vertical heat exchanger using an analytical method. Kavanaugh [10] constructed a test module for measuring pipe pressure loss and various heat losses in a geothermal heat pump system pipe design, thus describing the actual operation and yielding design values that can be utilized. It should be noted that Bose [8,9] and Kavanaugh [2] proposed the design criteria in the geothermal heat pump system, and provided the data of various examples of installed heat pump systems, thus accelerating the use of the geothermal heat pump [11].

Lund [12] showed a 58% growth in geothermal heat pump system usage after 1995, and that the growth was 9.8% annually, and around five million heat pump systems in use in America and European countries by the beginning of 2000s. Lund *et al.* [13] estimated the increase in the numbers of geothermal heat pump system being installed. Around 4.5 million geothermal heat pump systems were installed in America in 2005, with an increase of 0.5 million each year. The increase rates of installation were as follows: encapsulated type vertical heat exchanger, 46%; encapsulated type horizontal heat exchanger, 38%; open type system, 15%; and overall installation, 10%. In 2005, the number of 12 kW capacity geothermal heat pump systems installed had doubled compared to that in 2000 and by 2010, about 15 million geothermal heat pump systems had been installed.

Zeng *et al.* [14] analyzed borehole heat transfer in a vertical ground heat exchanger while utilizing single U-tube and double U-tube units. Hepbasli *et al.*

[15] conducted experimental research on geothermal heat pump performance with an encapsulated vertical ground heat exchanger. Bi *et al.* [16] analyzed the temperature distribution around a ground heat exchanger through theory and experiment. İnallı *et al.* [17] analyzed seasonal ground temperature distribution and system performance while varying the burying depth of the ground heat exchanger in the horizontal geothermal heat pump system. Esen *et al.* [18] analyzed heat pump system performance of the horizontal geothermal heat pump system in cases where the heat exchanger burying depth was 1.0 m and 2.0 m.

In this study, a fuel cell system was combined with a heat pump system and hybrid heat pump system currently in use. The fuel cell system transforms chemical energy directly to electrical energy, thus yielding higher efficiency compared to other electricity generating technologies. The cost is less than that of late-night electricity due to its capability to generate electricity on its own. Also, the waste heat from the fuel cell operation can be stored in the heat storage tank and used as a heat source, thus increasing the heat pump system coefficient of performance (COP) during heating.

## 2. THE SELECTION OF COMMUNITY BUILDING FOR NUMERICAL ANALYSIS

In this study, the object of numerical analysis is an actual building currently in operation. The specifications of the object community building are shown in Table 1. This community building has lower load change compared to other buildings, and has a large annual load, which is economically beneficial. Thus the building was selected as the object for numerical analysis. Also, to increase the building area to 33,057 $m^2$ and to observe the energy consumption variation according to regional change, the object building was fixed while the building location was varied, including Munmak, Daejeon, and Busan during numerical analysis.

**Table 1.** Specifications of the object community building.

| Item | Specifications |
|---|---|
| Object community building | Oak Valley Resort |
| Location | Munmak, Gangwon-do, Korea |
| Area ($m^2$) | 27,028 |
| Number of persons | Maximum: 2,260 |

# 3. NUMERICAL ANALYSIS OF GEOTHERMAL HEAT PUMP

## 3.1. Ground Source Heat Pump System

A schematic of the fuel cell driven ground source heat pump system in this study, which is the object of the numerical analysis, is shown in Figure 1.

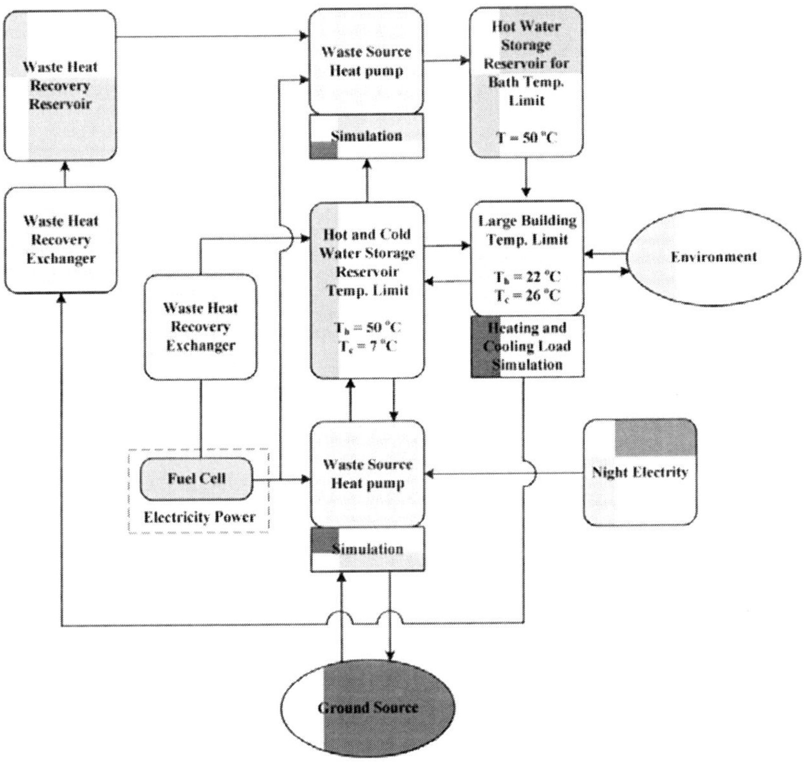


**Figure 1.** Schematic diagram of the fuel cell driven ground source heat pump heating and cooling system.

The heat pump system is a system combining a chilled water storage tank, a ground source heat pump, and a heat pump using bath water waste as a heat source. It consists of two heat pumps, while one heat pump can alternatively use the ground source and the waste heat as a source. In more detail, in the heat pump system, heat is stored in a chilled water storage tank with late-night electricity and heat from the fuel cell operation to be used during heating. Waste heat from the shower and bath water is stored in the hot water storage reservoir through the water source heat pump. Waste heat from the shower and bath water is stored in the hot water storage reservoir through the water source heat pump. In the case of cooling, the heat from the chilled water storage tank is pumped to the hot water storage reservoir using the water source heat pump so that both cold and hot heats are utilized simultaneously by a single heat pumping operation. This combined load system increases double the COP of the system. For numerical analysis of the heat pump system, the heating and cooling load calculation of the building, the heat pump capacity calculation, and the sub-program to predict ground temperature variation through the heat pump energy equilibrium equation are used to analyze two-dimensional heat transfer.

### 3.2. Fuel Cell System Waste Heat Recovery Apparatus

In a fuel cell system, waste heat recovery is more important than simply improving energy efficiency. According to the American DOE report [19], a polymer electrolyte membrane fuel cell system with 30% fuel-to-electricity efficiency can save up to 40% gas fuel when a waste heat recovery system with 65% efficiency is installed. In the actual fuel cell systems, water or air is used for cooling in order to maintain the optimal operation temperature, and pipelines are also already in the system. Thus, simple heat exchangers are added as shown in Figure 1 to store the heat generated during the fuel cell operation.

### 3.3. Dynamic Analysis of Community Building Load

In this study, the heat resistance-capacitance method [20,21] is used to calculate the heating and cooling load. In particular, the thermal-resistance capacitance method is widely used in the aviation industry. While a great deal of research has been carried out on the dynamic analysis of the building heating and cooling load such as DOE, BLAST, and TRNAYS [22,23,24], the program structure is complex and difficult to apply to building heating and cooling load calculation, and is also time-consuming. In this study, the entire building is considered in a one room equivalent model and a non-steady heat transfer equation is solved for the equivalent model to yield indoor temperature and wall temperature variation with time. Solar radiation load [25] due to solar radiation and permeability of solar radiation in the window are derived from the method proposed in ASHRAE Fundamentals [26]. Also, to obtain an infiltration air load, human body load, and other loads [27], numerical analysis was conducted while considering the object building's operation characteristics.

Lee *et al.* [28] applied thermal-resistance capacitance to office building energy consumption and the result showed the same accuracy as the dynamic analysis using the response factor method, with the error within 5% compared to the TRACLOAD code [22]. Lee [21] and Choi *et al.* [23] found that heat performance concurred well with the results from the experiments by using the thermal-resistance capacitance method, and they conducted a simulation of the continuous heat supply control in a high-rise apartment building.

In this study, a program is developed to easily analyze a heat pump heating and cooling system through dynamic heat load analysis using the thermal-resistance capacitance method.

### 3.4. Analysis of GSHP Cycle

In order to simulate the heat pump system, the cycles for the air source heat pump and water source heat pump must be simulated. In this study, the Engineering Equation Solver (EES) is used to simulate the heat pump cycle. Assumptions for the cycle simulation are as follows:

1) Temperature difference between condenser and water is ±5 °C in the heat pump heat storage tank or chilled cooling storage tank heat exchanger.
2) Temperature difference between the outdoors and the evaporator or condenser is ±10 °C for the evaporation and condensing in water-air heat exchange.
3) Procedures for simulating the heat pump with the EES code are as follows.
4) Determine the evaporator temperature and condenser temperature and then calculate the R-22 refrigerant heat pump system coefficient of performance (COP) with the EES program.
5) Ground brine temperature is altered automatically in response to the ground heat conduction condition until the total heat absorbed or emitted from the entire pipeline becomes 2RT. 2RT is the postulated heating or cooling capacity of a module ground pipe hole.

### 3.5. Ground Heat Exchange Analysis

Figure 2 shows a schematic of a ground source heat pump system in a geothermal heat pump system with a vertical ground heat exchanger. The single U-tube vertical ground heat exchanger consists of two tubes with flowing fluid, grout, and a borehole which is the boundary of the soil and grout. In the case of the heat pump cooling operation, discharged fluid flows through the pipeline and passes through the tube and grout, emitting heat to the soil, and returns to the heat pump. In reverse, during the heating operation, the fluid discharged from the heat pump flows through the tube and absorbs heat from the soil and returns to the heat pump.

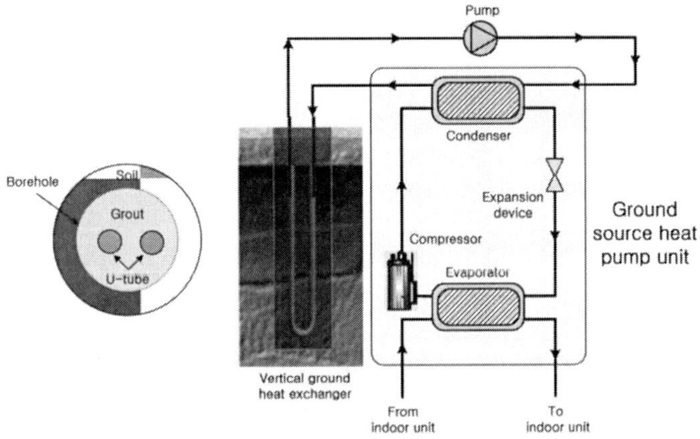


**Figure 2.** Schematic of a ground source heat pump system.

In this study, an equivalent tube is used to substitute the U-tubes, and the square calculation region is changed to a circular region, to simplify the simulation, as shown in Figure 3. Borehole pitch is assumed to be 4.5 m and borehole depth is 100 m. The properties of the underground soil and bentonite, which fills the borehole, were selected as shown in Table 2. Equation (1) is used to calculate the valid thermal conductivity in the soil. Water contained in the underground soil was assumed to be 35% and 75% [24].

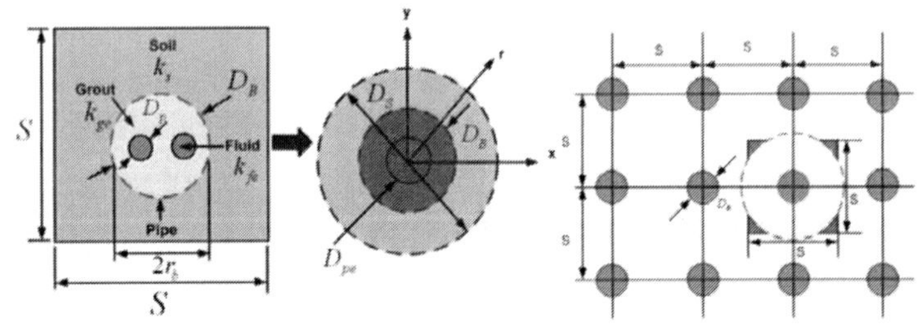


**Figure 3.** Transformation of a square domain to an equivalent concentric domain.

**Table 2.** Properties of the element of the ground heat exchanger.

| Material | Soil | Refrigerant (R-22) [27] | Grout (Bentonite) |
|---|---|---|---|
| $\rho$ (kg/m$^3$) | 2600 | 1133 | 1800 |
| $C_p$ (J/(kg K)) | 837 | 1332 | 500 |
| $k$ (W/(m K)) | 3.1 | 0.077 | 2.66 |

The valid thermal conductivity of the soil is:

$$\frac{1}{k_g} = \frac{1-\Phi}{k_s} + F \cdot \frac{\Phi}{(1-S_a)k_a + S_a k_w} \tag{1}$$

where $F = 0.5(\varepsilon + 0.014)^{0.25}$; $\varepsilon = S_a \cdot \Phi$. Here $k_s$, $k_a$, $k_w$ represent the thermal conductivity of soil particle, air, and water, respectively. $\Phi$ is underground soil porosity and $S_a$ is the degree of saturation. The single module for the ground heat exchanger is a square with the length of each side $S$ as shown in Figure 3. $D_B$ is the grout diameter and $D_p$ is the ground pipe diameter. The square region is substituted for the circular region with equivalent diameter $D_s$, while the two circular tubes with diameter $D_p$ are substituted for the equivalent tube with diameter $D_{pc}$. The equivalent diameter $D_{pc}$ is calculated as $2\sqrt{Dp}$, while the ground equivalent diameter $D_s$ is $2\sqrt{S}$.

Figure 4 shows the mesh generated to analyze ground temperature distribution. Fine mesh is used for regions where rapid temperature variation due to heat pump system operation is expected to occur. The TEACH code program was used for the ground heat transfer method [27] during simulation. A pipe depth of 100 m and a ground depth 150 m were analyzed, while the borehole pitch $S$ varied.

### 3.6. Governing Equation and Boundary Condition

The U-tubes are substituted for an equivalent tube as shown in Figure 3, and an assumption is made that the heat velocity originated from the equivalent tube and was transferred to the ground through the grout. The heat velocity has a positive value when the heat is transferred from a refrigerant tube to the ground, and has a negative value when heat is transferred from ground to refrigerant tube.

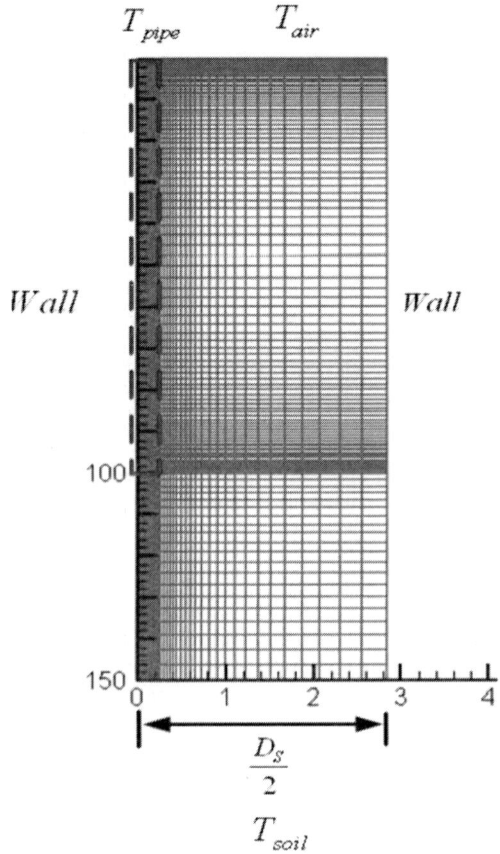


**Figure 4.** Transformation of a square domain to an equivalent concentric domain.

The heat conduction equation in the axis-symmetric coordinate is shown as follows [29]:

$$\rho C_p \frac{\partial T}{\partial t} = \frac{1}{r}\frac{\partial}{\partial r}\left( rk\frac{\partial T}{\partial r} \right) + \frac{\partial}{\partial z}\left( k\frac{\partial T}{\partial z} \right) \tag{2}$$

Heat conduction equation for grout is with $r_{fe} < r < r_b$:

$$\left( \rho C_p \right)_g \frac{\partial T}{\partial t} = \frac{k_g}{r}\frac{\partial}{\partial r}\left( r\frac{\partial T_g}{\partial r} \right) + k_g \frac{\partial^2 T_g}{\partial^2 z} \tag{3}$$

Here, $r_{fe}$ is radius of equivalent refrigerant tube; and $r_b$ is radius of grout, with $r_b < r < \infty$:

$$\frac{k_s}{r}\frac{\partial}{\partial r}\left(r\frac{\partial T_s}{\partial r}\right)+k_s\frac{\partial^2 T_s}{\partial^2 z}=\left(\rho C_p\right)_s\frac{\partial T_s}{\partial t} \tag{4}$$

Boundary condition at the outermost boundary of domain is:

$$\left.\frac{\partial T}{\partial r}\right|_{r=\frac{D_z}{2}}=0 \tag{5}$$

Boundary condition at the surface of refrigerant tube is:

$$\left.T\right|_{r=\frac{D_{pe}}{2}}=T_{pipe} \tag{6}$$

with an initial condition $T_{pipe}$ = 14.2 °C.

Ground surface boundary condition is:

$$\left.T_s\right|_{z=0}=T_{air} \tag{7}$$

Boundary condition for the bottom of the domain is:

$$\left.T_s\right|_{z=150}=T_{s,\infty} \tag{8}$$

Computational analysis is conducted for Equations (2)–(8) by discretization using finite volume method. Ground temperature is continuously varied according to depth, as time changes. However, after the depth reaches a certain value, the ground temperature variation maintains a constant annual temperature.

In this study, such a depth is defined as 150 m, and the temperature at the depth is assumed to be $T_{s,\infty}$ = 14.2 °C [24]. The ground heat exchanger is axis-symmetric, thus a two-dimensional analysis is conducted for computation analysis, while the boundary condition at the symmetric surface around the borehole is defined as zero gradient of temperature [29]. To analyze the ground heat conduction equation, the ground surface condition is needed along with the boundary condition. Thus, in this study, the regional annual climate data is used as the outdoor condition [30] and air is assumed to have the same properties as the outdoor condition [31].

### 3.7. Cost of Electricity and Fuel

A comparison of the late night electricity cost and the heat pump system operational cost is shown in Table 3. The cost of electricity generated from the fuel cell system is calculated using 46 won/kWh for hydrogen cost and 40% for fuel cell efficiency.

In this study, the energy cost is computed from the hydrogen cost and fuel cell efficiency, without taking into account the installation cost of the fuel cell. Also, the operational cost of the machines is excluded and only net energy cost is computed. The base price of late night electricity $cost_{base}$ is computed from the following equation. In Equation (9), the $COP_{bath}$ is the average of the hot water supply heat pump *COP*, while $COP_{heating}$ is the average of the heat pump heating *COP*:

$$cost_{base} = (0.746 \times (4.73 \times (\frac{bath\ heat\ pump\ capacity}{COP_{bath}} + \frac{cooling\ and\ heating\ heat\ pump}{COP_{heating}}) + fan\ power + circulating\ pump\ power) + light\ load) \times cost_b \times 12 \quad (9)$$

where $cost_b$ = basic cost.

### 3.8. Heat Pump Capacity and Heat Storage Tank Calculation

Heat pump capacity (HPC) is calculated from Equation (10) with the maximum daily load during a year obtained from the heating and cooling load program. The heat storage tank capacity is calculated from Equation (11):

$$HPC\,(kJ/hr) = PL(kJ/day) \times \frac{SR}{OT(hr)} \quad (10)$$

$$RC(m^3) = \frac{HPC(kJ/hr) \times NT(hr) \times SR - NL(kJ)}{RT(^{\circ}C) \times RE \times 1{,}000(kJ/m^3)} \quad (11)$$

Here, *PL* is maximum daily load; *SR* is excess rate; *OT* is heat source apparatus operating time; *NT* is operating time during night; *NL* is night load; *RT* is temperature difference due to heat storage tank usage, and *RE* is heat storage tank efficiency.

The temperature difference due to heat storage tank usage was assigned as 5 °C, with a 10 h operating time during the night, a heat storage tank efficiency of 95%, and an excess rate of 10%. The amount of waste water was assigned as 1.8 times that of daily heated water, while heat storage tank capacity was the amount of the daily heated water, and the waste water storage tank capacity was 1.8 times that of the heat storage tank.

# 4. RESULTS AND DISCUSSION

## 4.1. Regional Data of Ground Temperature Variation

In Korea, the heating load is larger than the cooling load, whereby more ground heat source is used during the winter than is stored heat during the summer; thus the ground temperature continuously decreases. Figure 5 shows the variation of ground mean temperature obtained from computational analysis. The difference of initial ground temperature according to the region is due to the fact that when the ground heat source is used, more heat is used than the amount stored in the ground heat source. Therefore the temperature difference in the ground is increased during use of the geothermal heat pump. Southern regions show a smaller decrease in ground temperature, while the northern regions show a larger variation in ground temperature. In southern regions, the heat load is small when heating with a ground heat source, making ground heat source recovery possible. However, in the northern regions, the amount of heat used is greater than the amount stored when heating with the ground heat source. Thus, after 1 year, the following ground temperature drop for each city is observed through computational analysis: in Busan, 1.5 °C; in Dajeon, 2.0 °C; and in Munmak 2.0 °C. Also, the increased use of the geothermal heat pump resulted in a greater decrease in temperature.

Thus, the most important issue in the geothermal heat pump system is how to compensate for the continual decrease of ground temperature. To resolve this problem, it is possible to control the system operation time and to store the heat source generated from summer cooling, as well as to use the heat source during winter heating. However, while this method can prevent ground temperature decrease, any long-term use of this system requires the use of multiple heat sources to deal with the continual decrease of ground temperature.

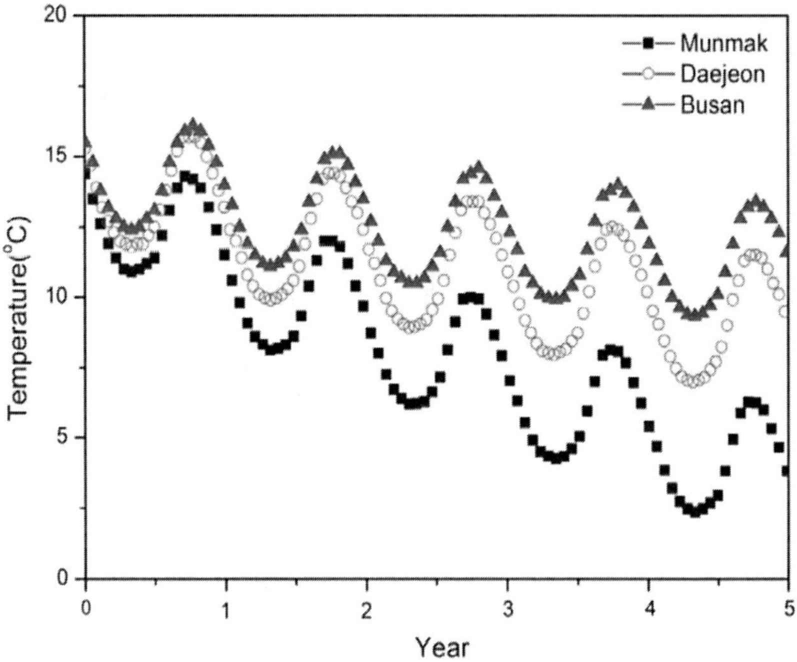


**Figure 5.** Variation of ground mean temperature.

## 4.2. Variation of Heat Pump System Performance According to Building Occupancy

Figure 6 shows the ground temperature variation according to occupancy in different regions. Heat generated from the human body differs as occupancy varies, and affects the building heat and cooling loads as well as the ground mean temperature, thus affecting overall heat pump system performance. As can be observed from Table 3, a reduced heat pump system performance can result from a ground temperature drop when the annual average occupancy is 60%. Thus, the adequate operation control is needed to maintain the ground mean temperature in different regions.

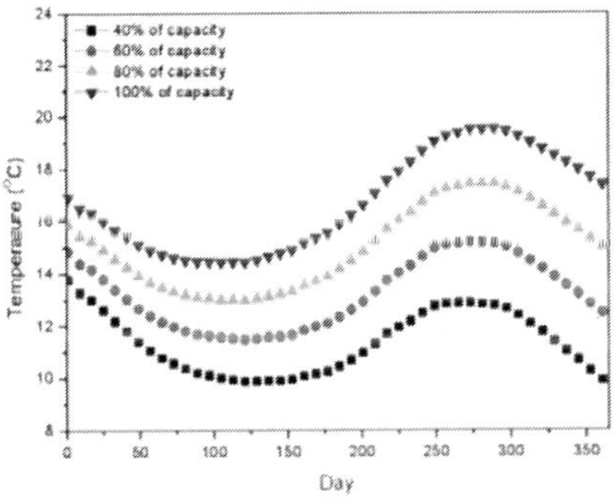


(a) Munmark

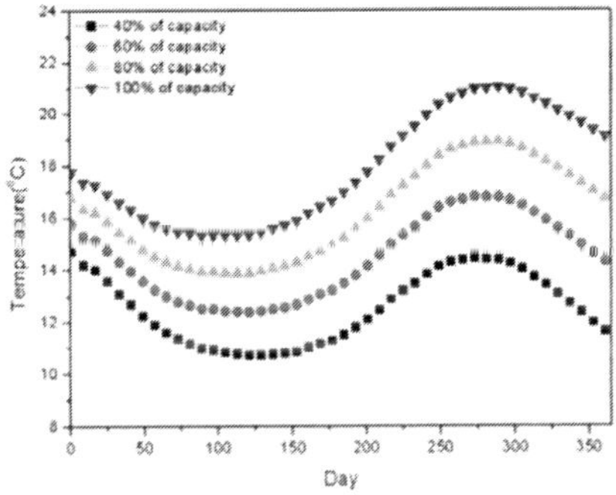


(b) Daejeon

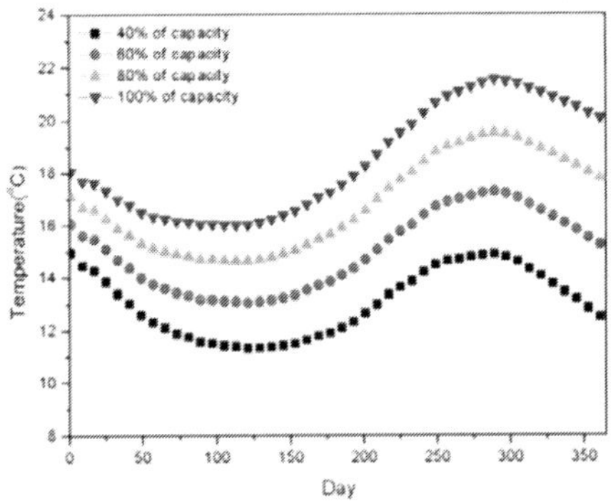


(c) Busan

**Figure 6.** Effect of occupancy on the ground mean temperature. (a) Munmak; (b) Daejeon; (c) Busan.

**Table 3.** Variation of mean ratio of number of people staying in the building to the maximum number of people staying in the building (total floor area 33,000 $m^2$, Heat recovery, Night-electricity driven Ground source heat pump system).

| Location | People living in building | Hourly maximum cooling load * | Hourly maximum heating load * | Annual maximum cooling load * | Annual maximum heating load * | Heating and cooling heat pump capacity * | Bath heat pump capacity (RT) | Total energy cost ** | COP (heating) | COP (cooling) |
|---|---|---|---|---|---|---|---|---|---|---|
| Munmak | 40% | −3,671,875 | 5,969,645 | $-3.92 \times 10^9$ | $11.30 \times 10^9$ | 1,731 | 267.6 | 11,363 | 3.60 | 6.57 |
| | 60% | −4,377,429 | 5,993,169 | $-5.29 \times 10^9$ | $10.75 \times 10^9$ | 1,681 | 267.6 | 16,552 | 3.75 | 6.12 |
| | 80% | −4,390,529 | 5,999,053 | $-6.85 \times 10^9$ | $10.46 \times 10^9$ | 1,668 | 267.6 | 16,572 | 3.84 | 5.68 |
| | 100% | −5,835,283 | 6,040,654 | $-8.53 \times 10^9$ | $9.58 \times 10^9$ | 1,620 | 267.6 | 16,611 | 4.40 | 4.52 |
| Daejeon | 40% | −4,274,281 | 5,125,525 | $-5.42 \times 10^9$ | $9.28 \times 10^9$ | 1,583 | 267.6 | 12,386 | 3.67 | 6.40 |
| | 60% | −4,903,655 | 5,126,583 | $-6.97 \times 10^9$ | $8.40 \times 10^9$ | 1,544 | 267.6 | 16,562 | 3.94 | 5.85 |
| | 80% | −5,754,941 | 5,128,469 | $-8.69 \times 10^9$ | $7.94 \times 10^9$ | 1,492 | 267.6 | 16,578 | 4.23 | 4.78 |
| | 100% | −6,495,485 | 5,130,355 | $-10.5 \times 10^9$ | $7.39 \times 10^9$ | 1,460 | 267.6 | 16,721 | 4.53 | 3.98 |
| Busan | 40% | −4,274,281 | 5,125,525 | $-5.80 \times 10^9$ | $9.28 \times 10^9$ | 1,498 | 267.6 | 11,363 | 3.68 | 6.40 |
| | 60% | −5,014,825 | 5,126,104 | $-7.48 \times 10^9$ | $8.15 \times 10^9$ | 1,434 | 267.6 | 11,731 | 3.94 | 4.93 |
| | 80% | −6,322,806 | 4,574,010 | $-9.28 \times 10^9$ | $6.26 \times 10^9$ | 1,631 | 267.6 | 16,351 | 4.01 | 4.88 |
| | 100% | −6,495,485 | 5,130,355 | $-10.5 \times 10^9$ | $6.22 \times 10^9$ | 1,788 | 267.6 | 17,070 | 4.54 | 3.98 |

* in kJ/h, ** in ten thousand won

### 4.3. Heat Pump System Performance Variation for Various Heat Sources

According to the results of this study, the air source heat pump using night time electricity and geothermal heat pump was modified so that the area was increased to 33,000 $m^2$ during analysis of the standard model, and 60% occupancy was assumed. The same building model was used for locations Munmak, Dajeon and Busan. Computational analysis was conducted for each case. Calculated data includes hourly maximum heating and cooling load, annual heating and cooling load, heat pump capacity, hot water supply heat pump capacity, total energy cost, COP for each heat pump system, size of hot water supply, size of return heat storage tank, size of heat storage tank, pump power, and fan power. The heating and cooling heat pump capacity is calculated from the hourly maximum heating or cooling load that has a higher absolute value.Table 4 and Table 5 show a comparison of air source and geothermal heat pump system performance using night time electricity and fuel cells. Observing the annual heating and cooling maximum load, Munmak resort has the highest heating load, followed by Dajeon and Busan. Busan has the highest cooling load, followed by Dajeon and the Munmak resort. In each case, the heat pump capacity is determined by either the maximum daily heating or cooling load, whichever is the larger.

Thus, Busan which has the largest maximum cooling load needs the largest heat pump capacity. Therefore, the base price for electricity rises. In determining the heat pump capacity, the optimal situation occurs when the amount of heating load and the amount of cooling load are similar. When either the heating load or cooling load is relatively large, the heat pump capacity and the size of the heat storage tank increase, thus increasing the cost of the initial installation and the cost of electricity. In Korea, the daily maximum heating load of the northern region is greater than that of the cooling load, which is the disadvantage in heat pump system application. Also, since the heating load is larger than the cooling load, the heat source used in winter is greater than the amount stored during the summer, thus resulting in a continual decrease of ground temperature. In southern regions, the ground temperature drop is less than in the Northern regions. In the ground source heat pump (GSHP), compensation of the continual increase or decrease of ground heat source is an important issue. Thus, in winter, heating the waste heat generated from the fuel cell operation can be used in system heating to yield a 1.5 COP increase in system performance compared to present ground heat source heat pump system. Electricity generated from the private power station can be supplied to the heat pump, saving energy costs.

**Table 4.** Heat pump system simulation results for heating and cooling a resort building (total floor area 33,000 m$^2$, Heat recovery, Night-electricity driven heat pump system).

| Location | Munmak | | Daejeon | | Busan | |
|---|---|---|---|---|---|---|
| Heat source | Air | Ground | Air | Ground | Air | Ground |
| Hourly maximum cooling load (kJ/h) | −4,390,529 | −4,390,529 | −4,903,655 | −4,903,655 | −6,322,806 | −6,322,806 |
| Hourly maximum heating load (kJ/h) | 5,999,053 | 5,999,053 | 5,126,583 | 5,126,583 | 4,574,010 | 4,574,010 |
| Annually maximum cooling load (kJ/year) | $-4.66 \times 10^9$ | $-4.66 \times 10^9$ | $-6.17 \times 10^9$ | $-6.17 \times 10^9$ | $-6.55 \times 10^9$ | $-6.55 \times 10^9$ |
| Annually maximum heating load (kJ/year) | $10.0 \times 10^9$ | $10.0 \times 10^9$ | $8.0 \times 10^9$ | $8.0 \times 10^9$ | $6.0 \times 10^9$ | $6.0 \times 10^9$ |
| Heating and cooling heat pump capacity (kJ/h) | 1,668 | 1,668 | 1,418 | 1,418 | 1,631 | 1,631 |
| Bath heat pump capacity (RT) | 267.6 | 267.6 | 267.6 | 267.6 | 267.6 | 267.6 |
| Total energy cost (ten thousand won) | 18,371 | 16,572 | 18,306 | 16,562 | 17,702 | 16,351 |
| COP (heating) | 3.16 | 3.84 | 3.27 | 3.94 | 3.42 | 4.01 |
| COP (cooling) | 3.84 | 4.68 | 3.69 | 4.85 | 3.90 | 4.93 |
| COP (bath) | 4.21 | 4.18 | 4.21 | 4.18 | 4.21 | 4.18 |
| Thermal storage reservoir for bath (t) | 511 | 511 | 511 | 511 | 511 | 511 |
| Waste heat recovery storage reservoir (ton) | 919 | 919 | 919 | 919 | 919 | 919 |
| Thermal storage reservoir for cooling and heating (t) | 2,209 | 2,209 | 1,879 | 1,879 | 2,161 | 2,161 |
| Pump power (heating and cooling) | 46.1 | 46.1 | 39.4 | 39.4 | 48.8 | 48.8 |
| Pump power (bath) | 6.2 | 6.2 | 6.2 | 6.2 | 6.2 | 6.2 |
| Fan power (ventilation) | 17.9 | 17.9 | 17.9 | 17.9 | 17.9 | 17.9 |

**Table 5.** Heat pump system simulation results for heating and cooling a resort building (total floor area 33,000 m$^2$, Heat recovery, Fuel-cell driven heat pump system).

| Location | Munmak | | Daejeon | | Busan | |
|---|---|---|---|---|---|---|
| Heat source | Air | Ground | Air | Ground | Air | Ground |
| Hourly maximum cooling load (kJ/h) | −4,390,529 | −4,390,529 | −4,903,655 | −4,903,655 | −6,322,806 | −6,322,806 |
| Hourly maximum heating load (kJ/h) | 5,999,053 | 5,999,053 | 5,126,583 | 5,126,583 | 4,574,010 | 4,574,010 |
| Annually maximum cooling load (kJ/year) | $-4.66 \times 10^9$ | $-4.66 \times 10^9$ | $-6.17 \times 10^9$ | $-6.17 \times 10^9$ | $-6.55 \times 10^9$ | $-6.55 \times 10^9$ |

**Table 5. (Continued)**

| Location | Munmak | | Daejeon | | Busan | |
|---|---|---|---|---|---|---|
| Heat source | Air | Ground | Air | Ground | Air | Ground |
| Annually maximum heating load (kJ/year) | $10.0 \times 10^9$ | $10.0 \times 10^9$ | $8.0 \times 10^9$ | $8.0 \times 10^9$ | $6.0 \times 10^9$ | $6.0 \times 10^9$ |
| Heating and cooling heat pump capacity (kJ/h) | 1,668 | 1,668 | 1,418 | 1,418 | 1,631 | 1,631 |
| Bath heat pump capacity (RT) | 267.6 | 267.6 | 267.6 | 267.6 | 267.6 | 267.6 |
| Total energy cost (ten thousand won) | 11,669 | 8,324 | 11,511 | 8,111 | 10,821 | 7,970 |
| COP (heating) | 4.62 | 5.34 | 4.68 | 5.44 | 4.73 | 5.51 |
| COP (cooling) | 3.84 | 4.68 | 3.69 | 4.85 | 3.90 | 4.93 |
| COP (bath) | 4.21 | 4.18 | 4.21 | 4.18 | 4.21 | 4.18 |
| Thermal storage reservoir for bath (t) | 511 | 511 | 511 | 511 | 511 | 511 |
| Waste heat recovery storage reservoir (t) | 919 | 919 | 919 | 919 | 919 | 919 |
| Thermal storage reservoir for cooling and heating (t) | 2,209 | 2,209 | 1,879 | 1,879 | 2,161 | 2,161 |
| Pump power (heating and cooling) | 46.1 | 46.1 | 39.4 | 39.4 | 48.8 | 48.8 |
| Pump power (bath) | 6.2 | 6.2 | 6.2 | 6.2 | 6.2 | 6.2 |
| Fan power (ventilation) | 17.9 | 17.9 | 17.9 | 17.9 | 17.9 | 17.9 |

### 4.4. Energy Cost

The cost of energy in the heat pump heating and cooling system is calculated as the sum of the electricity cost for the heat pump power, the hot water heat pump power, and the fan power. The total cost varies slightly in a heat storage heat pump system, but the variance of total energy cost according to location is small in Munmak, Dajeon, and Busan. In Busan, while the cooling energy cost is the largest, it has a small heating load energy cost. On the contrary, Munmak has a large heating load energy cost, but has a small cooling load energy cost. Thus, the sum of heating and cooling load energy cost was similar for Busan and Munmak. Table 6 shows the cost of electricity, gas, and the sum when absorption cooler-heat system is used instead of the heat pump system. Cases are classified by whether the public bath is installed in the resort and whether the waste heat is recovered. Cases are classified by whether the public bath is installed and whether the waste heat is recovered. In the cases when the public bath is installed and the waste heat is not recovered, total energy cost showed about a 0%–5% increase.

Table 7 shows a comparison of the total energy cost in a resort building when various heat sources are used. The large energy cost for gas absorption cooling-heating is due to an expensive fuel cost and general electricity cost, thus showing the largest energy cost. Energy cost difference according to location was slight in the air source and ground source heat pump system using night time electricity.

Thus fuel cell driven heat pump system with a high system performance coefficient and low unit price showed the lowest energy cost.

**Table 6.** Energy cost of gas absorption cooler-heater system for a resort building (total floor area = 33,000 m²).

| Location | Energy cost | Munmak | | Daejeon | | Busan | |
|---|---|---|---|---|---|---|---|
| | Bath existence | Non-exist | Exist | Non-exist | Exist | Non-exist | Exist |
| Heat recover | Absorption cooler-heater gas cost (ten thousand won) | 38.988 | 38,988 | 37.464 | 37,464 | 34.594 | 34.594 |
| | Absorption cooler-heater electric cost (ten thousand won) | 24.966 | 24,973 | 24.003 | 23.997 | 22.166 | 22.159 |
| | Absorption cooler-heater total cost (ten thousand won) | 63.954 | 63.961 | 61,467 | 61.416 | 56,760 | 56.753 |
| Non heat recovery | Absorption cooler-heater gas cost (ten thousand won) | 53.300 | 53.300 | 51,766 | 51.776 | 48,906 | 48.906 |
| | Absorption cooler-heater electric cost (ten thousand won) | 24,189 | 23.280 | 23.213 | 22.346 | 21.372 | 20,537 |
| | Absorption cooler-heater total cost (ten thousand won) | 77,489 | 76,580 | 74.989 | 74,122 | 70.281 | 69,443 |

**Table 7.** Comparison of operation costs.

| Heat pump heating and cooling system | Munmak | Daejeon | Busan |
|---|---|---|---|
| Gas absorption cooler-heater(ten thousand won) | 76.580 | 74,122 | 69,443 |
| Gas absorption cooler-heater + waste heat recovery (ten thousand won) | 63,961 | 61.461 | 56.753 |
| Night-electricity driven air source heat pump water storage (ten thousand won) | 18.371 | 18,306 | 17,702 |
| Night-electricity driven ground source heat pump water storage (ten thousand won) | 16,572 | 16.562 | 16.351 |
| Fuel cell driven air source heat pump water storage (ten thousand won) | 11,669 | 11.511 | 10.821 |
| Fuel cell driven ground source heat pump water storage (ten thousand won) | 8.324 | 8,111 | 7.970 |

# 5. CONCLUSIONS

In this study a ground source heat pump heating and cooling system driven by a fuel cell was computationally analyzed, in which heat pump system performance and economic feasibility were considered. Ground temperature ground the pipe is affected both by the variation of air temperature in the pipe brine temperature. The results are as follows:

(1) While the fuel cell driven heat pump system incurs a high initial installation cost, the cost is expected to decrease as fuel cell technology is developed. Also, night time electricity is predicted to increase, thus the fuel cell driven heat pump system will be competent to night time electricity driven heat pump system in future. Also the fuel cell system can be operated without limit in time and thus the size of heat storage tank can be decreased, which is an advantage over night time electricity driven heat pump system.

(2) In Korea, where the heat load is larger than the cooling load, a continuous operation of ground heat source heat pump system is bound to lead to a continual drop in ground temperature. The system is also affected by outer conditions such as occupancy. Therefore, the fuel cell driven system can increase COP by using waste heat generated from the system and minimize ground temperature drop.

(3) A fuel cell driven ground heat source heat pump system increased the COP heating by about 1.5 through waste heat recovery.

## ACKNOWLEDGMENTS

This work was supported by the Human Resources Development program (No. 20124010203250) of the Korea Institute of Energy Technology Evaluation and Planning (KETEP) grant funded by the Korea Government Ministry of Knowledge Economy and by a Korea University Grant (T1101411).

## REFERENCES

1. Kim, O.J.; Hong, Y.J.; Nam, L.W.; Kang, W.I.; Kong, Y.S. Cycle simulation of an air source heat pump using liquid injection. *Korea J. Air Cond. Ref. Eng.* **2000**, *12*, 244–250.
2. Kavanaugh, S.P.; Raffery, K. *Ground-Source Heat Pumps: Design of Geothermal Systems for Commercial and Institutional Buildings*; American Society of Heating, Refrigerating and Air-Conditioning Engineers (ASHRAE): Atlanta, New York, NY, USA, 1997.
3. Healy, P.F.; Ugursal, V.I. Performance and economic feasibility of ground source heat pumps in cold climate. *Int. J. Energy Res.* **1997**, *21*, 857–870.
4. Phetteplace, G.; Sullivan, W. Performance of a hybrid ground-coupled heat pump system. *Am. Soc. Heat. Ref. Air Cond. Eng. Tans.* **1997**, *104*, 763–770.

5. Bose, J.E.; Parker, J.D.; McQuiston, F.C. *Design/Data Manual for Closed-Loop Ground-Coupled Heat Pump Systems*; ASHRAE: Atlanta, New York, NY, USA, 1985.

6. Stoecker, W.F. *Procedures for Simulating the Performance of Components and Systems for Energy Calculations*; ASHRAE: Atlanta, New York, NY, USA, 1975.

7. Mei, V.C.; Baxter, V.D. Experimental study of direct-expansion ground coil heat exchangers. *Am. Soc. Heat. Ref. Air Cond. Eng. Tans.* **1990**, *96*, 634–642.

8. Bose, J.E. *Closed-Loop Ground-Coupled Heat Pump Design Manual*; Engineering Technology Extension; Oklahoma State University: Stillwater, OK, USA, 1984.

9. Bose, J.E. *Closed-Loop/Ground-Source Heat Pump Systems Installation Guide*; Engineering Technology Extension; Oklahoma State University: Stillwater, OK, USA, 1988.

10. Kavanaugh, S.P. Development of design tools for ground-source heat pump piping. *Am. Soc. Heat. Ref. Air Cond. Eng. Tans.* **1998**, *104*, 932–937.

11. Cane, D.; Morrison, A.; Christopher, J. *Operating Experiences with Commercial Ground-Source Heat Pump Systems*; ASHRAE: Atlanta, New York, NY, USA, 1998.

12. Lund, J.W. Direct-use of geothermal energy in the USA. *Appl. Energy* **2003**, *74*, 33–42.

13. Lund, J.W.; Freeston, D.H. World-wide direct uses of geothermal energy 2000. *Geothermics* **2001**, *30*, 29–68.

14. Zeng, H.; Diao, N.; Fang, Z. Heat transfer analysis of boreholes in vertical ground heat exchangers. *Int. J. Heat Mass Trans.* **2003**, *46*, 4467–4481.

15. Hepbasli, A.; Akdemir, O.; Hancioglu, E. *Experimental Study of A Closed Loop Vertical Ground Source Heat Pump Systems*; ASHRAE: Atlanta, New York, NY, USA, 2003.

16. Bi, Y.; Chen, L.; Wu, C. Ground heat exchanger temperature distribution analysis and experimental verification. *Appl. Therm. Eng.* **2002**, *22*, 183–189.

17. İnallı, M.; Esen, H. Experimental thermal performance evaluation of a horizontal ground-source heat pump system.*Appl. Therm. Eng.* **2004**, *24*, 2219–2232.

18. Esen, H.; İnallı, M.; Esen, M. Technoeconomic appraisal of a ground source heat pump system for a heating season in eastern Turkey. *Energy Convers. Manag.* **2004**, *47*, 1281–1297.

19. Houchins, C.; Kleen, G.J.; Spendelow, J.S.; Kopasz, J.; Peterson, D.; Garland, N.L.; Ho, D.L.; Marcinkoski, J.; Martin, K.E.; Tyler, R.; Papageorgopoulos, D.C. U.S. DOE progress towards developing low-cost, high performance, durable polymer electrolyte membranes for fuel cell applications. *Membranes* **2012**, *2*, 855–878.

20. Lee, C.S.; Choi, Y.D. Analysis of energy consumption of office building by thermal resistance-capacitance method.*Korea J. Air Cond. Ref. Eng.* **1997**, *9*, 1–13.

21. Lee, C.S. Development of a Building Energy Analysis Computer Program Using the Resistance-Capacity Method. Master Thesis, Korea University, Seoul, Korea, 2003.

22. Lee, J.H.; Choi, Y.D.; Cho, S.H. Comparison on Simulation Results of Load Analysis Program. In Proceedings of Society of Air-Conditioning and Refrigerating Engineers of Korea (SAREK 1999), Seoul, Korea, 30 November 1999; pp. 51–64.

23. Choi, Y.D.; Yoon, J.H.; Hong, J.K.; Lee, N.H.; Kang, D.H. Simulation of the thermal performance on an ondol house with hot water heating in consideration of radiation heat transfer. *J. Air Cond. Ref.* **1994**, *8*, 3–16.

24. Jeong, D.H.; Byun, J.K.; Choi, Y.D.; Cho, S.H. Study on the performance fuel cell driven compound source heat pump system to a large community building. *New Renew. Eng.* **2008**, *4*, 23–35.

25. *ASHRAE Handbook 1993 Fundamentals*; ASHRAE: Atlanta, New York, NY, USA, 1993.

26. *ASHRAE Handbook 1993 Fundamentals*; ASHRAE: Atlanta, New York, NY, USA, 1993.

27. Choi, Y.D.; Han, S.H.; Cho, S.H.; Kim, D.S.; Um, C.J. Study on the simulation of heat pump heating and cooling systems to hospital building. *Korea Soc. Mech. Eng.* **2008**, *32*, 275–282.

28. Shim, Y.B.; Park, K.J.; Jung, D.S.; Kim, J.S. Condensation heat transfer coefficients of R245fa on a plain tube. *Int. J. Air Cond. Ref.* **2007**, *19*, 555–562.

29. Patankar, S.V. *Numerical Heat Transfer and Fluid Flow*; McGraw-Hill: New York, NY, USA, 1980.

30. Choi, Y.D.; An, J.S.; Kim, W.K.; Park, H.S.; Jang, M.S. Analysis on the Thermal Performances of Cool Tube Systems for One Year. In Proceedings of The Korean Society of Mechanical Engineering, Pusan National University, Pusan, Korea, 17–19 April 1996; pp. 725–731.

31. Tromble, A.; Serres, L. Air earth exchanger study in real site experimentation and simulation. *Energy Build.* **1994**, *4*, 23–35.

# CHAPTER 5

# Heat Transfer and Geometrical Analysis of Thermoelectric Converters Driven by Concentrated Solar Radiation

Clemens Suter [1], Petr Tomeš [2], Anke Weidenkaff [2] and Aldo Steinfeld [1,3,*]

[1] *Department of Mechanical and Process Engineering, ETH Zurich, 8092 Zurich, Switzerland*
[2] *Solid State Chemistry and Catalysis, Empa, 8600 Duebendorf, Switzerland*
[3] *Solar Technology Laboratory, Paul Scherrer Institute, 5232 Villigen, Switzerland*

## ABSTRACT

A heat transfer model that couples radiation/conduction/convection heat transfer with electrical potential distribution is developed for a thermoelectric converter (TEC) subjected to concentrated solar radiation. The 4-leg TEC module consists of two pairs of *p*-type $La_{1.98}Sr_{0.02}CuO_4$ and *n*-type $CaMn_{0.98}Nb_{0.02}O_3$ legs that are sandwiched between two ceramic $Al_2O_3$ hot/cold plates and connected electrically in series and thermally in parallel. The governing equations for heat transfer and electrical potential are formulated, discretized and solved numerically by applying the finite volume (FV) method. The model is validated in terms of experimentally measured temperatures and voltages/power using a set of TEC demonstrator modules, subjected to a peak radiative flux intensity of 300 suns. The heat transfer model is then applied to examine the effect of the geometrical parameters (e.g. length/width of legs) on the solar-to-electricity energy conversion efficiency.

## KEYWORDS

thermoelectricity; thermoelectric converter; solar; heat transfer; radiation; modeling; optimization

## 1. INTRODUCTION

A thermoelectric converter (TEC) comprises *p*-type and *n*-type semiconductor legs sandwiched between two ceramic hot/cold plates and connected thermally in parallel and electrically in series [1,2,3]. The temperature gradient across the legs induces a voltage due to the Seebeck effect. The TEC performance is characterized by its figure-of-merit, $ZT = S^2T/(\rho\kappa)$. Due to the relatively low heat-to-electricity conversion efficiencies approaching 5% for $ZT \leq 1$, TECs have been mainly used in space applications. With the advent of novel functional ceramic materials, new high-temperature application areas are being considered, e.g. waste heat recovery and solar thermoelectric generation [4,5,6]. Previous heat transfer analyses considered compatibility factors [7,8,9] and heat conduction models [10,11]. In this paper, a FV-based heat transfer model of a TEC module is developed for simulating its thermal performance and analyzing the effect of the geometrical parameters. Coupled radiation/conduction/convection heat transfer with electrical potential distribution is considered for a TEC module directly exposed to concentrated solar radiation. The model is experimentally validated with measurements of temperature and voltages/power using a set of simplified 4-leg TEC modules that were directly irradiated. With these demonstrator ("proof-of-concept") modules, the direct conversion of high-temperature solar heat is demonstrated [12]. However, no attempt has yet been undertaken to optimize the design or materials of these TEC modules for maximum energy conversion efficiency. Neither the designing of middle/large scale applications nor the combination with other technologies, e.g. PV cells, has been considered yet.

## 2. EXPERIMENTAL

Figure 1 depicts a schematic of a 4-leg TEC module used in the experimental runs. Six demonstrator modules were fabricated with leg lengths $l = 4$, 5, and 10 mm (2 modules for each leg length). Each leg has a quadratic cross section of width $a = 4.5$ mm and a distance $d = 10$ mm from the neighboring leg. The *p*-type legs are made of $La_{1.98}Sr_{0.02}CuO_4$; the *n*-type legs are made of $CaMn_{0.98}Nb_{0.02}O_3$. These perovskite materials exhibit chemical and mechanical stability at high temperatures, but at the expense of having $ZT \sim 0.05$ [13]. The $LxLxb = 30 \times 30 \times 0.25$ mm absorber (hot) and cooling (cold) plates are made of $Al_2O_3$ with ~5% porosity. Additionally, the absorber plate is coated with graphite to augment its absorptivity.

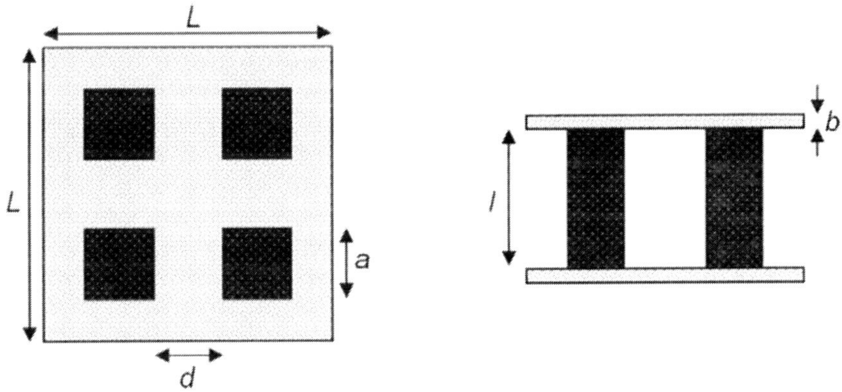


**Figure 1.** Top and front view of TEC module.

Experimentation was carried out at the ETH's High Flux Solar Simulator (HFSS): a high-pressure Ar arc close-coupled to an elliptical reflector that delivers an external source of intense thermal radiation to simulate the heat transfer characteristics of highly concentrating solar systems [14]. The solar flux concentration is characterized by the mean concentration ratio $\tilde{C}$, defined as $\tilde{C}$=Qsolar/(I· A), where $Q_{solar}$ is the solar power intercepted by a target of area *A*. The ratio $\tilde{C}$ is often expressed in units of "suns" when normalized to $I$ = 1 kW/m$^2$. The experimental set-up is shown schematically in Figures 2(a) and (b). Incident radiative fluxes were measured by a thermogage (with an accuracy of ±3%) [15], placed symmetrically to the TEC module at the focal plane. The TEC was exposed to a maximum mean solar concentration ratio of 300 suns. K-type thermocouples (tip = 0.5 mm, spatial accuracy = ±0.25 mm) were used to measure temperatures of the plates and of the hot end, middle, and cold end of the legs. Terminals were provided at the cold ends for measuring the voltage/power output of the module. The cold plate was attached to a water-circuit cooler at room temperature.

The temperature and solar radiative flux as a function of time are shown in Figure 3 for a representative experimental run using a module with leg length $l$ = 4 mm. The incident solar radiation was increased stepwise and held at constant level for 3-5-minute intervals. Due to the low thermal inertia and fast temperature response, steady-state conditions are assumed for each time interval. Maximum temperature was 625°C, at which graphite is no longer stable and starts to burn. For the same module ($l$ = 4 mm), Figure 4 shows the theoretical and measured voltage-power curves for incident solar radiative fluxes in the range q″solar = 1.8–10 W $cm^{-2}$, and for external loads with resistance in the range $R_{load}$ = 0.1-3.5 Ω. A parabola, which corresponds to the ideal voltage source with an internal resistance [16], is fitted through the data points. The maximum power is $P_{max}$ = 0.006, 0.015, 0.023, 0.031, 0.038 and 0.046 W for q″solar = 1.8, 2.9, 4.1, 5.4, 8.2 and 10 W $cm^{-2}$, respectively.

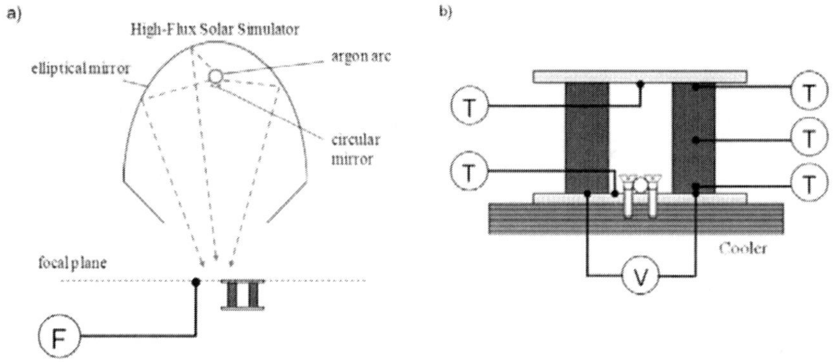


**Figure 2.** Schematic of the experimental setup at ETH's High Flux Solar Simulator. (a) the TEC module is placed at HFSS's focal plane; incident solar radiative fluxes measured by a thermogage (F). (b) position of type-K thermocouples (T) used to measure temperatures of the plates and of the hot end, middle, and cold end of the legs; terminals (V) provided at the cold ends for measuring the voltage/power output of the module. The cold plate was attached to a water-circuit cooler (denoted by screw fixation).

The measured temperature distribution for two tested modules with leg length $l$ = 10 mm is shown in Figure 5 for q″solar = 6 W $cm^{-2}$. As expected, the quasi linear profile indicates a predominant heat transfer by conduction across the legs. The abnormal behavior of 100 K temperature difference at the cold side is presumably due to the incorporation of the screw fixation (see Figure 2 (b)) causing different heat transfer rates.

*Efficiency* — The solar-to-power efficiency of the TEC module is defined as:

$$\eta = \frac{P_{\max}}{A_{abs} \cdot q''_{solar}} \tag{1}$$

where $P_{max}$ is the maximal power output and q″solar the mean solar radiative flux incident over the absorber surface$A_{absorber}$. For modules with leg lengths $l$ = 4, 5, and 10 mm the maximal power outputs $P_{max}$ are 45.6, 51.6 and 42.2 mW for q″solar = 9.9, 9.7, and 5.7 W cm$^{-1}$, respectively. The efficiency $\eta$ as a function of solar radiative flux is shown in Figure 6 for $l$ = 4, 5, and 10 mm. The curves are plotted up to the maximal solar flux of 9.9, 9.7 and 5.7 W cm$^{-1}$, respectively, for which $T_{hot}$ = 625°C is reached. Higher solar fluxes resulted in the burning of the graphite coating. The efficiency increases with q″solar as a result of the higher temperature difference across the legs, which in turn corresponds to a higher Carnot limitation [11]. In contrast, $\eta$ decreases with $T$ as re-radiation losses are proportional to $T^4$. Thus, an optimum q″solar for maximum $\eta$ is expected. For $l$ = 4 mm, $\eta_{max}$ = 0.065% at q″solar = 4 W cm$^{-2}$. For $l$ = 5 mm $\eta_{max}$ = 0.06% at q″solar = 8 W cm$^{-2}$. For $l$ = 10 mm,$\eta_{max}$ = 0.083% at q″solar = 4 W cm$^{-2}$.

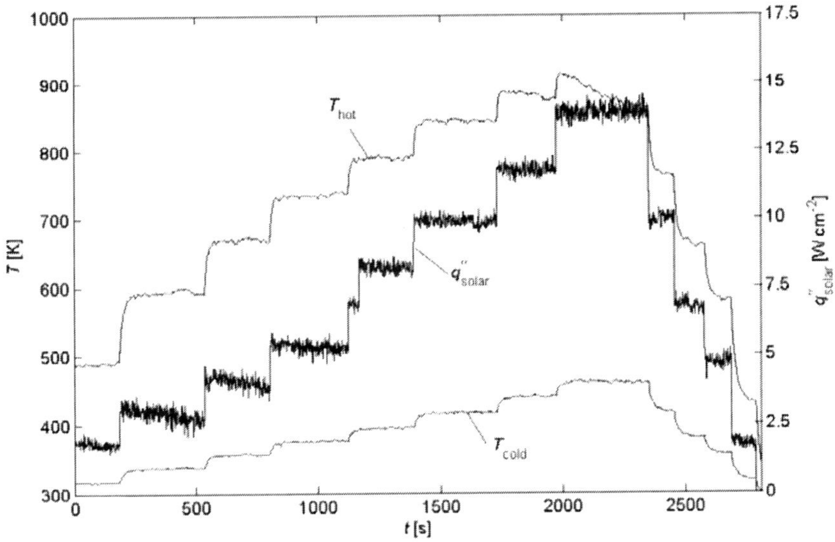


**Figure 3.** Temperature of hot and cold plates and solar radiative flux as a function of time during a representative experimental run for module with $l$ = 4 mm.

## 3. HEAT TRANSFER MODEL

A 2D steady-state heat transfer model is formulated. A cross section of the model domain, divided into $m \times n$ cells, is depicted in Figure 7. It contains the three major components: the absorber plate, one p- and one n-leg (P/N), and the space in-between. The domain is assumed to be infinitely long; therefore,

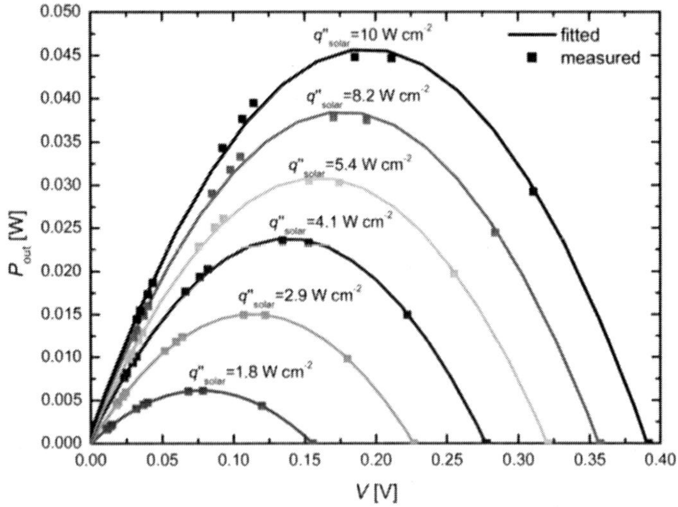


**Figure 4.** Fitted and measured voltage-power curves for incident solar radiative fluxes in the range q″solar = 1.8 – 10 W cm$^{-2}$, and for external loads with resistance in the range$R_{load}$ = 0.1-3.5 Ω for module with $l$ = 4 mm.

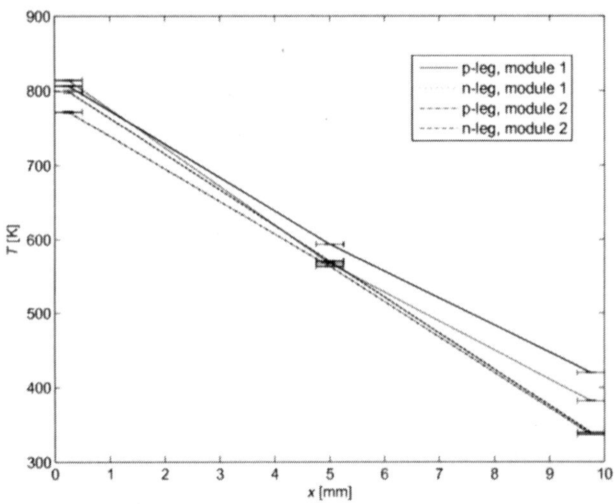


**Figure 5.** Temperature distribution along the p- and n-type legs for two modules with $l$= 10 mm. Error bars indicate spatial accuracy (±0.25 mm) of thermocouple placing.

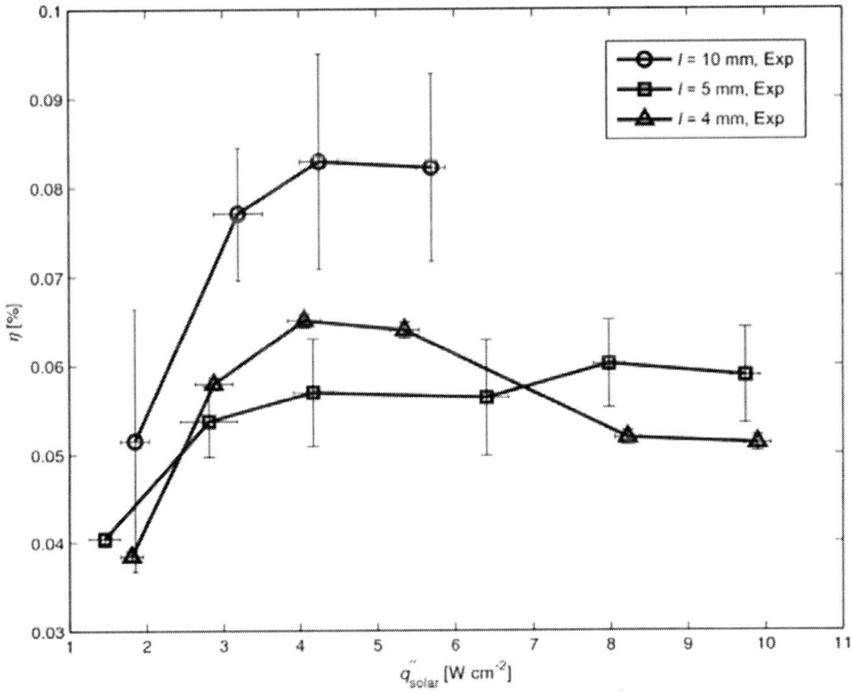


**Figure 6.** Efficiency $\eta$ as a function of solar radiative flux for three modules with $l$ = 4, 5, and 10 mm. Error bars indicating uncertainty of incident solar radiative flux and efficiency due to uncertain contact resistance.

periodic boundaries are set at the sides. The heat transfer modes considered are: (1) conduction in the complete domain, and (2) radiative heat transfer among all surfaces for two approaches: (a) assuming a semi-transparent absorber plate; (b) assuming an opaque absorber plate. It is further assumed: (*i*) the p/n solids are opaque, gray and diffuse scattering; (*ii*) gas phase is radiatively non-participating and its refractive index is equal to unity; (*iiia*) the absorber plate is radiatively participating with isotropic scattering and with temperature and wavelength independent extinction coefficient $\beta_{abs}$ and albedo $\omega_{abs}$; (*iiib*) the absorber plate is opaque, gray and diffuse scattering; (*iv*) convection is only considered from top of the hot plate; (*v*) open circuit voltage ($j = 0$). $P_{max}$and $\eta$ are calculated based on the matched load assumption, given by:

$$P_{max} = \frac{1}{4}\frac{V_{OC}^2}{R_{internal} + R_{contact}} \quad (2)$$

where $V_{OC}$ is the open circuit voltage, $R_{internal}$ the internal resistance of the TEC module, and $R_{contact}$ the contact resistance between legs and conduction strips.

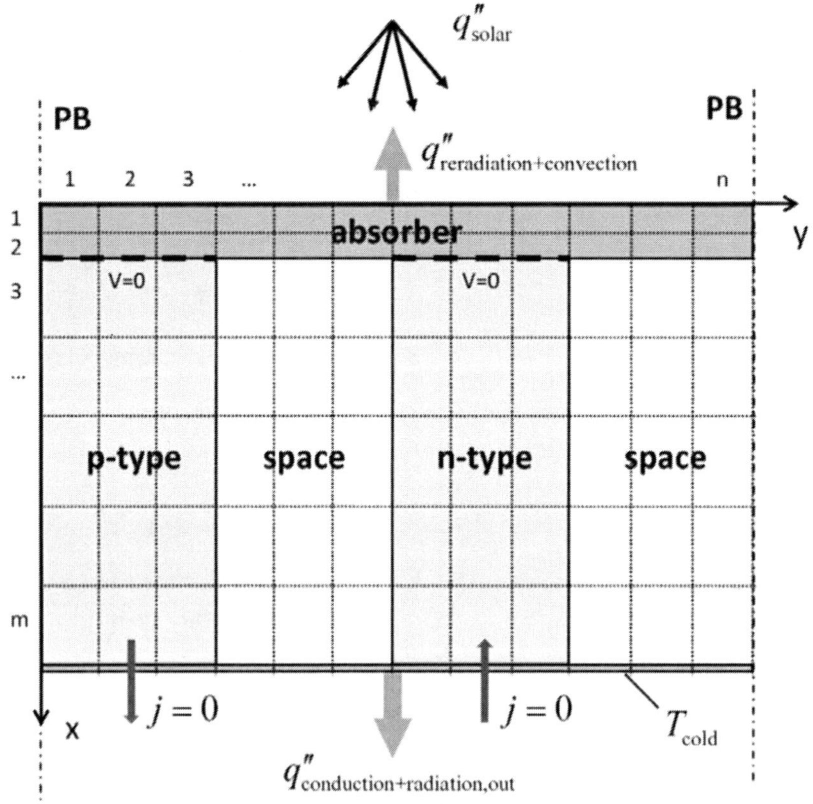


**Figure 7.** Scheme of the model domain (divided into m × n cells). Indicated are the boundary conditions.

*Conservation equations* — The steady-state energy conservation equation applied to the absorber for approach (a) is given by:

$$\nabla\left(\kappa_{absorber} \cdot \nabla T\right) - q'''_{radiation} = 0 \tag{3}$$

where $k_{absorber}$ is the absorber thermal conductivity and $q'''$radiation is the radiative volumetric heat source. For approach (b), $q'''$radiation = 0 and an additional boundary condition is necessary (see subchapter "boundary conditions"). The steady-state conservation equations applied to the legs (P/N domain) are given by:

$$\text{Energy: } \nabla\left(\kappa_{\text{leg}} \cdot \nabla T\right) + \underbrace{\rho_{\text{leg}} \cdot j^2}_{j=0} - \underbrace{T \cdot \frac{dS_{\text{leg}}}{dT} \cdot j \cdot \nabla T}_{j=0} = 0 \tag{4}$$

with thermal conductivity $k_{\text{leg}}$, electrical resistivity $\rho_{\text{leg}}$, current per area $j$, and Seebeck coefficient $S_{\text{leg}}$. Note that the two current terms cancel due to the open circuit condition:

$$\text{Current: } \nabla j = 0, \text{ where } j = -\frac{1}{\rho_{\text{leg}}}\left[\nabla\left(\underbrace{\frac{\mu_{\text{leg}}}{e_{\text{leg}}}}_{const} + V\right) + S_{\text{leg}} \cdot \nabla T\right] \tag{5}$$

with electrical resistivity $\rho_{\text{leg}}$, chemical potential $\mu_{\text{leg}}$, and the charge of charged particles of current $e_{\text{leg}}$ and Seebeck coefficient $S_{\text{leg}}$ [11]. Note that the gradient of $\frac{\mu_{\text{leg}}}{e_{\text{leg}}}$ cancels as this term is assumed constant.

For approach (a), the radiative heat transfer within the absorber plate is determined by the collision-based Monte Carlo (MC) method [17]. The radiative source term q‴radiation is approximated by:

$$q'''_{\text{radiation}} \approx \frac{N_{\text{absorbed}} \cdot q_{\text{ray}}}{\Delta V} - 2\beta_{\text{absorber}}\left(1 - \omega_{\text{absorber}}\right)\sigma T^4 \tag{6}$$

where $q_{\text{ray}}$ is the power carried by a single ray, $N_{\text{absorbed}}$ the number of rays absorbed within a control volume $\Delta V$, $\beta_{\text{absorber}}$the extinction coefficient, and $\omega_{\text{absorber}}$ its albedo. Thus, the net radiative flux q″radiation,net of the surfaces is calculated as:

$$q''_{\text{radiation,net}} \approx \frac{N_{\text{absorbed}} \cdot q_{\text{ray}}}{\Delta A} - \varepsilon_{\text{surface}}\sigma T^4 \tag{7}$$

where $q_{ray}$ is the power carried by a single ray, $N_{absorbed}$ the number of rays absorbed within a control surface $\Delta A$, and $\varepsilon_{surface}$ its emissivity. For approach (b), the net radiative flux q″radiation,net from inner surface elements is computed using the radiosity method (enclosure theory) [18], assuming p/n-type surfaces with emissivity $\varepsilon_P$, $\varepsilon_N$, respectively, and uncoated (white) surfaces from absorber and cold plates with emissivity $\varepsilon_{absorber}$. The corresponding system of equations is given by [17]:

$$\sum_{j=1}^{2(m_{P/N}+n_{space})} \left( \frac{\delta_{kj}}{\varepsilon_j} - F_{k-j} \frac{1-\varepsilon_j}{\varepsilon_j} \right) q''_{radiation,net,j} = \sum_{j=1}^{2(m_{P/N}+n_{space})} \left( \delta_{kj} - F_{k-j} \right) \sigma T_j^4$$

$$\text{for } k = 1 \ldots 2(m_{P/N} + n_{space}) \tag{8}$$

where $\delta$ is the Kronecker function, $m_{P/N}$ is the number of p/n-type elements in x-direction and $n_{space}$ the number of elements in y-direction. The view factors $F_{k-j}$ are calculated by applying reciprocity relations ($A_1F_{1-2}=A_2F_{2-1}$), enclosure criterion $(\sum_{j=1}^{N} F_{k-j} = 1)$, and tabulated view factors [19].

For simplicity, 2D geometry is considered. As the total absorber surface per leg must be the same for 3D and 2D geometries, the distance $d$ between the legs for 3D is transformed to $d^*$ for 2D. Similarly, the thermal conductivity in the direction along the plate is as adjusted as k∗absorber.

*Boundary conditions* — q″solar from the HFSS is assumed to be uniformly distributed. The heat losses from the absorber's top include re-radiation and free convection. Re-radiation is calculated by MC for approach (a), and by introducing a new boundary condition for approach (b): q″reradiation= εabsorber,coatedσT4. Free convection q″convection to the environment is calculated using a Nusselt correlation for a horizontal flat plate [20]:

$$\mathrm{Nu}_L = 0.54 \cdot \mathrm{Ra}_L^{1/4} \quad \left(10^4 \le Ra_L \le 10^7\right)$$

$$\mathrm{Ra}_L = \mathrm{Gr}_L \cdot \mathrm{Pr} = \frac{g\beta\left(T_{surface} - T_\infty\right)L^3}{\nu\alpha} \tag{9}$$

where Nu, Ra, Gr, and Pr are the Nusselt, Rayleigh, Grashof, and Prandtl numbers, $g$ the gravitational acceleration, $\beta$ the volumetric thermal expansion coefficient, ν the kinematic viscosity, α the thermal diffusity, $T_{surface}$ the surface temperature, $T_\infty$ the surroundings temperature, and $L$ the characteristic length (here the absorber width, $L = 30$ mm). The outgoing heat flux contains radiation losses q″radiation, out through the space to the cold plate as well as conduction losses q″conductionthrough the legs to the cold plate. q″radiation,out is either calculated by MC for approach (a), and by the radiosity method for approach (b).

*Numerical solution* — The finite volume technique (FV) is applied to discretize the governing equations (3) and (4) and solve the PDE system iteratively using the successive over-relaxation (SOR) method [21] implemented in FORTRAN. The algorithm is repeated until the convergence criterion:

$$\left| \frac{T_{i,j}^{\gamma} - T_{i,j}^{\gamma-1}}{T_{i,j}^{\gamma}} \right| \leq \varepsilon \qquad (10)$$

for all elements $i,j$ after γ iterations is satisfied, with $\varepsilon < 10^{-6}$ and the overall energy balance satisfied within 0.1%. After convergence, the potential distribution is calculated. A convergence study indicated optimal trade-off between accuracy and computational time with a grid containing 425 elements.

The difference between the $V_{OC}$ calculated by the two approaches (a) and (b) for analyzing the radiative heat transfer is shown in Figure 8 for $l =$ 10 mm. Different radiation properties ($\beta_{abs}$, $\omega$) of the absorber plate have been tested. For $\beta_{abs} \rightarrow \infty$ and $\omega \rightarrow 0$, no incoming radiation is transmitted through the absorber, and the solution obtained by approach (a) moves toward that for an opaque absorber obtained by approach (b). Since the absorber plate used in the measurements can be well approximated by an opaque surface, only approach (b) is applied in the analysis that follows.

# 4. MODEL VALIDATION

Validation is accomplished for the open circuit voltage $V_{oc}$, as this value is the most reliable magnitude to measure and is directly proportional to the mean temperature difference across the legs. The baseline parameter used for the model simulations are listed in Table 1.

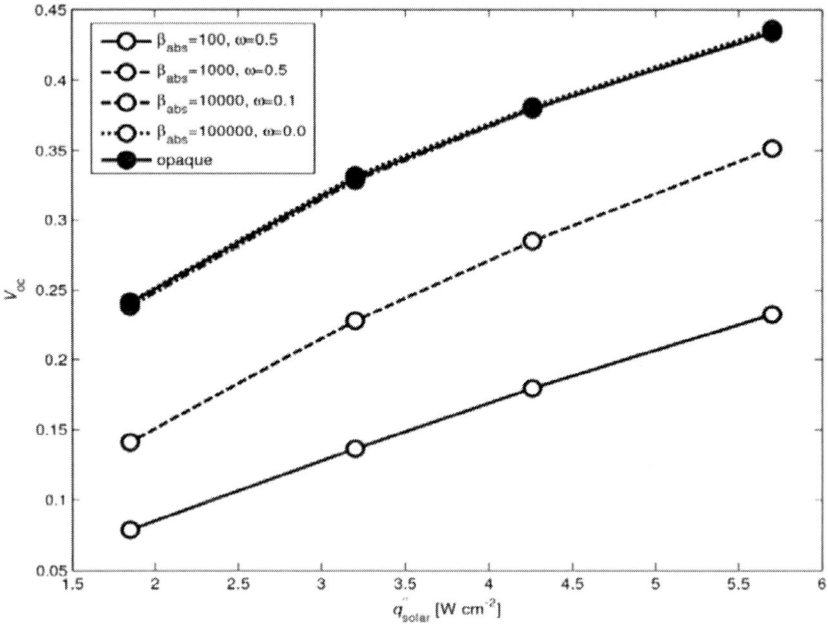


**Figure 8.** Simulated $V_{OC}$ as a function of solar radiative flux for $l = 10$ mm for approach (a) with $\beta_{absorber} = 100$, $\omega = 0.5$; $\beta_{absorber} = 1000$, $\omega = 0.5$; $\beta_{absorber} = 10000$, $\omega = 0.1$; $\beta_{absorber} = 100000$, $\omega = 0.0$ and for opaque approach (b).

**Table 1.** Baseline parameters.

| Parameter | Value | Unit | Source |
|---|---|---|---|
| $l$ | 4 - 15 | mm | measured/varied |
| $a$ | 3 - 6 | mm | measured/varied |
| $d$ | 1 - 10 | mm | measured/varied |
| $\varepsilon_{absorber,coated}$ | 0.95 | - | [22] |
| $\varepsilon_{absorber}$ | 0.3 | - | [22] |
| $\varepsilon_{P/N}$ | 0.7 | - | assumed |
| $\beta_{absorber}$ | 100 - 100000 | $m^{-1}$ | varied |
| $\omega$ | 0.0 - 0.5 | - | varied |
| $k^*_{absorber}$ | 250 | $W\,m^{-1}\,K^{-1}$ | assumed |
| $k_{absorber}$ | $1.78 \times 10^{-11}T^4 - 9.79 \times 10^{-8}T^3 + 2.02 \times 10^{-4}T^2 - 1.90 \times 10^{-1}T + 75.77$ | $W\,m^{-1}\,K^{-1}$ | [23] |
| $k_{P/N}$ | ~ 1 – 2.5/~ 1.75 – 3 | $W\,m^{-1}\,K^{-1}$ | measured |
| $S_{P/N}$ | ~ 120 – 260/~ -170 – -230 | $\mu V\,K^{-1}$ | measured |
| $\rho_{P/N}$ | ~ 0.025 - 0.05/~ 0.02 – 0.036 | $\Omega$ cm | measured |
| $T_\infty$ | 300 | K | assumed |
| $T_{cold}$ | 300 | K | assumed |
| $R_{contact}$ | 0.40-0.66 | $\Omega$ | [12]/assumed |

The experimentally measured and numerically calculated $V_{OC}$ are shown in Figure 9 for $l$ = 4, 5, and 10 mm. A reasonable good agreement is observed, except for the 4 mm case at high fluxes (q″solar > 8 W cm$^{-2}$), where the model predicts a 15% higher value. This discrepancy is attributed to the insufficient cooling of the cold plate at high fluxes, as evidenced by a rise of its temperature, which in turn caused higher absorber plate temperature and, consequently, higher re-radiation losses. Thus, the temperature difference across the legs is shifted to higher temperatures and reduced due to the higher re-radiation.

The numerically simulated solar-to-power efficiencies are shown in Figure 10 (together with the experimentally determined efficiencies from Figure 6), calculated using Equation (1) with maximal power output $P_{max}$ from Equation (2). The internal leg resistances $R_{internal}$ are calculated according to:

$$R_{internal} = \sum_i \rho_{leg,i} \cdot \frac{\Delta x}{a^2} \tag{11}$$

where $\rho_{leg,i}$ is the leg's temperature dependent electrical resistivity, $\Delta x$ the cell length in x-direction, $a$ the width of the leg and $i$ the index of summation over the number of cells in x-direction along the leg (see Figure 7). The mean contact resistance is assumed to be 0.53 ± 0.13 Ω for all cases, determined in [12]. The calculated values lie in the same range as the measured ones, expect for the 4 mm case at high fluxes (q″solar >8 W cm$^{-2}$) which result from the overestimated $V_{OC}$ (seeFigure 9).

The percentage of $Q_{solar}$ transferred by the different heat transfer modes is shown in Figure 11 for two cases; 1) q″solar = 6 W cm$^{-2}$ and $l$ = 10 mm leg length, and 2) q″solar = 10 W m$^{-2}$ and $l$ = 5 mm. In both cases, the heat losses by re-radiation and free convection from the absorber plate represent more than 70% of $Q_{solar}$. About 20% of $Q_{solar}$ is transferred by conduction through the legs, and <10% is lost by radiation to the cold plate.

Figure 12 shows the temperature distribution along a p-type leg of $l$ = 10 mm obtained for q″solar = 6 W m$^{-2}$. A comparable distribution is obtained for an n-type leg. The profile is linear, as corroborated by the experimental data (see Figure 5). Perpendicular to the length axis, the temperature is almost uniform, with a slightly higher temperature at the surface because of the radiative exchange among legs and plates. The small temperature gradient indicates that this radiative heat exchange is not predominant, as confirmed by the fact that <10% of $Q_{solar}$ is lost by radiation to the cold plate (see Figure 11).

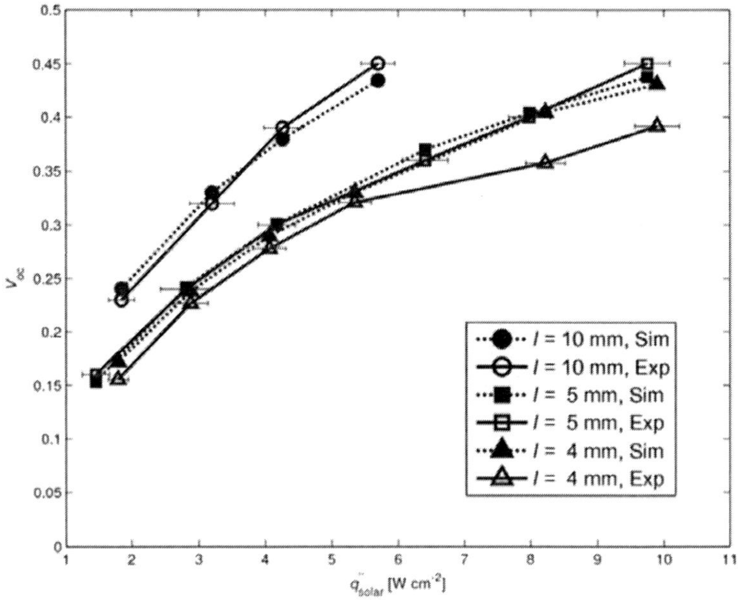


**Figure 9.** Simulated and experimental $V_{OC}$ as a function of solar radiative flux for $l = 4, 5, 10$ mm. Error bars indicating uncertainty of incident solar radiative flux.

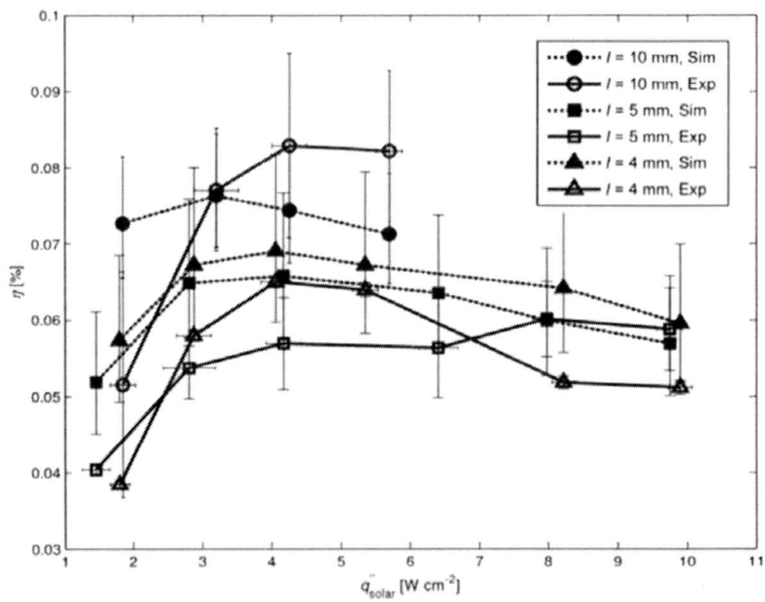


**Figure 10.** Simulated and experimental $\eta$ as a function of solar radiative flux for $l = 4, 5, 10$ mm (experimental data from Figure 6). Error bars for simulated data points indicating uncertainty of contact resistance.

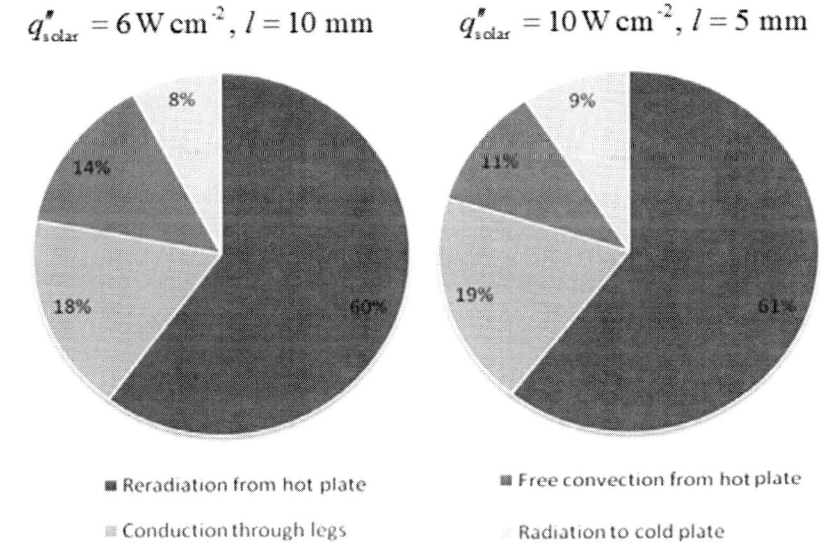


**Figure 11.** Percentage of $Q_{solar}$ transferred by the heat transfer modes.

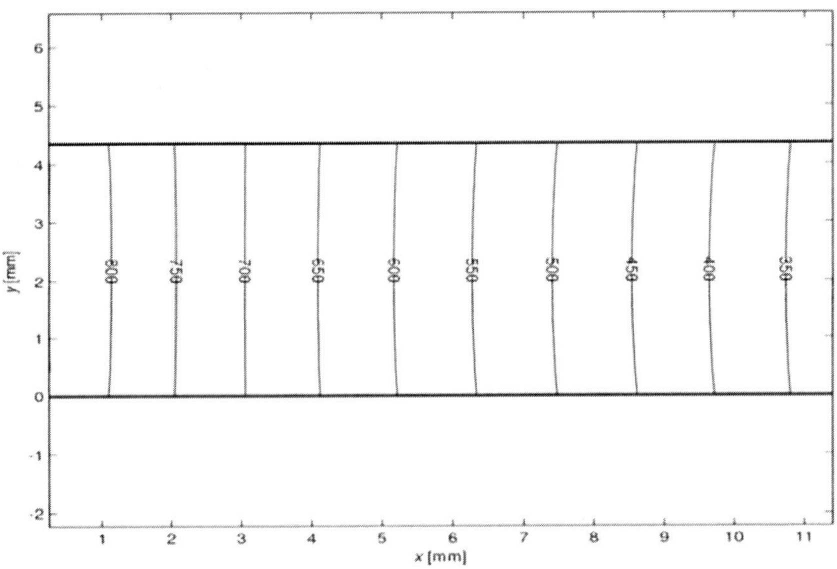


**Figure 12.** 2D temperature profile in p-type leg for $l$ = 10 mm at q″solar = 6 W m$^{-2}$.

## 5. EFFICIENCY

*Leg length* —The simulated dimensions of the modules are $l = 5$ - 15 mm, $a = 4.5$ mm, and plates with $LxLxb = 30 \times 30 \times 0.25$ mm. The cold plate temperature is set to 300 K. The contact resistance is $R_{Contact} = 0.55\ \Omega$. The solar radiative fluxes are varied in the range q"solar=2 - 10 W cm–2. The black coating of the absorber is assumed stable for all temperatures. The baseline parameter used for the model simulations are listed in Table 1.

Figure 13 shows the efficiency as a function of solar radiative flux for $l = 5$, 7.5, 10, 12.5 and 15 mm. The highest efficiency$\eta = 0.081\%$ is obtained for $l = 7.5$ mm at q"solar = 4 W cm$^{-2}$. Note that $l = 7.5$ mm is not optimal in the whole solar radiative flux range. For q"solar < 3 W cm$^{-2}$, $l = 10$ mm is most efficient, whereas for q"solar > 7 W cm$^{-2}$, $l = 5$ mm is most efficient. Thus, for increasing solar radiative fluxes, the optimal leg length decreases.

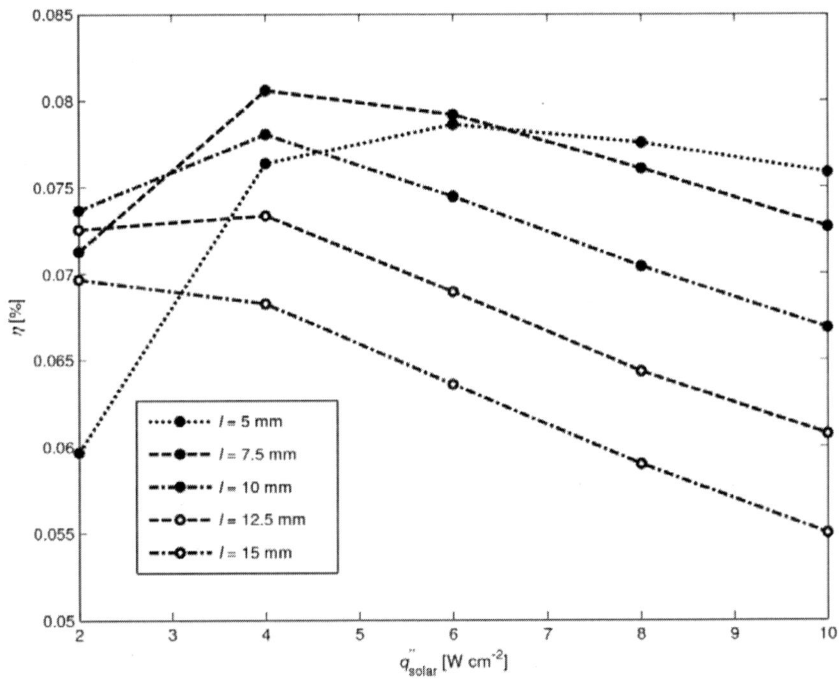


**Figure 13.** Efficiency as a function of solar radiative flux for $l = 5$ mm, $l = 7.5$ mm, $l = 10$ mm, $l = 12.5$ mm and $l = 15$ mm.

*Leg width and distance between neighboring legs* — For practical manufacturing purposes, it is assumed that the minimal leg width is $a = 3$ mm and the minimal distance $d$ = 1 mm. In Figure 14, the efficiencies are plotted as a function of solar radiative flux in the range q″solar = 2 – 20 W $cm^{-1}$ for a module with a leg length $l$ = 7.5 mm and for: (a) $d$ = 1 mm and $a$ = 3, 4.5, 6 mm, and (b) $a$ = 3 mm and $d$ = 1, 2, 3 mm. Highest efficiencies are obtained for $a$ = 3 mm in the whole solar flux range, and for $d$ = 1 mm in the range q″solar = 8 – 20 W $cm^{-2}$. The peak $\eta$ = 0.375% at q″solar = 20 W $cm^{-2}$ is obtained for $a$ = 3 mm and $d$= 1 mm, *i.e.* for the smallest leg width and distance between neighboring legs considered.

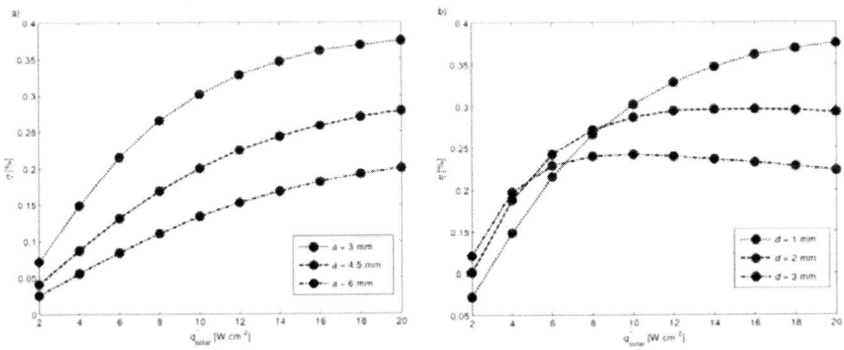


**Figure 14.** Efficiency as a function of solar radiative flux with $l$ = 7.5 mm for **(a)** $d$ = 1 mm and $a$ = 3, 4.5, 6 mm and for **(b)** $a$ = 3 mm and $d$ = 1, 2, 3 mm.

## 6. SUMMARY AND CONCLUSIONS

The conversion of high-temperature solar heat was demonstrated using a set of non-optimized TEC modules exposed to concentrated solar radiation. A 2D heat transfer numerical model of a TEC module has been implemented and validated based on experimental data. Two different approaches of modeling the radiation have been applied, namely the Monte Carlo method considering a semi-transparent absorber plate and the radiosity method considering only opaque surfaces. The absorber plate used in the measurements can be well approximated by an opaque surface. The heat transfer analysis for 4-leg modules with leg length $l$ = 5–10 mm and absorber plates of $L \times L \times b$ = 30 × 30 × 0.25 mm indicated that more than 70% of the incident solar power is lost due to re-radiation and free convection from the absorber, while 20% is conducted through the legs and <10% is lost by radiation to the cold plate. Heat conduction is the predominant mode of heat transfer across the legs, as corroborated by the measured and simulated linear temperature profiles across the legs. The optimal leg length of the 4-leg module with leg width $a$ = 4.5 mm and plates $L \times L \times b$ =

30 × 30 × 0.25 mm is $l$ = 7.5 mm, which results in a maximal efficiency of 0.081% at q″solar = 4 W $cm^{-2}$. Smaller leg width and distance between neighboring legs, namely, width $a$= 3 mm and gap $d$ = 1 mm, and leg length $l$ = 7.5 mm, leads to $\eta$ = 0.4%. For smaller dimensions than simulated here, the efficiency of this 4 leg module is expected to be 0.5%.

## NOMENCLATURE

| | |
|---|---|
| $A$ | surface, $m^2$ |
| $a$ | width, m |
| $b$ | thickness, m |
| $C^{\sim}$ | solar concentration ratio |
| $d$ | distance between legs, m |
| e | charge of electron/charged particles, Coul |
| $F_{k\text{-}j}$ | view factor from surface $k$ to $j$ |
| g | gravitational acceleration, m $s^{-2}$ |
| $h$ | heat transfer coefficient, W $m^{-2}$ $K^{-1}$ |
| $j$ | current density, A $m^{-2}$ |
| $k$ | thermal conductivity, W $m^{-1}$ $K^{-1}$ |
| $l$ | leg length, m |
| $L$ | absorber plate width, m |
| $N$ | number of Monte Carlo rays |
| $P$ | power output, W |

| | |
|---|---|
| $R$ | resistance, Ω |
| $q''$ | heat flux, W m$^{-2}$ |
| Qsolar | incident solar power, W |
| q″solar | incident solar radiative flux, W m$^{-2}$ |
| $q'''$ | volumetric heat source, W m$^{-3}$ |
| $S$ | Seebeck coefficient, V K$^{-1}$ |
| $t$ | time, s |
| $T$ | temperature, K |
| $V$ | voltage, V |
| $v$ | volume, m$^3$ |
| $x,y$ | coordinates |
| $\Delta V$ | control volume, m$^3$ |
| $\Delta A$ | control surface, m$^2$ |
| $\Delta x$ | cell length in x-direction, m |

## GREEK LETTERS

| | |
|---|---|
| $\alpha$ | thermal diffusivity, m$^2$ s$^{-1}$ |
| $\beta$ | extinction coefficient, m$^{-1}$ |
| $\beta$ | volumetric thermal expansion coeff, m$^3$ K$^{-1}$ |
| $\gamma$ | iteration step |

| | |
|---|---|
| $\delta_{kj}$ | Kronecker delta |
| $\varepsilon$ | total emissivity |
| $\varepsilon$ | error |
| $\eta$ | efficiency |
| $\mu$ | chemical potential, J mol$^{-1}$ |
| $\rho$ | electrical resistivity, Ω m |
| $\sigma$ | Stephan-Boltzmann constant, W m$^{-2}$ K$^{-4}$ |
| ν | kinematic viscosity, m$^2$ s$^{-1}$ |
| $\omega$ | scattering albedo |

## SUBSCRIPTS

| | |
|---|---|
| *i*, *j* | finite volume indices |
| max | maximum |
| *m*, *n* | numbers of finite volumes |
| N | n-type |
| oc | open circuit |
| P | p-type |
| ray | single ray in Monte Carlo |
| ∞ | surroundings |

# DIMENSIONLESS GROUP

| | |
|---|---|
| Gr | Grashof number |
| Nu | Nusselt number |
| Pr | Prandtl number |
| Ra | Rayleigh number |
| Re | Reynolds number |
| ZT | figure-of-merit |

# ABBREVIATIONS

| | |
|---|---|
| FV | Finite Volume |
| HFSS | high flux solar simulator |
| MC | Monte Carlo |
| N | n-type |
| OC | open circuit |
| P | p-type |
| PDE | Partial Differential Equation |
| SOR | successive over-relaxation |
| TEC | Thermoelectric converters |

## ACKNOWLEDGEMENTS

We thank P. Haueter and L. Schlumpf for technical support during the experimental campaign at ETH's High-Flux Solar Simulator, and M. Trottmann, P. Hinz, A. Haemmerli for the fabrication of the TEC demonstrator modules at EMPA.

## REFERENCES

1. Goldschmid, H.J.; Nolas, G.S. A review of the New Thermoelectric Materials. In 20th International Conference on Thermoelectrics, Beijing, China, June, 2001.
2. Min, G. Thermoelectric Module Design Theories. In *Thermoelectrics Handbook: Macro to Nano*; CRC Press: Boca Raton, FL, USA, 2006.
3. Rowe, D.M. General Principles and Basic Considerations. In *Thermoelectrics Handbook: Macro to Nano*; CRC Press: Boca Raton, FL, USA, 2006.
4. Riffat, S.B.; Ma, X. Thermoelectrics: a review of present and potential applications. *Appl. Therm. Eng.* **2003**, *23*, 913–935.
5. Tomeš, P.; Robert, R.; Bocher, L.; Trottmann, M.; Aguirre, M.H.; Weidenkaff, A.; Haueter, P.; Steinfeld, A.; Hejtmánek, J. Direct conversion of simulated solar radiation into electrical energy by a perovskite thermoelectric oxide module (TOM). *Proc. Mater. Sci. Technol. Confer. Exhibit.* **2008**, *1*, 429–435.
6. Weidenkaff, A.; Robert, R.; Aguirre, M.H.; Bocher, L.; Lippert, T.; Canulescu, S. Development of Thermoelectric Oxides for Renewable Energy Conversion Technologies. *Renewable Energy* **2008**, *33*, 342–347.
7. Snyder, G.J. Application of the compatibility factor to the design of segmented and cascaded thermoelectric generators. *Appl. Phys. Lett.* **2004**, *84*, 2436.
8. Snyder, G.J. Thermoelectric Power Generation: Efficiency and Compatibility. In *Thermoelectrics Handbook: Macro to Nano*; CRC Press: Boca Raton, FL, USA, 2006.
9. Snyder, G.J.; Ursell, T.S. Thermoelectric Efficiency and Compatibility. *Phys. Rev. Lett.* **2003**, *91*, 148301:1–148301:4.

10. Yamashita, O. Effect of linear temperature dependence of thermoelectric properties on energy conversion efficiency.*Energ. Conv. Manage.* **2008**, *49*, 3136–3169.

11. Hogan, T.P.; Shih, T. Modeling and Characterization of Power Generation Modules Based on Bulk Materials. In*Thermoelectrics Handbook: Macro to Nano*; CRC Press: Boca Raton, FL, USA, 2006.

12. Tomeš, P.; Trottmann, M.; Weidenkaff, A.; Suter, C.; Haueter, P.; Steinfeld, A. Thermoelectric oxide modules (TOMs) applied in direct conversion of simulated solar radiation into electrical energy. *Materials* **2010**. submitted.

13. Bocher, L.; Aguirre, M.H.; Logvinovich, D.; Shkabko, A.; Robert, R.; Trottmann, M.; Weidenkaff, A.; Quelle, A. $CaMn_{1-x}Nb_xO_3$ ($x \leq 0.08$) perovskite-type phases as promising new high-temperature n-type thermoelectric materials.*Inorg. Chem.* **2008**, *47*, 8077–8085.

14. Hirsch, D.; von Zedtwitz, P.; Osinga, T. A New 75 kW High-Flux Solar Simulator for High-Temperature Thermal and Thermochemical Research. *J. Sol. Energ. Eng.* **2003**, *125*, 117–120.

15. *Thermogage Vatell, 1000 Series*, http://www.vatell.com/thermogage.htm (December 2009).

16. Hagmann, G. *Grundlagen der Elektrotechnik*; Aula-Verlag: Wiebelsheim, Germany, 2009.

17. Modest, M.F. *Radiative Heat Transfer*, 2nd ed.; Academic Press: San Diego, CA, USA, 2003.

18. Siegel, R.; Howell, J. *Thermal Radiation Heat Transfer*; Taylor & Francis Inc.: New York, NY, USA, 2002.

19. A catalog of radiation heat transfer configuration factors. http://www.me.utexas.edu/~howell/tablecon.html(December 2009).

20. Incropera, F.P.; DeWitt, D.P. *Fundamentals of Heat and Mass Transfer*; John Wiley & Sons Inc.: Hoboken, NY, USA, 2002.

21. Patankar, S. *Numerical Heat Transfer and Fluid Flow*; Taylor & Francis: New York, NY, USA, 1980. Touloukian, Y.S. *Thermal Radiative Properties: Nonmetallic solids*; IFI/Plenum: New York, NY, USA, 1972.

22. MPDB 5.50. JAHM Software, Inc. North Reading, PA, USA, 1999.

# CHAPTER 6

# Forced Convection Heat Transfer Coefficient and Pressure Drop of Diamond-Shaped Fin-Array

Shigeki Hirasawa*, Atsushi Fujiwara, Tsuyoshi Kawanami, Katsuaki Shirai

*Department of Mechanical Engineering, Kobe University, Kobe, Japan*

## ABSTRACT

Forced convection cooling of fins on a high-temperature wall has been used to cool high-power electronic devices. We numerically calculated and experimentally measured the forced convection heat transfer coefficient and pressure drop of a diamond-shaped fin-array with water flow in this study, which had been reported to have a self-induced flip-flop flow phenomenon. Although the flip-flop flow phenomenon occurred in calculations, it was not observed in experiments. The heat transfer and pressure drop of the diamond-shaped fin-array could be estimated with equations for turbulent flow in tubes.

## KEYWORDS

Forced Convection, Heat Transfer Performance, Diamond-Shaped Fin, Flip-Flop Flow

# 1. INTRODUCTION

The heat density of high-load electronics, such as high-performance servers and the inverters used in vehicles, has been increasing as a result of higher processing speeds and higher power outputs. The method used to cool electronic devices depends on the level of heat generation and the operating environment, whereas available methods include natural convection cooling [1] , forced convection cooling, and heat pipe cooling [2] . Forced convection cooling is used most often because of its practical advantages in cost and performance. For these reasons, we have been developing forced convection cooling technologies to cool high-power electronic devices [3] . Pin-fin arrays have been studied to enhance forced convection heat transfer in a wide variety of applications, such as heat exchangers and electrical devices. Pin-fin arrays usually consist of a number of cylinders attached perpendicularly to a high-temperature wall with a coolant fluid passing a cross flow over the cylinders. Tsuzuki et al. [4] studied an S-shaped fin-array and reported a method of calculating heat transfer in the fin-array. Tanda [5] studied heat transfer and pressure drop in a diamond-shaped fin-array with a leading angle of 90°. Umeda et al. [6] [7] reported experimental results obtained from a self-induced flip-flop flow phenomenon in diamond-shaped cylinder bundles with a leading angle of 30°. The flip-flop flow oscillation promoted fluid mixing and had the potential to enhance convection heat transfer. However, they did not present experimental results for convection heat transfer in a diamond-shaped fin-array with flip-flop flow oscillation.

In this work, we numerically calculated and experimentally measured the forced convection heat transfer coefficient and pressure drop of a diamond-shaped fin-array with water passing a cross flow over the fin-array.

# 2. NUMERICAL CALCULATION METHOD

We used a 3-dimensional thermo-fluid computation code called "STAR-CCM+" [8] in this work to calculate flow velocity and temperature distributions in a water flow around a diamond-shaped fin-array. We used the finite volume method for the numerical calculations. The conditions of incompressible fluid, unsteady turbulent flow with the large eddy simulation (LES) model, constant properties, and discretization with polyhedralmesh were used in the calculations. Figure 1 outlines the calculation conditions. The length was 66.3 mm in the x direction, the width of the flow region was 20.7 mm in the y direction, and the thickness was 2 mm in the z direction. The diamond-shaped fins were in a staggered array with six rows in the x direction and five rows in the y direction. The leading angle of the diamond-shaped fin was 30°, their width was 2.07 mm in the y direction, and their length was 7.73 mm in the x direction.

The width of the grooves between the fins was w = 2 mm in the flow direction of 15° was different from that in the x direction. We only calculated the temperature distribution in the fluid between the spaces created by fins and did not calculate the temperature distribution in the fins. The main reason for this was to understand the convection heat transfer coefficient for basic boundary conditions with the constant wall temperature of fins. The surface temperature of the fins was $T_w$ = 40°C. The flow boundary condition at the surface of the fins had no-slip wall. The boundary condition for the outside wall of the flow region was symmetrical except for the inlet and outlet regions. The fluid was water with an inlet temperature of $T_a$ = 20°C, density of r = 998 kg/m$^3$, specific heat capacity of $C_p$ = 4180 J/(kg·K), viscosity of m = 8.89 × $10^{-4}$ (Pa·s), and thermal conductivity of l = 0.62 W/(m·K). The time step of unsteady calculation was 0.001 s, the number of iterations at each time step was 20, and the flow velocity and temperature distributions were obtained after more than 0.1 s with less than 0.2% change in the average heat flux from the outside wall of the fins. The transient local heat flux from the outside wall of the fins changed periodically because of the flip-flop flow phenomenon. The inlet velocities were $u_{in}$ = 1, 3 and 10 m/s, which corresponded to the Reynolds number of Re = (rud/m) = 8980, 26,900 and 89,800 using average the velocity in grooves u = $2u_{in}$, and the hydraulic diameter d = 2w of grooves with the width w between the fins.

## 3. CALCULATION RESULTS

Figure 2 has the calculated results for fluid flow velocity at a transient time ($t = t_0$) in the center cross section for the inlet velocity of $u_{in}$ = 3 m/s and Re = 26,900. The fluid flow caused separation behind the diamond-shaped fins and the flow pattern was not symmetrical in the y direction especially behind the fin-array near the outlet. Figure 3 shows the fluid flow velocity and temperature distributions at time ($t = t_0 - 0.02$ s), which is 0.02 s before from that for Figure 2. The separation and fluid flow behind the diamond-shaped fins are different from those in Figure 2 and there is a flip-flop flow phenomenon as was reported by Umeda et al. [6] [7] . Figure 4 shows the flow velocity distribution near the fin in the (3, 3)-row at time (a) $t = t_0 - 0.012$ s, (b) $t = t_0 - 0.008$ s, (c) $t = t_0 - 0.004$ s, and (d) $t = t_0$. We can see the change in the flow direction of the flip-flop flow phenomenon and the change in the flow direction does not occur at the same time for all fins.

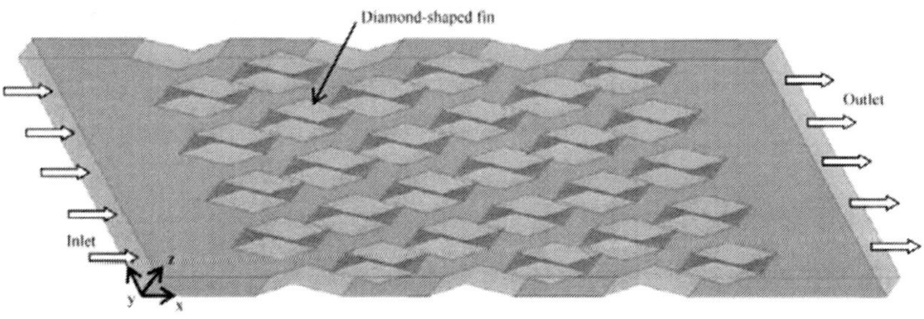


**Figure 1**. Calculation condition.

Figure 5 has the temperature distributions at time ($t = t_0$) in the center cross section. The temperature of the fluid increases along the flow direction due to convection heat transfer from the fins.

Figure 6 plots the local heat flux distribution q on the surface around the fin in the (3, 3)-row at time ($t = t_0$). The local heat heat flux q is high along the upstream side of the fin especially at its leading and flank edges. The local heat heat flux q is low along the downstream side of the fin because of the separation of fluid flow. Figure 7 plots the average heat flux $q_{ave}$ around the fins in the n-th row along the row in the x direction at time ($t = t_0$).

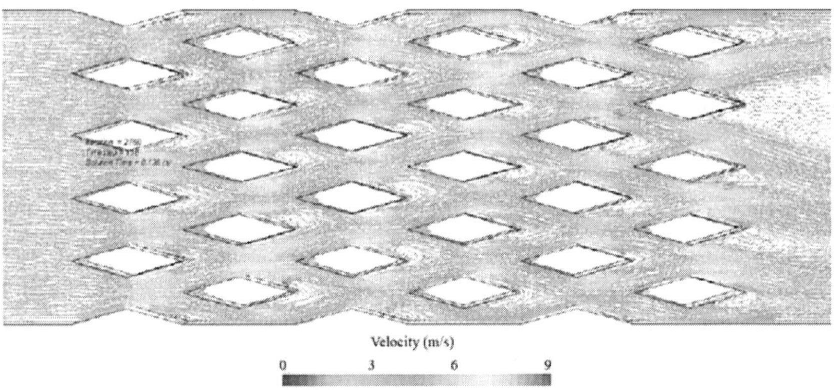


**Figure 2**. Flow velocity distribution ($t = t_0$, Re = 26,900).

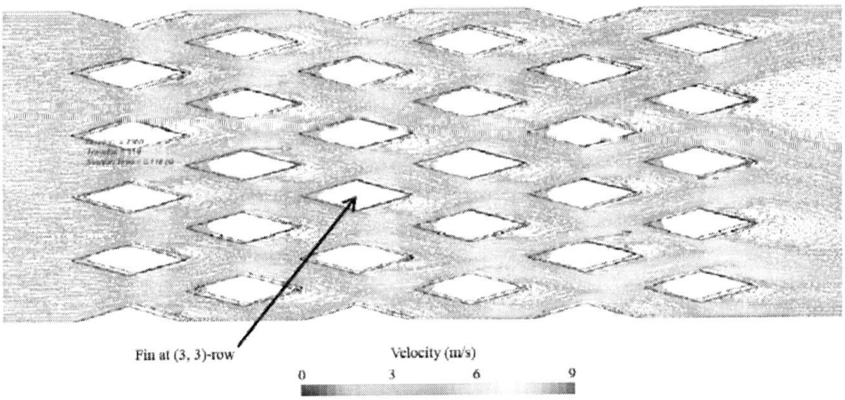


**Figure 3**. Flow velocity distribution (t = $t_0$ – 0.02s, Re = 26,900).

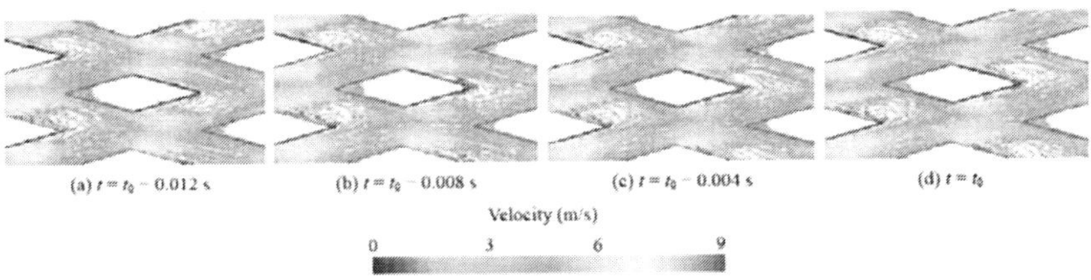


**Figure 4**. Flow velocity distribution near fin in (3, 3)-row (Re = 26,900).

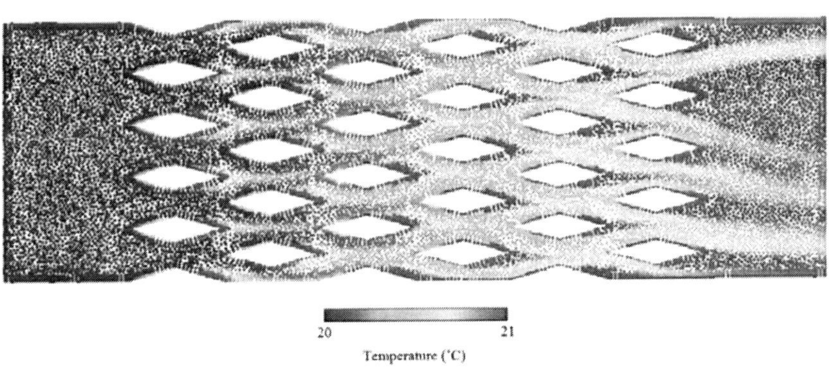


**Figure 5**. Temperature distribution (t = $t_0$, Re = 26,900).

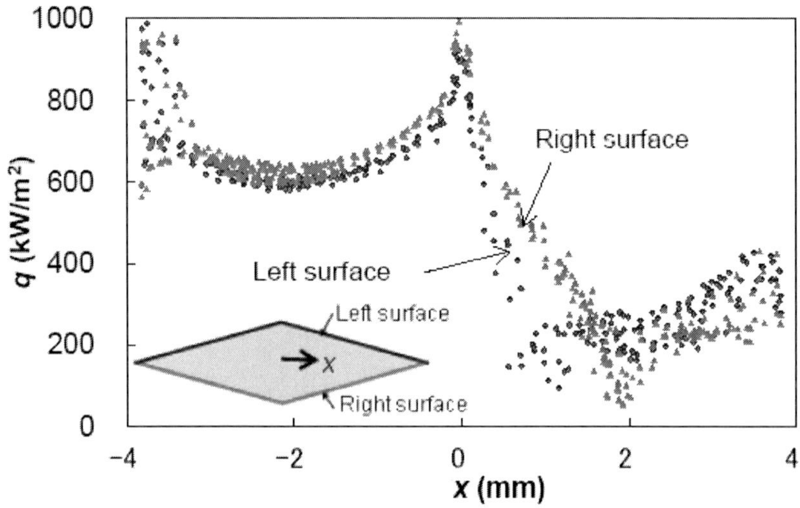


**Figure 6.** Local heat flux distribution around fin in (3, 3)-row ($t = t_0$, Re = 26,900).

The average heat flux $q_{ave}$ around the fins changes about 10% because of the flip-flop flow phenomenon. The average heat flux $q_{ave}$ around the fin in the first row is small because flow velocity is small on the upstream side of the fins.

Figure 8 and Figure 9 show the fluid flow velocity at a transient time in the center cross section for the inlet velocities of $u_{in}$ = 1 and 10 m/s (Re = 8980 and 89,800). The flow velocity distributions are different from those in Figure 2 ($u_{in}$ = 3 m/s, Re = 26,960) but the same flip-flop flow phenomena occur.

Average heat transfer coefficient $a\left(=q_{ave}/(T_w - T_a)\right)$ and Nusselt number Nu (=ad/l) were obtained from average heat flux $q_{ave}$ for all fins in the calculation region and the temperature difference between the wall of the fins $T_w$ = 40°C and inlet of fluid $T_a$ = 20°C, the hydraulic diameter d of grooves, and thermal conductivity l.

Figure 10 plots the relation between Reynolds number Re and Nusselt number Nu for our calculated results with the triangles "D". Nusselt number Nu for turbulent flow in a tube is known to be calculated with Equation (1) [9] .

$$\mathrm{Nu} = 0.023\,\mathrm{Re}^{0.8}\,\mathrm{Pr}^{0.4} \tag{1}$$

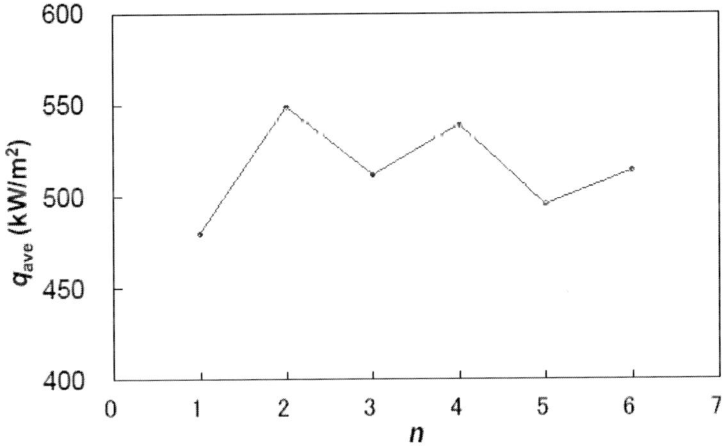


**Figure 7**. Average heat flux of fins along row ($t = t_0$, Re = 26,900).

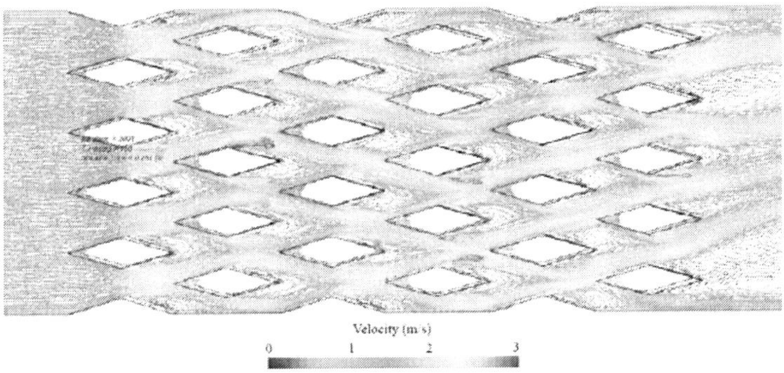


**Figure 8**. Flow velocity distribution (Re = 8980).

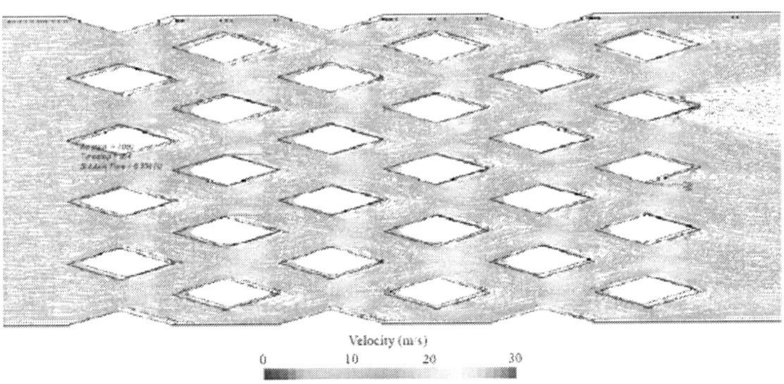


**Figure 9**. Flow velocity distribution (Re = 89,800).

Here, Pr is the Prandtl number of fluid (Pr = 6.0). The line in Figure 10 indicates the relation with Equation (1). Our calculated results agree with the relation in Equation (1).

Figure 11 plots the relation between Reynolds number Re and the pressure drop through fins DP for our calculated results with the triangles "D". Pressure drop DP for turbulent flow in a tube with the entrance and exit pressure loss can be calculated with Equation (2) [9] .

$$\Delta P = \left( k\frac{x}{d} + K_c \right)\frac{\rho u^2}{2} \tag{2}$$

Here, k is the friction coefficient with the function of the Reynolds number (k = 0.031, 0.024, and 0.013 at Re = 8980, 26,960, and 89,800 [9] ), $K_c$ is the sum of entrance and exit loss coefficients ($K_c$ = 1), r is fluid density, and u is the flow velocity. The line in Figure 11 plots the relation with Equation (2). Our calculated results agree with the relation in Equation (2). Therefore, we found that heat transfer and pressure drop in a diamond- shaped fin-array could be estimated with equations for turbulent flow in tubes though they had the flip-flop flow phenomenon.

## 4. EXPERIMENTAL APPARATUS

Figure 12 outlines the experimental apparatus. The test section was placed in an acrylic case (170 mm ′ 120 mm ′ 30 mm) and the excess spaces in the case were filled with dummy acrylic blocks. Water at a temperature of 15°C was supplied to the test section from a chiller after the flow rate was measured. Figure 13 shows the test fin-array. The diamond-shaped fin-array was made out of a copper block (70 mm ′ 70 mm ′ 15 mm) by cutting crossed grooves w = 2 mm in width, 2 mm in depth, and 4 mm in pitch. The leading angle of the diamond-shaped fins was 30°. The copper block was heated with three cartridge heaters (5 mm in diameter and a total heating rate of 200 W). The temperatures of the copper block and inlet water were measured with T-type thermocouples. The flow velocities u = 3.3 and 5 m/s in the grooves, which corresponded to the Reynolds number of Re = (rud/m) = 7410 and 11,200 using the hydraulic diameter d = w of rectangular grooves between the fins.

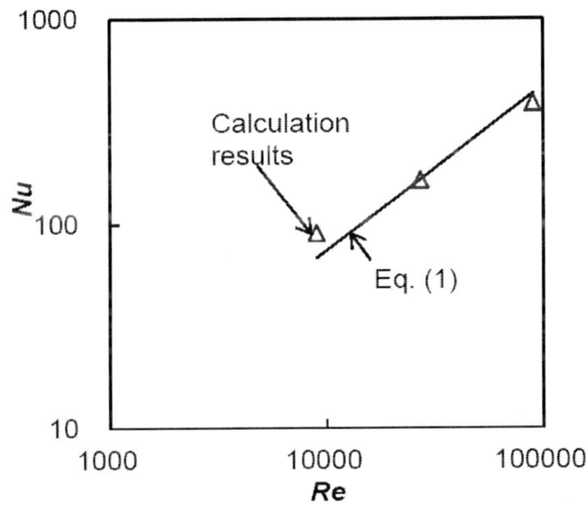


**Figure 10**. Calculation results of relation between Nusselt number and Reynolds number.

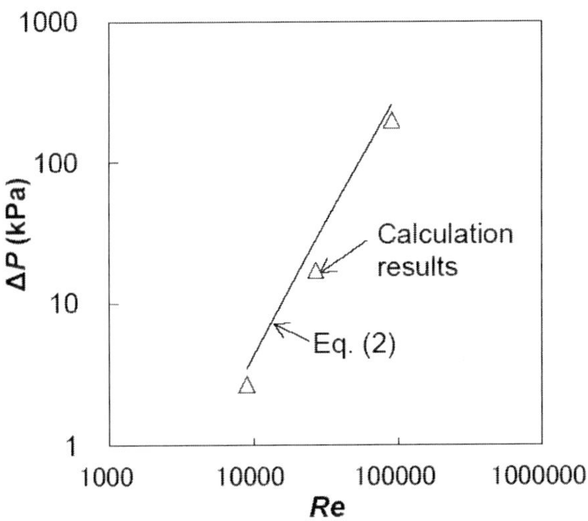


**Figure 11**. Calculation results of relation between pressure drop and Reynolds number.

## 5. EXPERIMENTAL RESULTS

The water flow pattern between the fins could be observed with tiny particles in water. The flip-flop flow phenomenon was not observed in our experiment under any experimental conditions. The reason for this is that the fluid space behind the fin-array was not enough to generate oscillation.Figure 14 plots the relation between Reynolds number Re and Nusselt number Nu in our experimental results with the circles "o". The line in Figure 14 indicates the relation with Equation (1) for turbulent flow in a tube. Our experimental results agreed with the relation in Equation (1) which was similar to the calculated results in Figure 10.

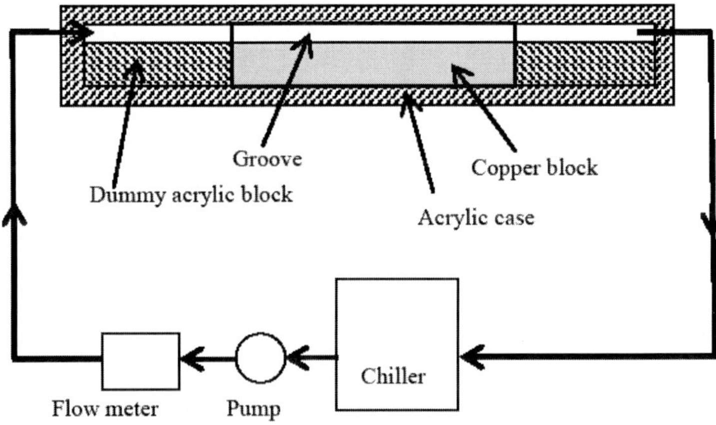


**Figure 12**. Experimental appatatus.

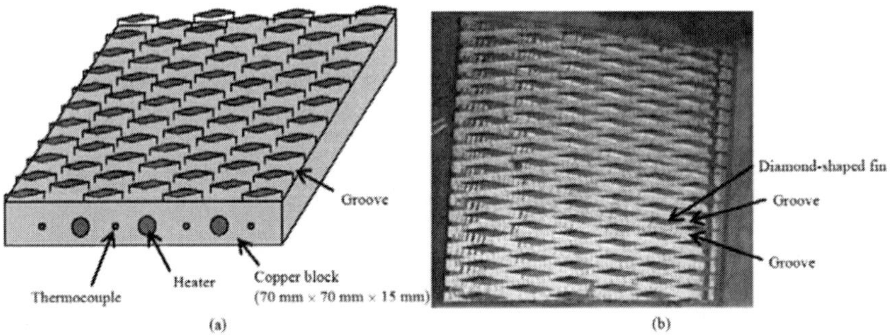


**Figure 13**. Test fin-array. (a) Schematic view; (b) Top view.

## 6. CONCLUSIONS

We numerically calculated and experimentally measured the forced convection heat transfer coefficient and pressure drop of a diamond-shaped fin-array with water flow. The following results were obtained.

1) The flip-flop flow phenomenon occurred in calculations but was not observed in experiments.

2) The heat transfer coefficient and pressure drop of the diamond-shaped fin-array could be estimated with equations for turbulent flow in tubes even though they had the flip-flop flow phenomenon.

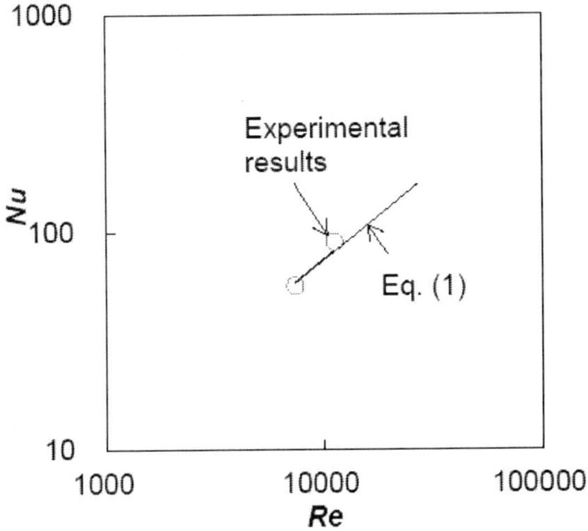


**Figure 14**. Experimental results of relation between Nusselt number and Reynolds number.

## REFERENCES

1. Nakazato, N., Hirasawa, S. and Mato, T. (1998) Natural Convection Cooling in Vertical Finned Plates in a Cabinet for Communication Equipment. IEICE Transactions on Electronics, 81, 421-426.

2. Toyoda, H. and Kondo, Y. (2013) Effect of Non-Condensable Gas Leakage on Long Term Cooling Performance of Loop Thermosyphon. Journal of Electronics Cooling and Thermal Control, 3, 131-135.

3. Atarashi, T., Tanaka, T., Tsubaki, S. and Hirasawa, S. (2014) Calculation Method for Forced-Air Convection Cooling Heat Transfer Coefficient of Multiple Lows of Memory Cards. Journal of Electronics Cooling and Thermal Control, 4, (Accepted to Be Published).

4. Tsuzuki, N., Utamura, M. and Ngo, T.L. (2009) Nusselt Number Correlations for a Microchannel Heat Exchanger Hot Water Supplier with S-Shaped Fins. Applied Thermal Engineering, 29, 3299-3308.

5. Tanda, G. (2001) Heat Transfer and Pressure Drop in a Rectangular Channel with Diamond-Shaped Elements. International Journal of Heat and Mass Transfer, 44, 3529-3541.

6. Umeda, S., Hasegawa, S. and Yang, W.J. (2005) Occurrence of Flip-Flop Flows in Diamond-Shaped Cylinder Bundles. Transaction of JSME B, 71, 2949-2955.

7. Umeda, S. and Yang, W.J. (2004) Effects of Mesh-Induced Upstream Turbulence on Flip-Flop Flow inside Diamond- Shaped Cylinder Bundles. Proceedings of ASMEHT-FED 2004 Conference, HT-FED2004-56140.

8. STAR-CCM+ Product Information (CD-Adapco Co.). http://www.cd-adapco.com/products/STAR-CCM_plus/index.html

9. Holman, J.P. (1989) Heat Transfer. McGraw-Hill Book Co., New York.

# CHAPTER 7

# Numerical Modeling of Non-Similar Mixed Convection Heat Transfer over a Stretching Surface with Slip Conditions

A. Subba Rao[1*], V. R. Prasad[1], N. Nagendra[1], K. V. N. Murthy[1], N. Bhaskar Reddy[2], O. Anwar Beg[3]

[1]*Department of Mathematics, Madanapalle Institute of Technology and Science, Madanapalle, India*
[2]*Department of Mathematics, Sri Venkateswara University, Tirupathi, India*
[3]*GortEngovation-Aerospace, Medical and Energy Engineering, Bradford, UK*

## ABSTRACT

In this paper, the heat transfer effect on the steady boundary layer flow of a Casson fluid past a stretching surface in the presence of slip conditions was analyzed. The stretching surface is maintained at a constant temperature. The boundary layer conservation equations, which are parabolic in nature, are normalized into non-similar form and then solved numerically with the well-tested, efficient, implicit, stable Keller-box finite difference scheme. The resulting equations are solved numerically by using the Kellerbox finite-difference method, and the expressions for velocity and temperature are obtained. They satisfy all imposed initial and boundary conditions and reduce to some well-known solutions for non-Newtonian fluids. Numerical results for velocity, temperature, skin friction and Nusselt number are shown in various graphs and discussed for embedded flow parameters. It is found that both velocity and temperature decrease with an increase of the Casson fluid parameter.

## KEYWORDS

Stretching Surface, Non-Newtonian Fluid, Slip Condition, Keller-Box Numerical Method, Heat Transfer, Skin Friction Coefficient

## 1. INTRODUCTION

Heat transfer in non-Newtonian fluids is an important research area due to its wide applications in food processing, petroleum production, cooling of an infinite metallic plate in a cooling bath and in many industries, for example polymers melt, polymer solutions employed in the plastic processing, a long thread traveling between a feed roll and a wind-up roll etc. Flow in the boundary layer on moving solid surface was historically first investigated by Sakiadis [1] [2] who observed that the boundary layer growth was in the direction of motion of the continuous solid surface and deviates from that of the classical Blasius flow past a flat plate. Erickson et al. [3] extended the Sakiadis problem to include blowing or suction at the moving surface and investigated its effects on the heat and mass transfer in the boundary layer. Bijjanal et al. [4] obtained closed-form similarity solutions for steady two-dimensional incompressible boundary layer flow caused by a stretching sheet with Non-Uniform Heat Source/Sink. Combined forced and free convection in boundary layers adjacent to a continuous horizontal sheet maintained at a constant temperature and moving with a constant velocity it was investigated numerically by Chen and Strobel [5] . Grubka and Bobba [6] have investigated the stretching sheet problem for a surface moving with linear velocity and with a variable surface temperature. Dutta et al. [7] studied numerically the Temperature Field in Flow over a Stretching Surface with Uniform Heat Flux. Chen and Char [8] investigated the effects of variable surface temperature and variable heat flux on the heat transfer characteristics of a linearly stretching sheet subject to blowing or suction. Several excellent studies of stretching flows in materials processing were presented by Karwe and Jaluria [9] [10] . Patil et al. [11] further analysed unsteady two-dimensional mixed convection flow along a vertical semi-infinite power law stretching sheet in a parallel free stream with a power-law temperature distribution. Nath et al. [12] analysed the three-dimensional, time dependent stretch surface flow. Ali and Al-Yousef [13] analysed mixed convection heat transfer from a uniformly stretching vertical surface with power function form for wall temperature. Partha et al. [14] described the effects of viscous dissipation on mixed convection heat transfer from an exponentially stretching surface.

Non-Newtonian transport phenomena arise in many branches of chemical and material processing engineering. Such fluids exhibit shear-stress-strain relationships which diverge significantly from the Newtonian (Navier-Stokes) model. Most non-Newtonian models involve some forms of modification to the momentum conservation equations. These include power-law, and thixotropic and viscoelastic fluids (Schowalter [15] ). Such rheological models however cannot simulate the microstructural characteristics of many important liquids including polymer suspensions, liquid crystal melts, physiological fluids, contaminated lubricants, etc. Several fluids in chemical engineering, multiphase mixtures, pharmaceutical formulations, china clay and coal in water, paints, synthetic lubricants, salvia, synovial fluid, jams, soups, jellies, marmalades, sewage sludge etc. are non-Newtonian. The constitutive relations for these kinds of fluids give rise to more complex and higher order equations than the Navier-Stokes equations. Considerable progress has been made on the topic by using different models of non- Newtonian fluids [16] - [25] . Previous studies indicate that not much has been presented yet regarding Casson fluid. This model (Casson [26] ; Nakamura et al. [27] ; Samir Kumar [28] ) in fact is a plastic fluid that exhibits shear thinning characteristics and that quantifies yield stress and high shear viscosity. Casson fluid model is reduced to a Newtonian fluid at a very high wall shear stress, when wall stress is much greater than yield stress. This fluid has good approximations for many substances such as biological materials, foams, molten chocolate, cosmetics, nail polish, some particulate suspensions etc. The boundary layer behaviour of viscoelastic fluid has technical applications in engineering such as glass fibre, paper production, manufacture of foods, the aerodynamic extrusion of plastic sheets, the polymer extrusion in a melt spinning process and many others.

Most of the existing studies on steady boundary layer flow and heat transfer with slip conditions are limited to the non-Newtonian fluid. The considered slip conditions especially are important in the non-Newtonian fluids such as polymer melts which often exhibit wall slip. This motivates us to consider the slip conditions in the present work for non-Newtonian fluids. More exactly, our aim is to investigate steady boundary layer flow and heat transfer of a Casson fluid past a stretching sheet with slip conditions. The equations of the problem are first formulated and then transformed into their dimensionless forms where the Keller box method is applied to find the exact solutions for velocity, temperature, Skin-friction and Nusselt number.

## 2. MATHEMATICAL ANALYSIS

We consider steady two-dimensional laminar mixed convection heat transfer flow along a stretching surface with partial slip. By applying two equal and opposing forces along the x-axis, the sheet is stretched with a speed proportional to the distance from the fixed origin x = 0 as shown in Figure 1. It is also assumed that the external electric field is zero and the electric field due to the polarization of chargers is negligible. The temperature is maintained at prescribed constant value. The fluid properties are assumed to be constant except the density variation in the buoyancy force term.

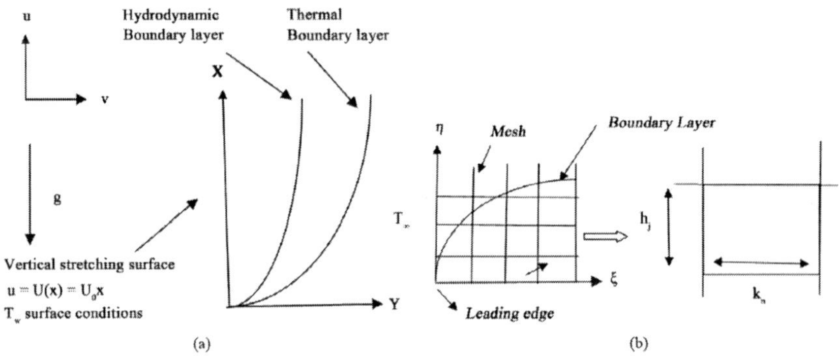


**Figure 1.** (a) Physical model and coordinate system; (b) Grid meshing and a Keller box computational cell.

The rheological equation of state for an isotropic flow of Casson fluid is (Nakamura et al. [27] ):

$$\tau_{ij} = \begin{cases} 2\left(\mu_B + \dfrac{p_y}{\sqrt{2\pi}}\right)e_{ij}, \pi \geq \pi_c \\ 2\left(\mu_B + \dfrac{p_y}{\sqrt{2\pi_c}}\right)e_{ij}, \pi < \pi_c \end{cases} \tag{1}$$

in which $\pi = e_{ij}e_{ij}$ and $e_{ij}$ represents the $(i, j)^{th}$ component of deformation rate, $\mu$ is the dynamic viscosity, $\pi$ denotes the product of the component of deformation rate with itself, $\pi_c$ shows a critical value of this product based on the non-Newtonian model, $\mu_B$ represents the plastic dynamic viscosity of non-Newtonian fluid and $p_y$ is the yield stress of fluid.

Under the usual Boussinesq and boundary layer approximations, the equations for mass continuity, (continuity/mass conservation) momentum and energy can be written in the following form:

$$\frac{\partial u}{\partial x}+\frac{\partial v}{\partial y}=0 \tag{2}$$

$$u\frac{\partial u}{\partial x}+v\frac{\partial u}{\partial y}=\nu\left(1+\frac{1}{\beta}\right)\frac{\partial^2 u}{\partial y^2}+g\Omega\left(T-T_\infty\right) \tag{3}$$

$$u\frac{\partial T}{\partial x}+v\frac{\partial T}{\partial y}=\alpha\frac{\partial^2 T}{\partial y^2} \tag{4}$$

where $u$ and $v$ are the velocity components in the $x$ - and $y$ - directions, $\nu$ is the kinematic viscosity of the conducting fluid, $\beta$ is the non-Newtonian Casson parameter, $\alpha$ is the thermal diffusivity, $T$ is the temperature respectively.

The boundary conditions are prescribed at the stretching surface and the edge of the boundary layer regime, respectively as follows:

$$\text{At } y=0,\ \ u=U(x)+N_0\left(1+\frac{1}{\beta}\right)\frac{\partial u}{\partial y},\ v=0,\ T=T_w+K_0\frac{\partial T}{\partial y}$$

$$\text{As } y\to\infty,\ \ u\to 0,\ \ T\to T_\infty \tag{5}$$

where $N_0$ is the velocity slip factor and $K_0$ is the thermal slip factor. For $N_0 = 0 = K_0$, one can recover the no- slip case. The stream function $\psi$ is defined by $u = \partial\psi/\partial y$ and $v = -\partial\psi/\partial x$, and therefore, the continuity equation is automatically satisfied. In order to write the governing equations and the boundary conditions in dimensionless form, the following non-dimensional quantities are introduced.

$$\psi = \nu \operatorname{Re}_x^{1/2} f(\xi,\eta),\ \xi = \frac{Gr_x}{\operatorname{Re}_x^2},\ \eta = \frac{y}{x}\operatorname{Re}_x^{1/2},$$

$$\Pr = \frac{\rho \nu c_p}{k},\ Gr = \frac{g\Omega(T_w - T_\infty)x^3}{\nu^2}, \operatorname{Re}_x = \frac{Ux}{\nu}$$

$$\theta(\xi,\eta) = \frac{T - T_\infty}{T_w - T_\infty},\ \beta = \mu_B \frac{\sqrt{2\pi_c}}{p_y} \tag{6}$$

where $\xi$ is the dimensionless stream wise coordinate, $f(\xi,\eta)$ is the dimensionless stream function, $\theta(\xi,\eta)$ is the temperature function, $\operatorname{Re}_x$ is the local Reynolds number, $Gr_x$ is the local thermal Grash of parameter, $\Omega$ is the coefficient of thermal expansion, $\rho$ is the density of the fluid, $Pr = \frac{\nu}{\alpha}$ is the Prandtl number, $T_\infty$ is the free stream temperature. The local mixed convection parameter $\xi$ is small near the leading edge where the forced convection dominates and large when the buoyancy force dominates the flow field. The stretching velocity of the surface obeys the relation:

$$U(x) = U_0(x) \tag{7}$$

where $U_0$ is a constant.

In view of Equations (6) and (7), Equations (2)-(4) reduce to the following coupled, nonlinear, dimensionless partial differential equations for momentum and energy for the regime

$$\left(1+\frac{1}{\beta}\right)f''' + ff'' - f'^2 + \xi\theta = \xi\left(f'\frac{\partial f'}{\partial \xi} - f''\frac{\partial f}{\partial \xi}\right) \tag{8}$$

$$\frac{\theta''}{\Pr} + f\theta' = \xi\left(f'\frac{\partial \theta}{\partial \xi} - \theta'\frac{\partial f}{\partial \xi}\right) \tag{9}$$

The transformed dimensionless boundary conditions are:

$$\begin{aligned} &\text{At } \eta = 0,\ f_w = 0,\ f' = 1 + \left(1+\frac{1}{\beta}\right)S_f f''(0),\ \theta = 1 + S_T\theta'(0) \\ &\text{As } \eta \to \infty,\ f' \to 0,\ \theta \to 0 \end{aligned} \tag{10}$$

In the above equations, the primes denote the differentiation with respect to $\eta$, the dimensionless transverse coordinate, and $\xi$ is the dimensionless tangential coordinate, $S_f = \frac{N_0 \mathrm{Re}_x^{1/2}}{x}$ and $S_T = \frac{K_0 \mathrm{Re}_x^{1/2}}{x}$ are the non-dimensional velocity and thermal slip parameters respectively. Here we assumed the typical values $K_0 = 0.5$, $N_0 = 0.25$ for finding the non-dimensional velocity and thermal slip parameters.

The engineering design quantities of physical interest include the skin-friction coefficient and Nusselt number, which are given by:

$$\frac{1}{2}\mathrm{Re}_x^{1/2} C_f = \left(1+\frac{1}{\beta}\right) f''(\xi,0) \tag{11}$$

$$\mathrm{Re}_x^{-1/2}\, Nu = -\theta'(\xi,0) \tag{12}$$

## 3. NUMERICAL SOLUTION

In this study the efficient Keller-Box implicit difference method has been employed to solve the general flow model defined by Equations (8)-(9) with boundary conditions (10). Therefore a more detailed exposition is presented here. This method, originally developed for low speed aerodynamic boundary layers by Keller [29] , and has been employed in a diverse range of coupled heat transfer problems. These include Ramachandra Prasad et al. [30] , Rao et al. [31] and Beg et al. [32] .

Essentially 4 phases are central to the Keller Box Scheme.

These are

a) Reduction of the N$^{th}$ order partial differential equation system to N first order equations;

b) Finite Difference Discretization;

c) Quasilinearization of Non-Linear Keller Algebraic Equations;

d) Block-tridiagonal Elimination of Linear Keller Algebraic Equations.

Phase a: Reduction of the N$^{th}$ order partial differential equation system to N first order equations

Equations (8)-(9) subject to the boundary conditions (10) are first written as a system of first-order equations. For this purpose, we reset Equations (8)-(9) as a set of simultaneous equations by introducing the new variables u, v and t:

$$f' = u \tag{13}$$

$$f'' = v \tag{14}$$

$$\theta' = t \tag{15}$$

$$\left(1+\frac{1}{\beta}\right)v' + fv - u^2 + \xi s = \xi\left(u\frac{\partial u}{\partial \xi} - v\frac{\partial f}{\partial \xi}\right) \tag{16}$$

$$\frac{1}{\Pr}t' + ft = \xi\left(u\frac{\partial s}{\partial \xi} - t\frac{\partial f}{\partial \xi}\right) \quad (17)$$

In terms of the dependent variables, the boundary conditions become:

$$\text{At } \eta = 0: u = 1 + \left(1 + \frac{1}{\beta}\right)f''(0),\ f = 0,\ s = 1$$

$$\text{As } \eta \to \infty: u \to 0,\ \ s \to 0 \quad (18)$$

Phase b: Finite difference discretization

A two dimensional computational grid is imposed on the x-η plane as sketched in Figure 2. The stepping process is defined by:

$$\eta_0 = 0, \eta_j = \eta_{j-1} + h_j, j = 1,2,\cdots,J,\ \eta_J = \eta_\infty \quad (19)$$

$$\xi^0 = 0, \xi^n = \xi^{n-1} + k_n, n = 1,2,\cdots,N \quad (20)$$

where $k_n$ and $h_j$ denote the step distances in the ξ and η directions respectively.

If $g_j^n$ denotes the value of any variable at $\left(\eta_j, \xi^n\right)$, then the variables and derivatives of Equations (13)-(17) at $\left(\eta_{j-1/2}, \xi^{n-1/2}\right)$ are replaced by:

$$g_{j-1/2}^{n-1/2} = \frac{1}{4}\left(g_j^n + g_{j-1}^n + g_j^{n-1} + g_{j-1}^{n-1}\right), \quad (21)$$

$$\left(\frac{\partial g}{\partial \eta}\right)_{j-1/2}^{n-1/2} = \frac{1}{2h_j}\left(g_j^n - g_{j-1}^n + g_j^{n-1} - g_{j-1}^{n-1}\right), \quad (22)$$

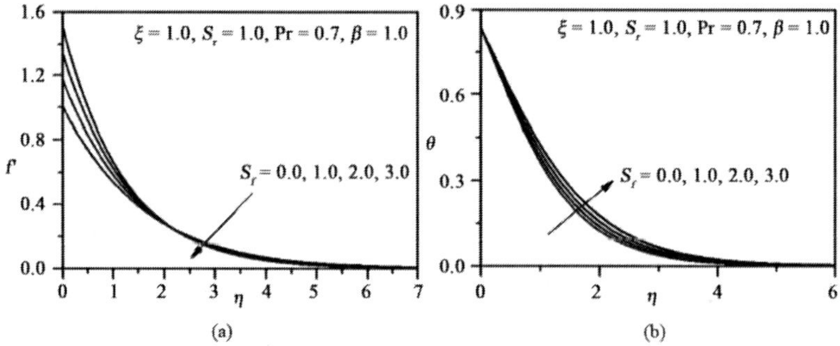


**Figure 2.** (a) Influence of $S_f$ on the velocity; (b) Influence of $S_f$ on the temperature.

$$\left(\frac{\partial g}{\partial \xi}\right)_{j-1/2}^{n-1/2} = \frac{1}{2k_n}\left(g_j^n - g_{j-1}^n + g_j^{n-1} - g_{j-1}^{n-1}\right), \tag{23}$$

We now state the finite-difference approximation of Equations (13)-(17) for the mid-point $\left(\eta_{j-1/2}, \xi^n\right)$, below

$$h_j^{-1}\left(f_j^n - f_{j-1}^n\right) = u_{j-1/2}^n, \tag{24}$$

$$h_j^{-1}\left(u_j^n - u_{j-1}^n\right) = v_{j-1/2}^n, \tag{25}$$

$$\left(1+\frac{1}{\beta}\right)\left(v_j - v_{j-1}\right) - \frac{h_j}{4}(1+\alpha)\left(u_j + u_{j-1}\right)^2 + \frac{\alpha h_j}{2} v_{j-1/2}^{n-1}\left(f_j + f_{j-1}\right)$$
$$+\frac{(1+\alpha)h_j}{4}\left[\left(f_j + f_{j-1}\right)\left(v_j + v_{j-1}\right)\right] - \frac{\alpha h_j}{2} f_{j-1/2}^{n-1}\left(v_j + v_{j-1}\right) + \frac{\xi h_j}{2}\left(s_j + s_{j-1}\right) = \left[R_1\right]_{j-1/2}^{n-1} \tag{26}$$

$$h_j^{-1}\left(\theta_j^n - \theta_{j-1}^n\right) = t_{j-1/2}^n, \tag{27}$$

$$\frac{1}{\Pr}\left(t_j - t_{j-1}\right) + \frac{(1+\alpha)h_j}{4}\left[\left(f_j + f_{j-1}\right)\left(t_j + t_{j-1}\right)\right] - \frac{\alpha h_j}{4}\left[\left(u_j + u_{j-1}\right)\left(s_j + s_{j-1}\right)\right]$$
$$+\frac{\alpha h_j}{2} s_{j-1/2}^{n-1}\left(u_j + u_{j-1}\right) - \frac{\alpha h_j}{2} u_{j-1/2}^{n-1}\left(s_j + s_{j-1}\right) - \frac{\alpha h_j}{2} f_{j-1/2}^{n-1}\left(t_j + t_{j-1}\right)$$
$$+\frac{\alpha h_j}{2} t_{j-1/2}^{n-1}\left(f_j + f_{j-1}\right) = \left[R_2\right]_{j-1/2}^{n-1} \tag{28}$$

where we have used the abbreviations

$$\alpha = \frac{\xi^{n-1/2}}{k_n} \tag{29}$$

$$\left[R_1\right]_{j-1/2}^{n-1} = -h_j\left[\left(1+\frac{1}{\beta}\right)\left(\frac{v_j - v_{j-1}}{h_j}\right) + (1-\alpha)\left(f_{j-1/2}v_{j-1/2}\right) - (1-\alpha)\left(u_{j-1/2}\right)^2 + \xi\left(s_{j-1/2}\right)\right] \tag{30}$$

$$\left[R_2\right]_{j-1/2}^{n-1} = -h_j\left[\frac{1}{\Pr}\left(\frac{t_j - t_{j-1}}{h_j}\right) + (1-\alpha)\left(f_{j-1/2}t_{j-1/2}\right) + \alpha\left(u_{j-1/2}s_{j-1/2}\right)\right] \tag{31}$$

The boundary conditions are

$$f_0^n = u_0^n = 0,\ \theta_0^n = 1,\ u_J^n = 0,\ \theta_J^n = 0 \tag{32}$$

Phase c: Quasilinearization of non-linear Keller algebraic equations

If we assume $f_j^{n-1}, u_j^{n-1}, v_j^{n-1}, s_j^{n-1}, t_j^{n-1}$ to be known for $0 \le j \le J$, Equations (24)-(28) are a system of 5J + 5 equations for the solution of 5J + 5 unknowns $f_j^n, u_j^n, v_j^n, s_j^n, t_j^n$, $j = 0, 1, 2, \cdots, J$. This non-linear system of algebraic equations is linearized by means of Newton's method as explained in Keller [29] and Prasad et al. [33] .

Phase d: Block-tridiagonal elimination of linear Keller algebraic equations

The linear system (24)-(28) can now be solved by the block-elimination method, since they possess a block- tridiagonal structure. Commonly, the block-tridiagonal structure consists of variables or constants, but here, an interesting feature can be observed, namely that it consists of block matrices. The complete linearized system is formulated as a block matrix system, where each element in the coefficient matrix is a matrix itself. Then, this system is solved using the efficient Keller-box method. The numerical results are affected by the number of mesh points in both directions. After some trials in the η-direction (radial coordinate) a larger number of mesh points are selected whereas in the ξ direction (tangential coordinate) significantly less mesh points are utilized. $\eta_{max}$ has been set at 10 and this defines an adequately large value at which the prescribed boundary conditions are satisfied. $\xi_{max}$ is set at 3.0 for this flow domain. Mesh independence is therefore achieved in the present computations. The computer program of the algorithm is executed in MATLAB running on a PC. The method demonstrates excellent stability, convergence and consistency, as elaborated by Keller [29] and this system is developed by Cebeci and Bradshaw [34] .

# 4. RESULTS AND DISCUSSIONS

Comprehensive solutions have been obtained and are presented in Figures 2-7. The numerical problem comprises 2 independent variables (x, h), 2 dependent fluid dynamic variables $(f,\theta)$ and 5 thermo physical and body force control parameters $\Pr, S_f, S_T, \beta, \xi$ . In the present computations, the following default parameters are prescribed (unless otherwise stated): Pr = 0.71, $S_f$ = 0.5, $S_T$ = 1.0, b = 1.0, x = 1.0.

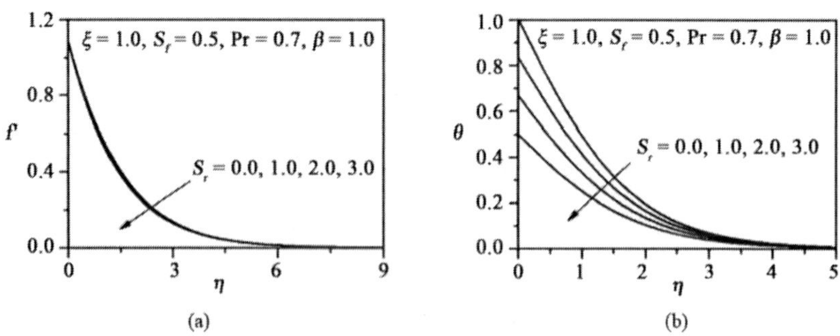


**Figure 3.** (a) Influence of $S_T$ on the velocity; (b) Influence of $S_T$ on the temperature.

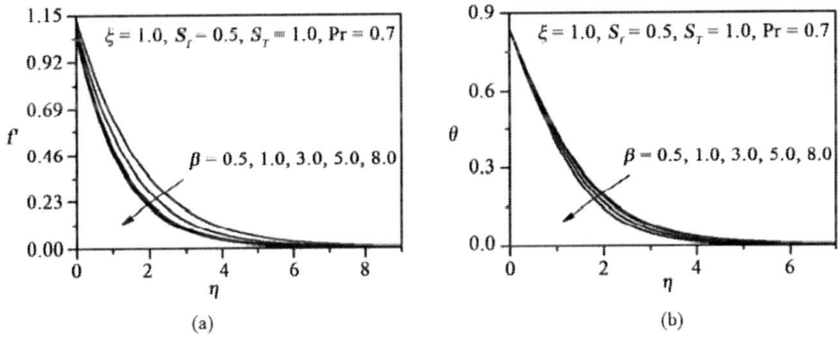


**Figure 4**. (a) Influence of β on the velocity; (b) Influence of β on the temperature.

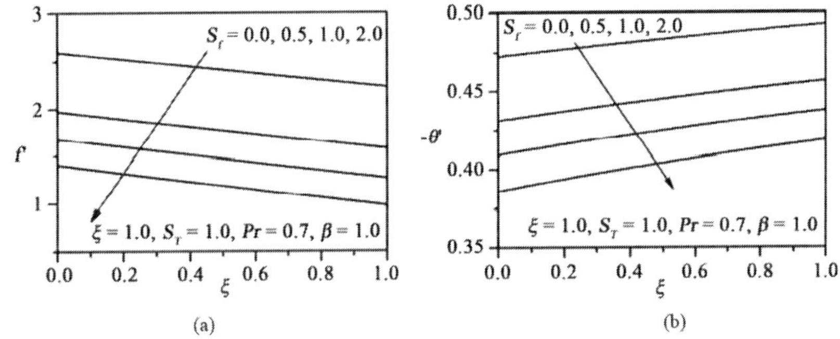


**Figure 5**. (a) Effect of $S_f$ on the Skin-friction coefficient results; (b) Effect of $S_f$ on the Nusselt number results.

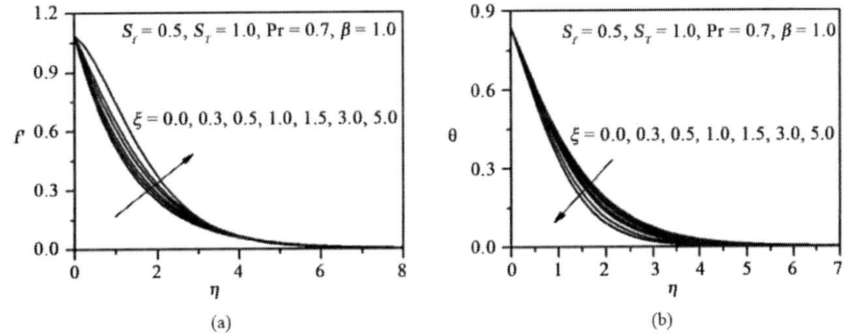


**Figure 6**. (a) Influence of ξ on the velocity; (b) Influence of ξ on the temperature.

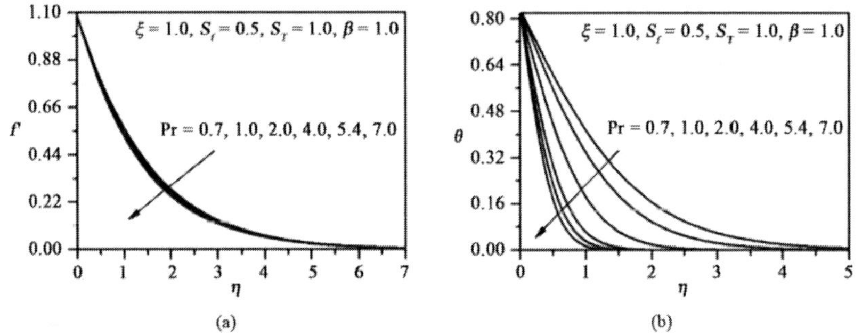


**Figure** 7. (a) Influence of Pr on the velocity; (b) Influence of Pr on the temperature.

In Figure 2(a) and Figure 2(b), the influence of velocity slip parameter on velocity and temperature is illustrated. In Figure 2(a) the dimensionless velocity component at the wall reduces with increase in slip parameter and hence there will be a decrease in the boundary layer thickness. The velocity profiles damped out a bit slower for the high amount of slip parameters, because of an interception which exhibits among them. Figure 2(b) indicates that an increase in slip parameter tends to increase temperature in the flow field. By increasing $S_f$, thermal boundary layer thickness enhances.

The variation of velocity and temperature with the transverse coordinate (h), over the thermal slip parameter $S_T$ is illustrated in Figure 3(a) and Figure 3(b). The response of velocity is much more consistent than for the case of changing velocity slip parameter, it is strongly decreased for all locations in the radial direction. The peak velocity accompanies the case of no thermal slip ($S_T = 0$). The maximum deceleration corresponds to the case of strongest thermal slip ($S_T = 3$). Temperatures (Figure 3(b)) are also strongly depressed with increasing thermal slip. The maximum effect is observed at the wall. Further into the free stream, all temperature profiles converge smoothly to the vanishing value. Figure 4(a) and Figure 4(b) depict the effect of the Casson fluid parameter β on velocity and temperature. Actually, with an increase in non-Newtonian Casson parameter (β), it produces resistance in the fluid flow. An increase in β implies a decrease in yield stress [illegible] of the Casson fluid and increase in the value of plastic dynamic viscosity $\mu_B$ this effect creates resistance in the flow of fluid. It is further noted that velocity decreases as the Casson fluid parameter $\beta$ increases. In Figure 4(b) it is shown that the effect of $\beta$ causes decreases in temperature. The effect of velocity slip parameter $S_f$ on stretching surface shear stress [illegible], local Nusselt number variation $\left(-\theta'\right)$ are presented in Figure 5(a) and Figure 5(b). In consistency with the earlier graphs described for velocity evolution, with an increase in $S_f$, wall shear stress is consistently reduced i.e. the flow is decelerated along the stretching surface. The impact of wall slip is therefore significant on the boundary layer characteristics of Casson flow from a surface. With an increasing $S_f$, the local Nusselt number is also considerably decreased and profiles are generally monotonic decays. Maximum local Nusselt number always arises at the stretching surface and is minimized with proximity to the greater distance from the stretching surface. In both Figure 5(a) and Figure 5(b), skin friction coefficient and local Nusselt number are maximized for the case of no-slip i.e. $S_f = 0$. In Figure 6(a) and Figure 6(b), the variation of velocity and temperature fields with different $\xi$ values is shown. Close to the stretching surface, velocity $(f')$ is found to be maximized closer to the stretching surface and minimized with progressive distance away from it i.e. the flow is decelerated with increasing $\xi$. However further from the wall, a marked acceleration in the flow is generated with greater distance from the surface i.e. velocity values are higher for higher values of $\xi$. Temperature q is found to noticeably decrease through the boundary layer with increasing $\xi$ values; as such the fluid regime is cooled most efficiently at the stretching surface and heated increasingly as we progress around the stretching surface periphery upwards. The effect of Prandtl number (Pr) on the primitive flow variables of velocity and temperature is shown in Figure 7(a) and Figure 7(b). Prandtl number signifies the ratio of viscous diffusion to thermal diffusion in the boundary layer regime. With greater Pr values, viscous diffusion rate exceeds thermal diffusion rate. An increase in Pr from 0.7 through 1.0, 2.0, 4.0, 5.4 to 7.0, strongly depresses velocities (Figure 7(a)) in the regime. For $Pr < 1$, thermal diffusivity exceeds momentum diffusivity i.e. heat will diffuse faster than momentum. For $Pr = 1.0$, both the viscous and energy diffusion rates will be the same as will the thermal and velocity boundary layer thicknesses. With increasing Pr values, temperature as shown in Figure 7(b), is markedly reduced throughout the boundary layer.

To validate the present solutions, we compare the present model with the earlier Newtonian model of Merkin [35] and we observe that an excellent agreement between the previous results as shown in Table 1.

## 5. CONCLUSIONS

In this study, numerical solutions have been presented for flow and heat transfer of Casson fluid from a permeable isothermal stretching surface with partial slip.

The model has been developed to simulate food stuff transport processes in industrial manufacturing operations. A robust, extensively-validated and implicit finite difference numerical scheme has been implemented to solve the transformed and dimensionless velocity and thermal boundary layer equations, subject to physically realistic boundary conditions. The results in summary have shown that, when increasing the velocity slip parameter, velocity, skin friction and Nusselt number decrease, but the temperature increases. A significant finding of this study is that flow separation can be controlled by increasing the value of Casson fluid parameter as well as by increasing Prandtl number.

**Table 1.** Values of the local heat transfer coefficient for various values of $\xi$ with $\beta \to \infty$, .

| $\xi$ | $-\theta'(\xi,0)$ | |
|---|---|---|
| | Merkin [35] | Present results |
| 0.0 | 0.4212 | 0.4198 |
| 0.2 | 0.4204 | 0.4201 |
| 0.4 | 0.4182 | 0.4191 |
| 0.6 | 0.4145 | 0.4149 |
| 0.8 | 0.4093 | 0.4098 |
| 1.0 | 0.4025 | 0.4034 |
| 1.2 | 0.3942 | 0.3945 |
| 1.4 | 0.3843 | 0.3847 |
| 1.6 | 0.3727 | 0.3729 |
| 1.8 | 0.3594 | 0.3598 |
| 2.0 | 0.3443 | 0.3448 |
| 2.2 | 0.3270 | 0.3272 |
| 2.4 | 0.3073 | 0.3079 |
| 2.6 | 0.2847 | 0.2852 |
| 2.8 | 0.2581 | 0.2588 |
| 3.0 | 0.2252 | 0.2254 |
| $\pi$ | 0.1963 | 0.1964 |

The current study has been confined to steady-state flow i.e. ignored transient effects and neglected thermal radiation heat transfer effects [36] . Generally, very stable and accurate solutions are obtained with the present finite difference code and it is envisaged that other non-Newtonian flows will be studied using this methodology in the future, including Maxwell upper convected fluids [37] , and couple stress fluids [38] . These aspects are also of relevance to rheological food processing simulations and will be considered in future investigations.

## ACKNOWLEDGEMENTS

The authors are grateful to the reviewers for giving their constructive comments for improving this article. The work is supported by the University Grants Commission-SERO. The authors are thankful to UGC-NEWDELHI, S.V University, Tirupati and management of MITS, Madanapalle.

## REFERENCES

1. Sakiadis, B.C. (1961) Boundary-Layer Behavior on Continuous Solid Surfaces: I. Boundary-Layer Equations for Two- Dimensional and Axisymmetric Flow. AIChE Journal, 7, 26-28.
2. Sakiadis, B.C. (1961) Boundary-Layer Behavior on Continuous Solid Surfaces: II. Boundary-Layer Equations on a Continuous Flat Surface. AIChE Journal, 7, 221-225.
3. Erickson. L.E., Fan. L.T. and Fox, V.G. (1966) Heat and Mass Transfer on Moving Continuous Flat Plate with Suction or Injection. Industrial Engineering Chemistry Fundamentals, 5, 19-25.
4. Gireesha, B.J., Roopa, G.S. and Bagewadi, C.S. (2011) Boundary Layer Flow of an Unsteady Dusty Fluid and Heat Transfer over a Stretching Sheet with Non-Uniform Heat Source/Sink. Scientific Research, 3, 726-735.
5. Chen, T.S. and Strobel, F.A. (1980) Buoyancy Effects in Boundary Layer Adjacent to a Continuous, Moving Horizontal Flat Plate. Journal of Heat Transfer, 102, 170-172.
6. Grubka, L.J. and Bobba, K.M. (1985) Heat Transfer Characteristics of a Continuous, Stretching Surface with Variable Temperature. ASME J. Heat Transfer, 107, 248-250.
7. Dutta, B.K., Roy, P. and Gupta, A.S. (1985) Temperature Field in Flow over a Stretching Surface with Uniform Heat Flux. International Communications in Heat and Mass Transfer, 12, 89-94.
8. Chen, C.K. and Char, M.I. (1988) Heat Transfer of a Continuous Stretching Surface with Suction or Blowing. Journal of Mathematical Analysis and Applications, 135, 568-580.
9. Karwe, M.V. and Jaluria, Y. (1988) Fluid Flow and Mixed Convection Transport from a Moving Plate in Rolling and Extrusion Processes. ASME J. Heat Transfer, 110, 655-661.

10. Karwe, M.V. and Jaluria, Y. (1991) Numerical Simulation of Thermal Transport Associated With a Continuously Moving Flat Sheet in Materials Processing. ASME J. Heat Transfer, 113, 612-619.

11. Patil, P.M., Roy, S. and Pop, I. (2010) Unsteady Mixed Convection Flow over a Vertical Stretching Sheet in a Parallel Free Stream with Variable Wall Temperature. International Journal of Heat and Mass Transfer, 53, 4741-4748.

12. Rajeswari, V., Kumari, M. and Nath, G. (1993) Unsteady Three-Dimensional Boundary Layer Flow Due to a Stretching Surface. Actamechanica, 98, 123-141.

13. Ali, M. and Al-Yousef, F. (2002) Laminar Mixed Convection Boundary Layers Induced by a Linearly Stretching Permeable Surface. International Journal of Heat and Mass Transfer, 45, 4241-4250.

14. Partha, M.K., Murthy, P.V.S.N. and Rajasekhar, G.P. (2005) Effect of Viscous Dissipation on the Mixed Convection Heat Transfer from an Exponentially Stretching Surface. Heat and Mass Transfer, 41, 360-366.

15. Schowalter, W.R. (1978) Mechanics of Non-Newtonian Fluids. Pergamon Press, New York.

16. Rana, P. and Bhargava, R. (2012) Flow and Heat Transfer of a Nanofluid over a Nonlinearly Stretching Sheet: A Numerical Study. Communications in Nonlinear Science and Numerical Simulation, 17, 212-226.

17. Nazar, M., Fetecau, C., Vieru, D. and Fetecau, C. (2010) New Exact Solutions Corresponding to the Second Problem of Stokes for Second Grade Fluids. Nonlinear Analysis: Real World Applications, 11, 584-591.

18. Fetecau, C., Hayat, T., Zierep, J. and Sajid, M. (2011) Energetic Balance for the Rayleigh—Stokes problem of an Oldroyd-B fluid. Nonlinear Analysis: Real World Applications, 12, 1-13.

19. Wang, S.W. and Tan, W.C. (2008) Stability Analysis of Double-Diffusive Convection of Maxwell Fluid in a Porous Medium Heated from Below. Physics Letters A, 372, 3046-3050.

20. Tan, W.C. and Xu, M.Y. (2004) Unsteady Flows of a Generalized Second Grade Fluid with the Fractional Derivative Model between Two Parallel Plates. Acta Mechanica Sinica, 20, 471-476.

21. Zhang, Z.Y., Fu, C.J., Tan, W.C. and Wang, C.Y. (2007) On Set of Oscillatory Convection in a Porous Cylinder Saturated with a Viscoelastic Fluid. Physics of Fluids, 19, 98-104.

22. Rashidi, M.M., Chamkha, A.J. and Keimanesh, M. (2011) Application of Multi-Step Differential Transform Method on Flow of a Second Grade Fluid over a Stretching or Shrinking Sheet. American Journal of Computational Mathematics, 6, 119-128.

23. Ali, N., Hayat, T. and Asghar, S. (2009) Peristaltic Flow of Maxwell Fluid in a Channel with Compliant Walls. Chaos, Solitons & Fractals, 39, 407-416.

24. Attia, H.A. and Seddeek, M.A. (2007) On the Effectiveness of Uniform Suction or Injection on Two Dimensional Stagnation-Point Flow towards a Stretching Surface with Heat Generation. Chemical Engineering Communications, 194, 553-564.

25. Hussain, M., Hayat, T., Asghar, S. and Fetecau, C. (2010) Oscillatory Flows of Second Grade Fluid in a Porous Space. Nonlinear Analysis: Real World Applications, 11, 2403-2414.

26. Casson, N. (1959) In Reheology of Dipersed System. Peragamon Press, Oxford.

27. Nakamura, M. and Sawada, T. (1988) Numerical Study on the Flow of a Non-Newtonian Fluid through an Axisymmetric Stenosis. Journal of Biomechanical Engineering, 110, 137-143.

28. Samir Kumar, N. (2013) Analytical Solution of MHD Stagnation-Point Flow and Heat Transfer of Casson Fluid over a Stretching Sheet with Partial Slip. ISRN Thermodynamics, 2013, Article ID: 108264.

29. Keller, H.B. (1970) A New Difference Method for Parabolic Problems. In: Bramble, J., Ed., Numerical Methods for Partial Differential Equations, Academic Press, New York, 327-350.

30. Prasd, V.R., Vasu, B. and Beg, O.A. (2011) Thermo-Diffusion and Diffusion-Thermo Effects on Boundary Layer Flows. LAP Lambert Academic Publishing GmbH & Co. KG, Saarbrücken.

31. Rao, A.S., Prasad, V.R., Reddy, N.B. and Bég, O.A. (2013) Heat Transfer in a Casson Rheological Fluid from a Semi-infinite Vertical Plate with Partial Slip. Heat Transfer-Asian Research, 44, 272-291.

32. Bég, O.A., Prasad, V.R., Vasu, B., Reddy, N.B., Li, Q. and Bhargava, R. (2011) Free Convection Heat and Mass Transfer from an Isothermal Sphere to a Micropolar Regime with Soret/Dufour Effects. International Journal of Heat and Mass Transfer, 54, 9-18.

33. Prasad, V.R., Rao, A.S., Reddy, N.B., Vasu, B. and Beg, O.A. (2013) Modelling Laminar Transport Phenomena in a Casson Rheological Fluid

from a Horizontal Circular Cylinder with Partial Slip. Proceedings of the Institution of Mechanical Engineers, Part E: Journal of Process Mechanical Engineering, 227, 309-326.

34. Cebeci, T. and Bradshaw, P. (1984) Physical and Computational Aspects of Convective Heat Transfer. Springer, New York.
35. Merkin, J.H. (1977) Free Convection Boundary Layers on Cylinders of Elliptic Cross Section. Journal of Heat Transfer, 99, 453-457.
36. Prasad, V.R., Vasu, B., Prashad, D.R. and Bég, O.A. (2012) Thermal Radiation Effects on Magneto-Hydrodynamic Heat and Mass Transfer from a Horizontal Cylinder in a Variable Porosity Regime. Journal of Porous Media, 15, 261- 281.
37. B´eg, O.A. and Makinde, O.D. (2011) Viscoelastic Flow and Species Transfer in a Darcian High-Permeability Channel. Journal of Petroleum Science and Engineering, 76, 93-99.
38. Kairi, R.R. and Murthy, P.V.S.N. (2012) Effect of Melting on Mixed Convection Heat and Mass Transfer in a Non-Newtonian Fluid Saturated Non-Darcy Porous Medium. Journal of Heat Transfer, 134, Article ID: 042601.

# CHAPTER 8

# Optimal Design of Magnetohydrodynamic Mixed Convection Flow in a Vertical Channel with Slip Boundary Conditions and Thermal Radiation Effects by Using an Entropy Generation Minimization Method

Mohamad Yaghoub Abdollahzadeh Jamalabadi [1,2], Jae Hyun Park [3,*] and Chang Yeop Lee [4]

[1] Ship *Engineering Department, School of Mechanical Engineering, Maritime University of Chabahar, Chabahar 99717-56499, Iran*
[2] *Graduate School of Mechanical and Aerospace Engineering, Gyeongsang National University, Jinju, Gyeongnam 660-701, Korea*
[3] *Department of Aerospace and Software Engineering and Research Center for Aircraft Parts Technology, Gyeongsang National University, Jinju, Gyeongnam 660-701, Korea*
[4] *Thermochemical Energy System R&BD Group, Korea Institute of Industrial Technology, Cheonan-si, Chungcheongnam-do 331-822, Korea*

## ABSTRACT

Investigation of the effect of thermal radiation on a fully developed magnetohydrodynamic (MHD) convective flow of a Newtonian, incompressible and electrically conducting fluid in a vertical microchannel bounded by two infinite vertical parallel plates with constant temperature walls through a lateral magnetic field of uniform strength is presented. The Rosseland model for the conduction radiation heat transfer in an absorbing medium and two plates with slip-flow and no-slip conditions are assumed. In addition, the induced magnetic field is neglected due to the assumption of a small magnetic Reynolds number. The non-dimensional governing equations are solved numerically using Runge-Kutta–Fehlberg method with a shooting technique. The channel is optimized based on the Second Law of Thermodynamics by changing various parameters such as the thermal radiation parameter, the temperature parameter, Hartmann number, Grashof to Reynolds ratio, velocity slip length, and temperature jump.

## KEYWORDS

thermal radiation; slip boundary; mixed convection; entropy minimization; magnetohydrodynamic (MHD)

## 1. INTRODUCTION

Heat transfer in free and mixed convection in vertical channels takes place in many applications such as the design of cooling systems for electronic devices [1], chemical processing equipment [2], microelectronic cooling [3], and solar energy [4], *etc.* There is some research that deals with the assessment of the temperature and velocity fields for the vertical fully developed flow without considering the effect of thermal radiation [5,6], but heat transfer by simultaneous radiation and convection is important in various cases including combustion flows [7], furnaces [8], high-temperature reactors [9] and heat exchangers [10], combustion [11], solar energy [12], thermal radiative loading [13], and many others [14–16]. In addition, the significance of radiative heat transfer is found in the industrial processes and magnetohydrodynamic (MHD) flow [17]. MHD is the field of fluid mechanics that deals with the dynamics of an electrically conducting fluid under the influence of a magnetic field. MHD has gained considerable importance because of its wide ranging applications in fluid flow control and propulsion [18,19].

Slip effects under boundary conditions and thermal radiation in the fluid is the topic of many researches [20]. Industrial application of this combination in microchannels is in the flow of metal vapors [21,22], plasma chemical vapor deposition (CVD) [23], electrohydrodynamic mixers [24], rhombus microchannels [25], and laser-driven radiative shock experiments [26]. As stated in common textbooks [27] the slip flow regime is valid when $0.01 \leq Kn \leq 0.1$. In the slip flow regime the flow is dense enough to be considered a continuum but the no-slip boundary condition is not valid. In this regime a sub-layer on the order of one mean free path, known as the Knudsen layer, starts to become dominant between the bulk of the fluid and the wall surface. In the slip flow regime the flow is ruled by the Navier Stokes equations, and rarefaction properties are simulated by the limited slip at the wall using Maxwell's velocity slip and von Smoluchowski's temperature jump boundary conditions. However the slip boundary condition is not limited to compressible fluid flow and as shown by experiments for common Newtonian liquids such as isopropanol, *n*-hexadecane, water, ethanol, and toluene in microchannels with heights of approximately 350 nm–5 μm, it is demonstrated that the behaviour is not the same as expected by the no-slip boundary condition, and the slip boundary condition should be applied [28].

When fluid flow is confined by small systems such as micro/nanochannels, some discontinuity in velocity and temperature profiles might happen at the interface of the fluid and solid surface [28–30]. Ulmanella and Ho [28] experimentally observed the velocity for various micro-sized channels, which is a function of shear rate, type of liquid and surface morphology. Bocquet and Barrat [29] described a possibility of temperature jump along with velocity slip. They modeled velocity slip and temperature jump by introducing velocity slip length and temperature slip length terms, respectively. The velocity slip and temperature jump relations should be used as the boundary conditions for liquid-solid interface of micro/nanosized channel in other references [27].

One of the potential applications of current research is the design of collectors for thermal storage [30], such as in thermo-syphon solar water heaters [31]. As shown in Figure 1, a vertical channel acting as a collector of the thermal energy contains a moving phase change material heated by the solar radiation. This kind of configuration is commonly found in solar collectors for thermal storage applications and thermo-syphon solar water heaters. The working fluid is forced by a pump to transfer thermal energy from the collector to the thermal storage tank. The designed pump should create enough pressure gradients through the channel to ensure stable movement of the phase change against the wall friction and the natural convection flow caused by temperature difference between two walls of the channel. Here the right wall is assumed to be at a higher temperature than the left wall because it is heated by solar radiation and the left wall, the cold heat source, is at the melting point of the phase change material. The purpose of a solar thermal collector is to absorb as much sunlight as possible and a number of studies have been focused on increasing the absorptivity of the fluid medium in the solar collector. Recently, researchers have found that the absorptivity can be significantly enhanced by using a nanofluid that is a mixture of liquid and suspended nanoparticles [32]. By seeding the nanoparticles in the base fluid, the extinction coefficient of the medium increases over 100 $cm^{-1}$ [32]. It is apparently an optically thick medium even for microchannels whose characteristic length is on the microscale and therefore the optically thick medium approximation is applicable to the analysis of radiative transfer. In addition for the application of metal gas vapors and plasma in microchannels [21–23] the radiative absorption coefficient of the medium is large enough to assume a thick optical medium. The radiative heat transfer through such media can be simplified by the Rosseland diffusion approximation [4]. Other than the other applications [33–39], the other trend of analysing flow and heat transfer is to apply a Second Law of Thermodynamics analysis, and its design-related concept of entropy generation minimization. Entropy generation is associated with thermodynamic irreversibility, which presents in all flow and heat transfer developments. Bejan [40,41] founded theoretical work on entropy generation in flow systems. He showed that by minimizing the entropy, the efficiency of a thermal system could be enhanced based on the Second Law of Thermodynamics. Thereafter, several authors [42–51] have analyzed different problems to study the entropy effects on thermal systems and to find ways to minimize them.

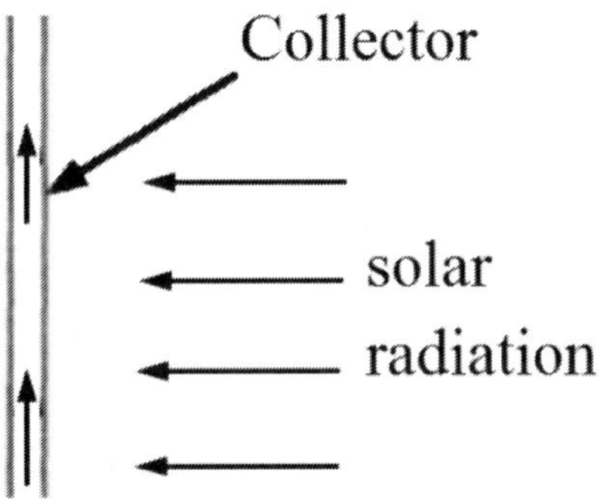


**Figure 1.** Thermal storage thermo-syphon solar water heaters.

Considering all the above, the goal of the current research was to optimize the steady state fully developed mixed convection incompressible flow in a vertical microchannel such that the walls of the channels are subjected to different wall temperatures and the Rosseland approximation model. The effects of thermal radiation coefficient, the thermal parameter, Prantdl number, Reynolds number, Hartmann number, Grashof number, Hartmann number, specific heat ratio and Knudsen number on entropy generation are investigated analytically.

## 2. GOVERNING EQUATIONS

In order to understand the fundamental basics of the fully developed laminar flow through an infinitely long vertical channel, a simple configuration as shown in Figure 2 is considered. An optically thick incompressible fluid is confined by two parallel planar walls that are separated by a distance of 2L. The two infinite vertical parallel plates of the channel are non-permeable, but there are slip-flow conditions on both of the plates. Since the infinitely long geometry is assumed, the flow velocity u is only a function of the y-direction across the channel and the origin of the system is placed at the middle point. The left and right walls (y = ±L) are maintained at the uniform temperatures of $T_L$ and $T_R$, respectively. The temperature difference of the two plates is also assumed to be high enough to induce heat transfer due to radiation. A uniform magnetic field of strength B is applied perpendicular to the plates of the channel. The magnetic Reynolds number is presumed small enough consequently that the induced magnetic field is neglected. In the steady state, this system can be described by momentum and energy equations employing Boussinesq approximation for buoyancy force (All fluid properties are assumed to be constant except that of the influence of density variation with temperature is considered only in the body force term). By ignoring the horizontal component of the velocity in the continuity equation, the simplified momentum equation for fully developed velocity (u(y)) is:

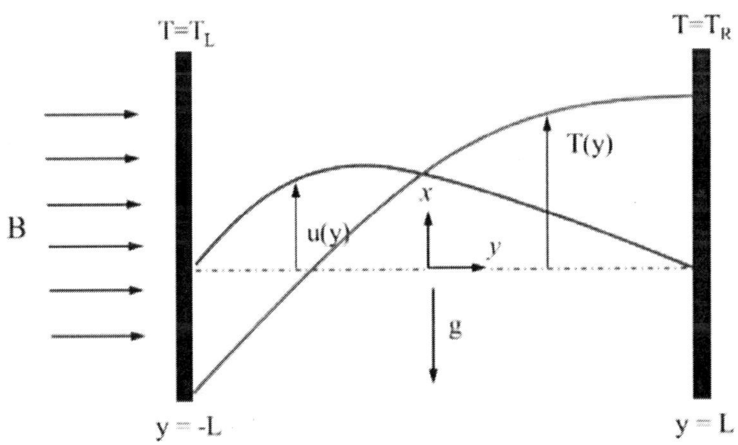


**Figure 2.** Schematic of the system.

$$\frac{d^2u}{dy^2} = \frac{1}{\mu}\frac{dp}{dx} + \frac{\sigma u B^2}{\mu} - \frac{\rho_0 g\beta\left(T - T_{ref}\right)}{\mu} \quad (1)$$

and the energy equation for thermal radiative absorbance of solar radiation in phase change material (radiative heat flux transport in a gray medium) is based on the Rosseland approximation (diffusion of radiative heat flux) which is useable when the medium is optically thick, and is recommended for use in problems where the optical thickness is greater than 3 [4]:

$$\frac{d^2T}{dy^2} = -\frac{4\sigma}{3k\chi}\frac{d^2T^4}{dy^2} \tag{2}$$

where dp/dx is a pressure gradient in the channel in the flow direction, *i.e.*, x-direction, which is perpendicular to the y-direction (constant everywhere), T = T(y) is the medium temperature, k is the thermal conductivity of the fluid, β is the thermal expansion coefficient, μ is the dynamic viscosity, ρ is the fluid density, and $T_{ref}$ is a constant reference temperature defined as $T_{ref} = (T_L + T_R)/2$. σ is the Stefan-Boltzmann constant and $\chi$ is the mean absorption coefficient of the medium. Since the channel size is on a microscale, the velocity slip and the temperature jump need to be considered as the boundary conditions at the solid walls:

$$u(y=L) = l_v\left(\frac{du}{dy}\right)_{y=L} \tag{3}$$

$$u(y=-L) = l_v\left(\frac{du}{dy}\right)_{y=-L} \tag{4}$$

$$T(y=L) = T_R + l_T\left(\frac{dT}{dy} + \frac{4\sigma}{3k\chi}\frac{dT^4}{dy}\right)_{y=L} \tag{5}$$

$$T(y=-L) = T_R + l_T\left(\frac{dT}{dy} + \frac{4\sigma}{3k\chi}\frac{dT^4}{dy}\right)_{y=-L} \tag{6}$$

where $l_v$ is the velocity slip length and $l_T$ is the temperature slip length, which is referred to as "temperature jump length" hereafter. From the Maxwell theory for gases the mean free path of the molecules is equal to velocity slip length and can be estimated from the $l = 3\mu/2\rho a$ where "a" is the sound speed. Also the temperature slip length from the Maxwell theory is equal to $l_T = \frac{2C_p l_v}{C_p + C_v}\frac{C_p \mu}{k}$. There is no analytical formula for liquids and many experimental studies have been performed [33–39] reporting a wide range of slip lengths, ranging from micrometers to nanometers (including no-slip). Furthermore, molecular dynamics simulations suggest slip lengths of less than 100 nm [33–35], which were confirmed by experiments using near-surface particle tracking [36,37]. Hydrophobic surfaces do appear to introduce a discernible but small slip length of approximately 10–50 nm [38,39]. The modelling of velocity slip and temperature jump at a fluid-solid interface is not easy due to the complexity of the phenomena themselves. However, in this study the simplest models of Equations (3)–(6) are applied, which resemble those for fluid-solid interface, because the purpose of the current study is to investigate the physical essence rather than the quantitative details. Nevertheless, the correct physical principle from the simplest velocity slip and temperature jump models could be extracted even though they could have clear limitation in predicting the accurate quantities.

In thermodynamics, entropy is a measure of the number of specific ways in which a thermodynamic system may be arranged (a measure of disorder). According to the Second Law of Thermodynamics the entropy of an isolated system never decreases; such a system will spontaneously evolve toward thermodynamic equilibrium, the configuration with maximum entropy. The current channel which is not isolated and has the irreversible processes decrease in entropy provided they increase the entropy of its environment by at least that same amount (increase the combined entropy of the system and its environment). The entropy is a state function and its change related to the initial and final state.

The irreversibility in channel flow of a fluid has two components of energy and momentum. Consequently, entropy production may occur as a result of fluid friction and heat transfer in the direction of finite temperature gradients. Following Bejan [40,41], the volumetric rate of entropy generation can be expressed as:

$$\dot{S}_g''' = \frac{k}{T^2}\left(\frac{dT}{dy}\right)^2 + \frac{\mu}{T}\left(\frac{du}{dy}\right)^2 \qquad (7)$$

The first term in Equation (7) describes the heat transfer irreversibility and the second term represents the local entropy generation rate due to fluid friction, respectively. Energy minimization methods [41,42] seek the maximum exergy of a system (the maximum useful work possible during a process that brings the system into equilibrium with a heat reservoir). In the current system the description of the exergy is the potential of the fluid in channel to cause a change as it achieves equilibrium with its wall at constant temperature. In the present system, exergy is destroyed because the process involves a temperature change. This destruction is proportional to the entropy increase of the system together with its surroundings, so the minimization of the entropy generation is equal to the maximization of the energy that is available to be used in the system.

In order to clarify the physical essence, in this study the non-dimensionalized variables are introduced. The non-dimensional of main variables are:

$$X = \frac{x}{L} \tag{8}$$

$$Y = \frac{y}{L} \tag{9}$$

$$U(Y) = \frac{u(y)}{u_m} \tag{10}$$

$$\theta(Y) = \frac{T - (T_L + T_R)/2}{T_R - T_L} \tag{11}$$

$$P = \frac{pL}{\mu u_m} \tag{12}$$

and the non-dimensional parameters which arise are:

$$Gr = \frac{g\beta(T_R - T_L)L^3}{2\nu^2} \tag{13}$$

$$\mathrm{Re} = \frac{u_m L}{\nu} \tag{14}$$

$$R_d = \frac{\sigma(T_R - T_L)^3}{3k\chi} \tag{15}$$

$$\theta_R = \frac{(T_L + T_R)/2}{T_R - T_L} \tag{16}$$

$$Ha = BL\sqrt{\frac{\sigma}{\mu}} \tag{17}$$

$$\lambda_v = \frac{l_v}{L} \tag{18}$$

$$\lambda_T = \frac{l_T}{L} \tag{19}$$

$$N_s = \frac{L^2 \dot{S}_g'''}{k} \tag{20}$$

$$Br = \frac{\mu}{k(T_R - T_L)}(u_m)^2 \tag{21}$$

$$Be = \frac{\left(\frac{d\theta}{dY}\right)^2}{\left(\frac{d\theta}{dY}\right)^2 + Br\left(\frac{dU}{dY}\right)^2} \tag{22}$$

where the Gr is the Grashof number, Re is the Reynolds number, $R_d$ is the radiation parameter, Ha is the Hartmann number (the ratio of electromagnetic force to the viscous force in the absence of the Hall effect, electrical and polarization effects), and $\theta_R$ is the temperature parameter, Be is the Bejan number (in the context of thermodynamics,the ratio of heat transfer irreversibility to total irreversibility due to heat transfer and fluid friction; in the context of fluid mechanics, the dimensionless pressure drop along a channel), Br is the ratio between heat produced by viscous dissipation and heat transported by molecular conduction (the ratio of viscous heat generation to external heating), and $N_s$ is the dimensionless form of local entropy generation rate in Equation (7). From Equation (22), it is obvious that the Bejan number ranges from $0 < Be < 1$. While $Be = 0$ represents the limit case of fluid friction dominated irreversibility, $Be = 1$ corresponds to the limit case of heat transfer dominated irreversibility. The contribution of both heat transfer and fluid friction to entropy production in the flow system is the same when $Be = 0.5$.

As the term $u_m = \frac{1}{2}\int_{-L}^{L} u(y)\,dy$ is a constant $(\int_{-1}^{1} U(Y)\,dY = 2$ can be used to find the pressure gradient). Then, by multiplying Equation (1) in $L^2/u_m$ and Equation (2) in $2L^2/(T_R - T_L)$, respectively, and considering dimensionless parameters in (7)–(18), the non-dimensional momentum and energy equations as a system of ordinary coupled differential equations can be rewritten as:

$$\frac{d^2U}{dY^2} = Ha^2U - \frac{Gr\theta}{\mathrm{Re}} + \frac{dP}{dX} \tag{23}$$

$$\frac{d^2}{dY^2}\left[\theta + R_d(\theta + \theta_R)^4\right] = 0 \tag{24}$$

In addition, the non-dimensional form of the boundary conditions in the equations from (3) to (6) can be rewritten as:

$$U(Y=1) = \lambda_v\left(\frac{dU}{dY}\right)_{Y=1} \tag{25}$$

$$U(Y=-1) = \lambda_v\left(\frac{dU}{dY}\right)_{Y=-1} \tag{26}$$

$$\theta(Y=1) = 1 + \lambda_T\frac{d}{dY}\left[\theta + R_d(\theta + \theta_R)^4\right]_{Y=1} \tag{27}$$

$$\theta(Y=-1)=-1+\lambda_T\frac{d}{dY}\left[\theta+R_d(\theta+\theta_R)^4\right]_{Y=-1} \tag{28}$$

The components of dimensionless entropy generation are:

$$S_U=\frac{1}{(\theta+\theta_R)}\left(\frac{dU}{dY}\right)^2 \tag{29}$$

$$S_\theta=\frac{1}{(\theta+\theta_R)^2}\left(\frac{d\theta}{dY}\right)^2 \tag{30}$$

## 3. RESULTS AND DISCUSSION

The problem of optimal design of a steady state mixed convection MHD flow in a vertical channel filled with radiative absorbing medium is solved under the no-slip and slip-flow conditions at the channel walls in the presence of a crosswise uniform magnetic field. The Newtonian, incompressible and electrically conducting fluid is assumed. With the aim of acquiring a comprehensive vision of the physical problem, estimations have been accomplished to get the flow and temperature field with Hartmann number (Ha), radiation parameter ($R_d$), mixed convection parameter (Gr/Re), thermal parameter ($\theta_R$), temperature jump ($\lambda_T$), and velocity slip ($\lambda_v$) asshown in Figures 3–8. The non-dimensional governingEquations (23)–(28) are solved numerically using Runge–Kutta–Fehlberg method [52] with shooting technique [53]. The set of simultaneous first order differential equations of equivalent initial-value problem ($Z' = f(Z,Y)$) are constructed by the vector $Z = [U\ U'\ \theta\ \theta']$ and the first guess of the initial value is assumed as $Z(Y = -1) = [0\ 0\ -1\ 0]$.

Figure 3 presents the effect of thermal radiation parameter on the various aspect of thermal and fluid flow of the system for the Grashof number to Reynolds number ratio of one hundred, and $\lambda_T = 0.01$, $\lambda_v = 0.01$, $\theta_R = 10$, Ha = 0.1. For the case of no the thermal radiation effect ($R_d = 0$), the temperature profile is linear ($\theta = Y$). By increasing the thermal radiation coefficient from $10^{-4}$ to $10^{-2}$ the dimensionless temperature gradient increased, especially near the left wall as shown inFigure 3a, which leads to an increase of the heat transfer component of the entropy. Figure 3b shows the wavy shape of the velocity component of the entropy profile. By increasing the thermal radiation coefficient, the dimensionless velocity profile becomes flattened. The minimum velocity component of the entropy value occurs at the peak value of the velocity or zero shear stress points. Furthermore, the maximum velocity component of the entropy value occurs at the walls. The effects of thermal radiation on the total heat transfer and viscous parts of the entropy are illustrated in Figure 3c,d, respectively. By increasing $R_d$, the heat transfer part of the entropy increases dramatically while the viscous part of the entropy decreases. Although the entropy generation due to thermal effects increases, it is two orders of magnitude ($10^{-1}$ *vs.* $10^{1}$) smaller with respect to entropy generated by frictional effects.

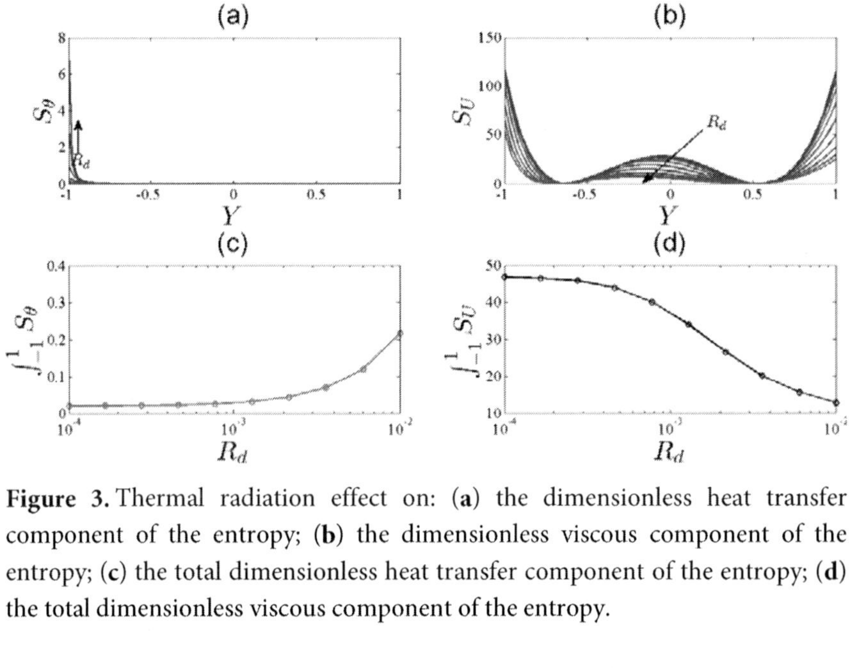


**Figure 3.** Thermal radiation effect on: (**a**) the dimensionless heat transfer component of the entropy; (**b**) the dimensionless viscous component of the entropy; (**c**) the total dimensionless heat transfer component of the entropy; (**d**) the total dimensionless viscous component of the entropy.

The effect of the Gr/Re parameter is demonstrated for the $R_d = 10$, and $\lambda_T = 0.01$, $\lambda_v = 0.01$, $\theta_R = 10$, Ha = 0.1 in Figure 4. As seen in Figure 4a, by increasing the Gr/Re from 1 to 30 the $S_\theta$ is constant because the dimensionless temperature is not changed. As shown the $S_\theta$ is maximum at the left wall and decreases smoothly to the right wall. Figure 4b illustrates the influence of the mixed convection parameter Gr/Re, on the dimensionless velocity component of the entropy profiles. As seen by increasing Gr/Re the velocity component of the entropy increases. The maximum again appears at the walls. By increasing of Gr/Re, the dimensionless velocity profile maximum increases. As illustrated in Figure 4c the Gr/Re has no effect on total $S_\theta$, but $S_u$ is increased by increases of Gr/Re (see Figure 4d), so a lower Gr/Re is better from the Second Law of Thermodynamics perspective.

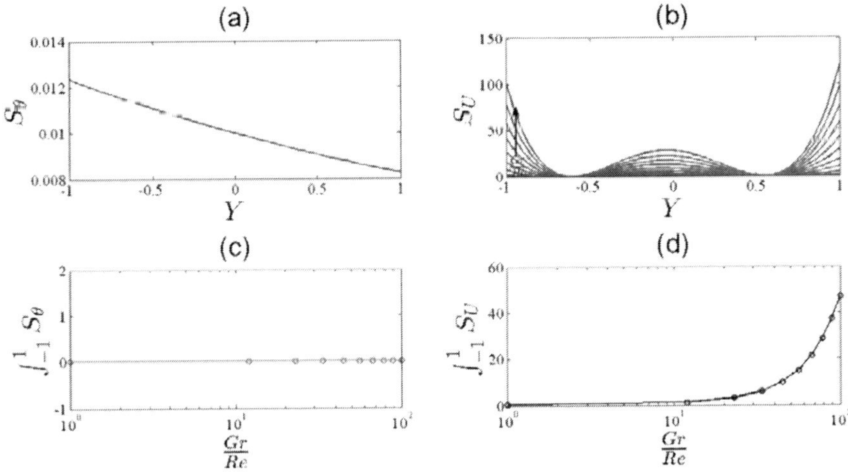


**Figure 4.** Grashof number to Reynolds number ratio effect on: (**a**) the dimensionless heat transfer component of the entropy; (**b**) the dimensionless viscous component of the entropy; (**c**) the total dimensionless heat transfer component of the entropy; (**d**) the total dimensionless viscous component of the entropy.

The effect of the Hartmann number is demonstrated in Figure 5. As seen in Figure 5a, by increasing Ha from 5 to 15 for the $R_d = 10$, $\lambda_T = 0.01$, $\lambda_v = 0.01$, $\theta_R = 10$, and Gr/Re = 50, because the dimensionless temperature profile is not dependent on the Hartmann number the $S_\theta$ not changed. As seen the $S_\theta(Y)$ is approximately linear and has a higher value at the left wall and a lower value at the right wall. Figure 5b shows the distribution of $S_u$ caused by the presence of the magnetic field.

The existence of the magnetic field generates a resistive force akin to the drag force that acts in the opposite direction of the fluid motion, thus making the velocity of the fluid decrease. By increasing the Hartmann number, the peaks of the dimensionless velocity profile are chamfered so the velocity gradient inside the channel is decreased which leads to lower viscous heating, so the general effect of the magnetic parameter is to decrease the velocity component of the entropy generation. The effect of the Hartmann number on the total $S_\theta$ is exemplified in Figure 5c. As stated before the Hartmann number has no effect on the total $S_\theta$, but as there is a minimum on $S_u$, for the total entropy generation there is a minimum at Ha = 8.

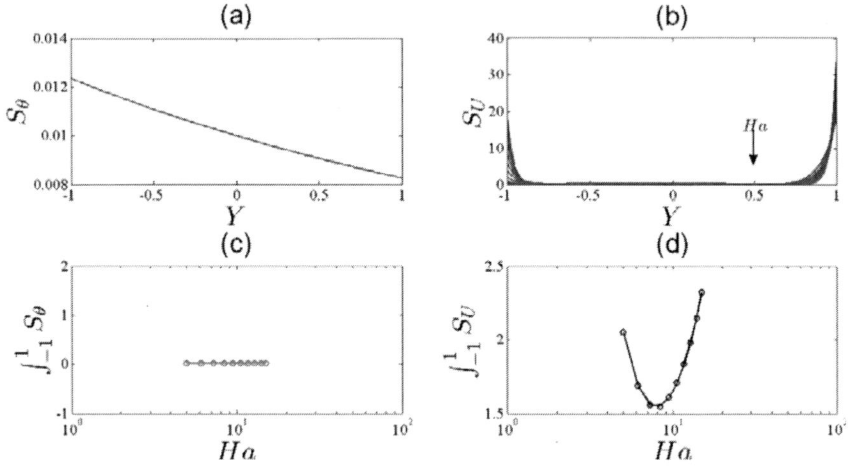


**Figure 5.** Hartmann number effect on: (**a**) the dimensionless heat transfer component of the entropy, (**b**) the dimensionless viscous component of the entropy, (**c**) the total dimensionless heat transfer component of the entropy, (**d**) the total dimensionless viscous component of the entropy.

The effect of the thermal parameter is demonstrated for the $R_d = 0.01$, Gr/Re = 100, $\lambda_T = 0.01$, $\lambda_v = 0.01$, and Ha = 0.1 in Figure 6. As shown in Figure 6a by increasing $\theta_R$ from 5 to 9 the dimensionless heat transfer entropy generation is increased, especially near the left wall. Figure 6b shows the wavy shape of the velocity entropy generation component which has a critical point inside and two peaks at the boundaries. As observed, by the increase of $\theta_R$, the dimensionless entropy generation component velocity profile decreases. The effect of $\theta_R$ on the total heat transfer entropy generation is illustrated in Figure 6c. By increasing $\theta_R$, the total heat transfer entropy generation has a parabolic shape with a minimum at $\theta_R = 6.6$. Also the total viscous entropy generation decreases with the increase of $\theta_R$ (see Figure 6d).

In Figure 6, the major contribution to the total entropy is due to the viscous dissipation and this term always decreases with the temperature parameter. Generally, when both the heat transfer entropy generation and viscous entropy generation are summed and displayed on a single curve, the decrease in the viscous entropy generation is dominant and the minimum does not exist. In Figure 7 both terms are added to show the dominance of the decrease in the viscous entropy generation and that there is no minimum on it.

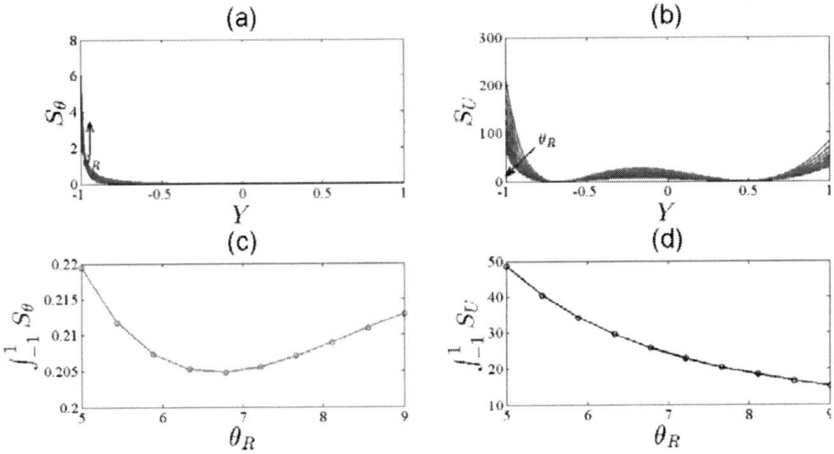


**Figure 6.** Thermal parameter effect on: (**a**) the dimensionless heat transfer component of the entropy; (**b**) the dimensionless viscous component of the entropy; (**c**) the total dimensionless heat transfer component of the entropy; (**d**) the total dimensionless viscous component of the entropy.

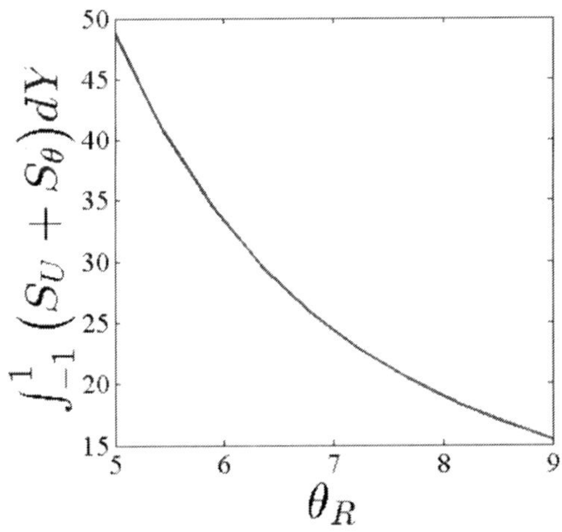


**Figure 7.** Thermal parameter effect on summation of the total dimensionless heat transfer component of the entropy and the total dimensionless viscous component of the entropy.

The effect of velocity slip ratio is demonstrated in Figure 8 by increasing $\lambda_v$ from 0 to 0.05 for the $R_d = 0.01$, and $\lambda_T = 0.01$, Ha = 0.1, Gr/Re = 50, $\theta_R = 0.1$. The $\lambda_v$ is equal to the Knudsen number for gas flow in the channel. It can be noticed fromFigure 8a that for different values of $\lambda_v$ the dimensionless heat transfer entropy generation rate is not changed. Further, inFigure 8b it is observed that the viscous entropy generation rate is maximum at the right wall, while the sinus shape of the velocity profile is not altered with an increase of $\lambda_v$. By increasing $\lambda_v$, the maximum dimensionless viscous entropy generation rate profiles increase to some extent. As anticipated from Figure 8a the $\lambda_v$ has no influence on the total dimensionless heat transfer entropy generation rate, while it is clearly seen from these Figure 8d that the total viscous entropy generation rate increases with an increase of $\lambda_v$, so the lower velocity slip is better for maximizing the exergy of the system.

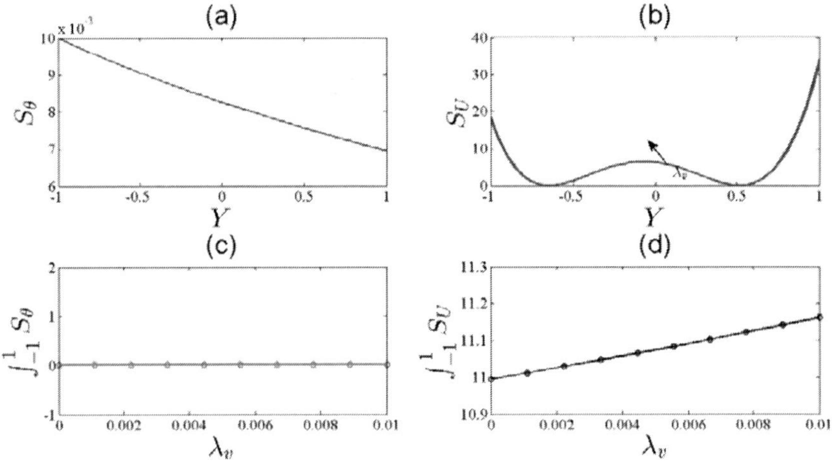


**Figure 8.** Velocity slip effect on: (**a**) the dimensionless heat transfer component of the entropy; (**b**) the dimensionless viscous component of the entropy; (**c**) the total dimensionless heat transfer component of the entropy; (**d**) the total dimensionless viscous component of the entropy

The effect of temperature slip is demonstrated in Figure 9. As depicted in Figure 9a, by increasing $\lambda_T$ from 0 to 0.1 the magnitude of the dimensionless heat transfer entropy generation rate distribution shifted downward. Figure 9b shows the sine-like shape of the velocity entropy generation rate distribution with fixed values of all other parameters as the $R_d = 0.01$, and $\lambda_v = 0.01$, Ha = 0.1, Gr/Re = 50, $\theta_R = 0.1$. By increasing $\lambda_T$, the dimensionless velocity profile is not changed meaningfully, although an increase of $\lambda_T$ causes the increase of $S_u$ at the middle of the channel and a decrease at the left boundary. For the gas flow in a vertical channel the temperature slip is proportional to the velocity slip length ratio or Knudsen number. By increasing of the gas Prantdl number and decreasing the specific heat ratio, the $\lambda_T$ increases. The effect of $\lambda_T$ on the total heat transfer entropy generation rate is illustrated in Figure 9c. By increasing $\lambda_T$, the

$S_\theta$ decreases. Furthermore the friction entropy generation rate decreases as $\lambda_T$ increases, as shown in Figure 9d.

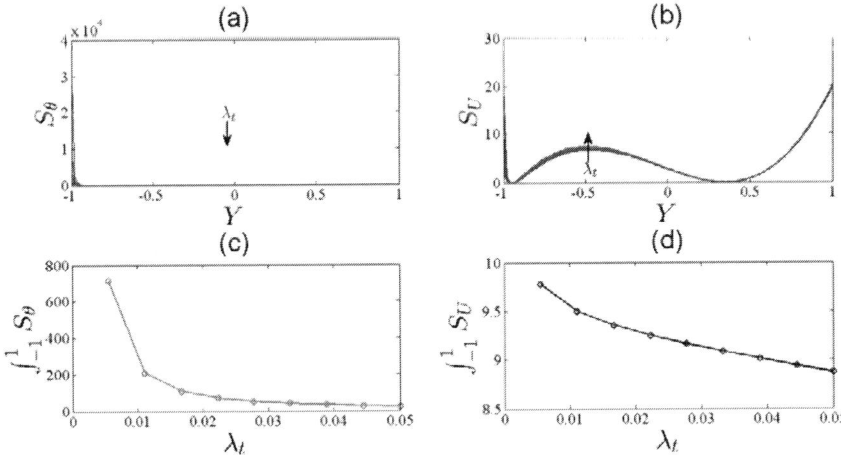


**Figure 9.** Temperature slip effect on: (**a**) the dimensionless heat transfer component of the entropy; (**b**) the dimensionless viscous component of the entropy; (**c**) the total dimensionless heat transfer component of the entropy; (**d**) the total dimensionless viscous component of the entropy

## 4. CONCLUSIONS

In this study, the optimization of the steady flow in a filled vertical microchannel with thermal radiation has been thoroughly investigated. The results can be summarized as follows:

1. By increasing $R_d$, the heat transfer part of the entropy increases dramatically while the viscous part of the entropy decreases.
2. By increasing $\theta_R$, the total heat transfer entropy generation has a minimum at $\theta_R = 6.6$. Also the total viscous entropy generation decreases as $\theta_R$ increases.
3. Lower velocity slip, Grashof number to Reynolds number ratio are better for maximize the exergy of the system.
4. Higher temperature slip is better for maximizing the exergy of the system.
5. The optimum Hartmann number is about 8.

## ACKNOWLEDGMENTS

This research was supported by the Basic Science Research Program through the National Research Foundation of Korea (NRF) funded by the Ministry of Education, Science and Technology (NRF-2012R1A1A1042920).


**PACS Codes:** 05.70.-a ; 44.40.+a ; 47.55.pb; 47.35.Tv; 47.65.-d; 83.50.Lh

## AUTHOR CONTRIBUTIONS

Mohamad Yaghoub Abdollahzadeh Jamalabadi and Jae Hyun Park designed the research with theoretical formulations. Mohamad Yaghoub Abdollahzadeh Jamalabadi performed the numerical simulations. Mohamad Yaghoub Abdollahzadeh Jamalabadi and Jae Hyun Park together analyzed the data and wrote the manuscript for initial submission. Chang Yeop Lee had critical contribution in revising the manuscript including the preparation of responses and English correction. All authors have read and approved the final manuscript.

## REFERENCES


1. Barletta, A. Analysis of combine forced and free flow in a vertical channel with viscous dissipation and isothermal-isoflux boundary conditions. *J. Heat Transf.* **1999**, *121*, 349–356.
2. Barletta, A. Fully developed mixed convection and flow reversal in a vertical rectangular duct with uniform wall heat flux. *Int. J. Heat Mass Transf.* **2002**, *45*, 641–654.
3. Boulama, K.; Galanis, N. Analytical solution for fully developed mixed convection between parallel vertical plates with heat and mass transfer. *J. Heat Transf.* **2004**, *126*, 381–388.
4. Modest, M.F. *Radiative Heat Transfer*, 2nd ed; Academic Press: New York, NY, USA, 2003.
5. Baytaş, A.C.; Liaqat, A.; Groşan, T.; Pop, I. Conjugate natural convection in a square porous cavity. *Heat Mass Transf.***2001**, *37*, 467–473.
6. Hadjadj, A.; Ben-Nasr, O.; Shadloo, M.S.; Chaudhuri, A. Effect of wall temperature in supersonic turbulent boundary layers: A numerical study. *Int. J. Heat Mass Transf.* **2015**, *81*, 426–438.

7. Kandlikar, S.G. High flux heat removal with microchannels—A roadmap of challenges and opportunities. *Heat Transf. Eng.* **2005**, *26*, 5–14.

8. Zhao, C.Y.; Lu, T.J. Analysis of microchannel heat sinks for electronics cooling. *Int. J. Heat Mass Transf.* **2002**, *45*, 4857–4869.

9. Liu, D.; Garimella, S.V. Investigation of liquid flow in microchannels. *J. Thermophys. Heat Transf.* **2004**, *18*, 65–72.

10. Judy, J.; Maynes, D.; Webb, B.W. Characterization of frictional pressure drop for liquid flows through microchannels.*Int. J. Heat Mass Transf.* **2002**, *45*, 3477–3489.

11. Abdollahzadeh Jamalabadi, M.Y. Effect of fuel inject angle on non-premixed combustion of air/methane mixtures in vertical cylinder. *Int. J. Multidiscip. Res. Dev.* 2014, 1. Available online: http://allsubjectjournal.com/vol1/issue5/PartA/pdf/13.1.pdf accessed on 11 February 2015.

12. Kandlikar, S.G.; Schmit, D.; Carrano, A.L.; Taylor, J.B. Characterization of surface roughness effects on pressure drop in microchannels. *Phys. Fluids.* **2005**, *17*.

13. Abdollahzadeh Jamalabadi, M.Y.; Ghasemi, M.; Hamedi, M.H. Two-dimensional simulation of thermal loading with horizontal heat sources. *Proc. Inst. Mech. Eng. C* **2012**, *226*, 1302–1308.

14. Abdollahzadeh Jamalabadi, M.Y.; Ghasemi, M.; Hamedi, M.H. Numerical investigation of thermal radiation effects on open cavity with discrete heat sources. *Int. J. Numer. Methods Heat Fluid Flow* **2013**, *23*, 649–661.

15. Abdollahzadeh Jamalabadi, M.Y. Experimental investigation of thermal loading of a horizontal thin plate using infrared camera. *J. King Saud Univ. Eng. Sci.* **2014**, *26*, 159–167.

16. Abdollahzadeh Jamalabadi, M.Y.; Park, J.H. Thermal radiation, joule heating, and viscous dissipation effects on MHD forced convection flow with uniform surface temperature. *Open J. Fluid Dyn.* **2014**, *4*, 125–132.

17. Abdollahzadeh Jamalabadi, M.Y. Analytical study of Magnetohydrodynamic propulsion stability. *J. Mar. Sci. Appl.***2014**, *13*, 281–290.

18. Shahidian, A.; Ghassemi, M.; Khorasanizade, S.; Abdollahzade, M.; Ahmadi, G. Flow Analysis of non-Newtonian blood in a magnetohydrodynamic pump. *IEEE Trans. Magn.* **2009**, *45*, 2667–2670.

19. Abdollahzadeh Jamalabadi, M.Y.; Park, J.H. Electro-magnetic ship propulsion stability under gusts. *Int. J. Sci. Basic Appl. Res.* **2014**, *1*, 421–427.

20. Abdollahzadeh Jamalabadi, M.Y.; Park, J.H.; Lee, C.Y. Thermal radiation effects on the onset of unsteadiness of fluid flow in vertical microchannel filled with highly absorbing medium. *Therm. Sci.* **2014**.

21. Field, J.E.; Hahn, K.H.; Harris, S.E. Observation of electromagnetically induced transparency in collisionally broadened lead vapor. *Phys. Rev. Lett.* **1991**, *67*, 3062–3065.

22. Lemoff, B.E.; Yin, G.Y.; Gordon, C.L., III; Barty, C.P.J.; Harris, S.E. Demonstration of a 10-Hz femtosecond-pulse-driven XUV laser at 41.8 nm in Xe IX. *Phys. Rev. Lett.* **1995**, *74*, 1574–1577.

23. Kadowaki, M.; Yoshizawa, H.; Mori, S.; Suzuki, M. Plasma CVD on the inner surface of a microchannel. *Thin Solid Films* **2006**, *506*, 123–127.

24. Kumar, V.; Paraschivoiu, M.; Nigam, K.D.P. Single-phase fluid flow and mixing in microchannels. *Chem. Eng. Sci.***2011**, *66*, 1329–1373.

25. Aghanajafi, C.; Vandadi, V.; Shahnazari, M.R. Investigation of convection and radiation heat transfer in rhombus microchannels. *Int. J. Res. Rev. App. Sci.* **2010**, *3*, 167–176.

26. Leibrandt, D.R.; Drake, R.P.; Reighard, A.B. A validation test of the flux-limited diffusion approximation for radiation hydrodynamics. *Astrophys. J.* **2005**, *626*, 616–625.

27. Karniadakis, G.; Beskok, A.; Aluru, N. *Microflows and Nanoflows: Fundamentals and Simulation*; Springer: New York, NY, USA, 2003.

28. Ulmanella, U.; Ho, C.M. Molecular effects on boundary condition in micro/nanoliquid flows. *Phys. Fluids.* **2008**, *20*.

29. Bocquet, L.; Barrat, J.-L. Flow boundary conditions from nano- to micro-scales. *Soft Matt* **2007**, *3*, 685–693.

30. Horvath, J.A.; Webb, R.N. Experimental study of radiation absorption by microchannels of varying aspect ratios. *Solar Energ.* **2011**, *85*, 1035–1040.

31. Hossain, M.S.; Saidur, R.; Fayaz, H.; Rahim, N.A.; Islam, M.R.; Ahamed, J.U.; Rahman, M.M. Review on solar water heater collector and thermal energy performance of circulating pipe. *Renew. Sustain. Energ. Rev.* **2011**, *15*, 3801–3812.

32. Taylor, R.A.; Phelan, P.E.; Otanicar, T.P.; Adrian, R.; Prasher, R. Nanofluid optical property characterization: towards efficient direct absorption solar collectors. *Nanoscale Res. Lett.* **2011**, *6*.

33. Thompson, P.A.; Troian, S.M. A general boundary condition for liquid flow at solid surfaces. *Nature* **1997**, *389*, 360–362.

34. Barrat, J.-L.; Bocquet, L. Large slip effect at a nonwetting fluid-solid interface. *Phys. Rev. Lett.* **1999**, *82*, 4671–4674.

35. Cieplak, M.; Koplik, J.; Banavar, J.R. Boundary conditions at a fluid-solid surface. *Phys. Rev. Lett.* **2001**, *86*, 803–806.

36. Guasto, J.S.; Breuer, K.S. High-speed quantum dot tracking and velocimetry using evanescent wave illumination. *Exp. Fluids.* **2009**, *47*, 1059–1066.

37. Bouzigues, C.I.; Tabeling, P.; Bocquet, L. Nanofluidics in the Debye layer at hydrophilic and hydrophobic surfaces.*Phys. Rev. Lett.* **2008**, *101*.

38. Huang, P.; Guasto, J.S.; Breuer, K.S. Direct measurement of slip velocities using three-dimensional total internal reflection velocimetry. *J. Fluid Mech.* **2006**, *566*, 447–464.

39. Huang, P.; Breuer, K.S. Direct measurement of slip length in electrolyte solutions. *Phys. Fluids.* **2007**, *19*.

40. Bejan, A. Second-law analysis in heat transfer and thermal design. *Adv. Heat Transfer* **1982**, *15*, 1–58.

41. Bejan, A. *Entropy Generation Minimization*; CRC: Boca Raton, FL, USA, 1996.

42. Makinde, O.D.; Eegunjobi, A.S. Entropy generation in a couple stress fluid flow through a vertical channel filled with saturated porous media. *Entropy* **2013**, *15*, 4589–4606.

43. Eegunjobi, A.S.; Makinde, O.D. Combined effect of buoyancy force and Navier slip on entropy generation in a vertical porous channel. *Entropy* **2012**, *14*, 1028–1044.

44. Ozkol, I.; Komurgoz, G.; Arikoglu, A. Entropy generation in the laminar natural convection from a constant temperature vertical plate in an infinite fluid. *J. Power Energ* **2007**, *221*, 609–616.

45. Mahmud, S.; Fraser, R.A. Mixed convection-radiation interaction in a vertical porous channel: Entropy generation.*Energy* **2003**, *28*, 1557–1577.

46. Tasnim, S.H.; Shohel, M.; Mamum, M.A.H. Entropy generation in a porous channel with hydromagetic effect. *Exergy: Int. J.* **2002**, *2*, 300–308.

47. Chauhan, D.S.; Kumar, V. Heat transfer and entropy generation during compressible fluid flow in a channel partially filled with porous medium. *Int. J. Energ. Tech.* 2011. Available online: http://www.as-se.org/ijet/paperInfo.aspx?ID=19088 accessed on 13 February 2015.

48. Chen, S.; Liu, Z.H.; Bao, S.; Zheng, C.G. Natural convection and entropy generation in a vertically concentric annular space. *Int. J. Therm. Sci.* **2010**, *49*, 2439–2452.

49. Chen, S. Entropy generation of double-diffusive convection in the presence of rotation. *Appl. Math. Comput.* **2011**, *217*, 8575–8597.

50. Bianco, V.; Manca, O.; Nardini, S. Second law analysis of $Al_2O_3$-water nanofluid turbulent forced convection in a circular cross section tube with constant wall temperature. *Adv. Mech. Eng.* **2013**, *2013*.

51. Bianco, V.; Manca, O.; Nardini, S. Entropy generation analysis of turbulent convection flow of $Al_2O_3$–water nanofluid in a circular tube subjected to constant wall heat flux. *Energ. Conver. Manage.* **2014**, *77*, 306–314.

52. Hairer, E.; Nørsett, S.; Wanner, G. *Solving Ordinary Differential Equations I: Nonstiff Problems*; Springer: Berlin/Heidelberg, Germany, 1993.

53. Press, W.H.; Teukolsky, S.A.; Vetterling, W.T.; Flannery, B.P. *Numerical Recipes: The Art of Scientific Computing*; Cambridge University Press: New York, NY, USA, 2007.

# CHAPTER 9

# Numerical Modelling of Non-Similar Mixed Convection Heat and Species Transfer along an Inclined Solar Energy Collector Surface with Cross Diffusion Effects

Osman Anwar Bég[1], Ahmed Bakier[2], Ramachandra Prasad[3], Swapan Kumar Ghosh[4]

[1]*Biomechanics and Energy Research, Aerospace Engineering, Sheffield Hallam University, Sheffield, England*
[2]*Department of Mathematics, Assiut University, Assiut, Egypt*
[3]*Department of Mathematics, Madanapalle Institute of Technology and Science, Madanapalle, India*
[4]*Department of Mathematics, Narajole Raj College, West Bengal, India*

## ABSTRACT

An analysis is performed to study thermo-diffusion and diffusion-thermo effects on mixed convection heat and mass transfer boundary layer flow along an inclined (solar collector) plate. The resulting governing equations are transformed and then solved numerically using the local nonsimilarity method and Runge-Kutta shooting quadrature. A parametric study illustrating the influence of thermal buoyancy parameter (x), Prandtl number (Pr), Schmidt number (Sc), Soret number (Sr), Dufour number (Du) and concentration-tothermal-buoyancy ratio parameter, N, on the fluid velocity, temperature and concentration profiles as well as on local skin-friction, Nusselt and Sherwood numbers is conducted. For positive inclination angle of the plate (g = 70 degrees), flow velocity (f′) is strongly increased i.e. accelerated, with thermal buoyancy force parameter (x), in particular closer to the plate surface; further into the boundary layer, x has a much reduced effect. Conversely temperature (q) and concentration (f) is decreased with increasing thermal buoyancy parameter, x. For negative plate inclination, the flow is accelerated whereas for positive inclination it is decelerated i.e. velocity is reduced. Conversely with negative plate inclination both the temperature and concentration in the boundary layer is reduced with the opposite apparent for positive inclination. Increasing Prandtl number strongly reduces temperature in the regime whereas an increase in Schmidt number boosts temperatures with temperature overshoots near the plate surface for Sc = 3 and 5 (i.e. for Sc > 1). Concentration is reduced continuously throughout the boundary layer, however, with increasing Schmidt number. A positive increase in concentration-to-thermal-buoyancy ratio parameter, N, significantly accelerates the flow in the domain, whereas negative N causes a deceleration. A velocity overshoot is also identified for N = 20, at intermediate distance from the plate surface. Negative N (thermal and concentration buoyancy forces oppose each other) induces a slight increase in both fluid temperature and concentration, with the reverse observed for positive N (thermal and concentration buoyancy forces assisting each other). Increasing Dufour number respectively causes a rise in temperature and a decrease in concentration, whereas an increase in Soret number cools the fluid i.e. reduces temperature and enhances concentration values. In the absence of Soret and Dufour effects, positive N causes a monotonic increase in local Nusselt number, $Nu_x \mathrm{Re}_x^{-1/2}$ with x Cosg; for N = -1 the local Nusselt number remains constant for all values of parameter, x Cosg. Local Sherwood number, [illegible] is boosted considerably with higher Schmidt numbers and also with positive N values. The computations in the absence of Soret and Dufour effects correlate accurately with the earlier study by Chen et al. (1980).

## KEYWORDS

Heat and mass transfer, inclined plate, Soret/Dufour effects, buoyancy, Numerical, Solar energy

## 1. INTRODUCTION

Combined heat and mass transfer from inclined surfaces finds numerous applications in solar energy systems, geophysics, materials processing etc. Many studies have appeared concerning natural and also mixed convection flows.

Kierkus [1] studied isothermal inclined plate natural convection boundary layer flow using a perturbation method for a Prandtl number of 0.7. Fujii and Imura [2] studied experimentally free convection from an inclined plate. Chen et al. [3] analyzed combined heat and mass transfer in mixed convection along vertical inclined plates. Uniform surface flux effects on inclined plate thermal convection were reported by Armaly et al. [4]. The combined buoyancy and wall transpiration effects on mixed convection along an inclined plate were studied by Lee and Hsu [5]. Wickern [6] examined the laminar boundary layer flow over an arbitrarily sloping plate showing that for opposing buoyancy forces singular as well as regular behaviour may prevail. Further studies have been communicated by Yan and Soong [7] in the context of evaporating inclined boundary layer convection and by Sheu and Lin [8] in combusting inclined flat plate flows. All the studies neglected Soret and/or Dufour effects. Such effects are significant when density differences exist in the flow regime. The Soret or thermo-diffusion effect refers to mass flux produced by a temperature gradient and was first reported by Ludwig [9] and Soret [10], these studies showing an increase (decrease) of salt concentration at the cold (hot) end of a tube filled with salty water. The Dufour effect refers to heat flux produced by a concentration gradient. The Soret and Dufour effects are usually minor and can be neglected in simple models of coupled heat and mass transfer. According to Platten and Legros [11], the mass fraction gradient established under the effect of thermal diffusion is very small. However, it has a disproportionately large influence on hydrodynamic stability relative to its contribution to the buoyancy of the fluid. They also argued that in most liquid mixtures the Dufour effect is negligible, but this may not be the case in gases. Mojtabi and Charrier-Mojtabi [12] have confirmed this by noting that in liquids the Dufour coefficient is an order of magnitude smaller than the Soret effect. Many investigations of Soret/Dufour effects on convection flows have been presented. Twefik and Yang [13] experimentally investigated Helium mass transfer via a porous horizontal cylinder and also confirmed the major influence of Soret and Dufour phenomena on transpiration-cooled boundary layer flows. Sparrow et al. [14] studied convection along a horizontal cylinder for a Helium-air system. Sparrow et al. [15] further conducted experiments using Hydrogen, Helium, Carbon Dioxide and Freon-12, injected via a horizontal porous cylinder, showing the strong influence of Soret effects on the flow regime for injected gases with a lower density than air (i.e. Helium and Hydrogen). Abreu et al. [16] used the Adomian's polynomial method to study both forced and free convection laminar boundary layer heat and mass transfer with Soret and Dufour effects. Hort et al. [17] analyzed in detail the thermal convection and mass transfer in binary gas mixtures with Dufour effects. Coelho and Silva Telles [18] studied the Graetz problem with thermo-diffusion effects with conduction in the axial and transverse directions plus longitudinal advection. Weaver and Viskanta [19] reported numerical results for the Soret/Dufour effects on free convection heat and mass transfer in a cavity, indicating that species interdiffusion reduces the overall heat transfer, but increases the mass transfer through the cavity. They also showed that heat transfer due to Dufour effects is comparable to that by heat conduction and that the total mass flux through the cavity due to Soret diffusion can reach 10% to 15%. Further excellent studies of Soret and/or Dufour effects on mixed-free-forced convection, have been communicated by Kafoussias and Williams [20] for temperature-dependent viscosity and for the case of a Darcian porous regime by Anghel et al. [21]. Postelnicu [22] studied the influence of chemical reaction on heat and mass transfer from vertical surfaces in porous media considering Soret and Dufour effects. Bég et al. [23] presented the first computational study of micropolar convection boundary layers in nonlinear porous media with Soret and Dufour cross-diffusion effects, employing a finite element method with a focus on suspension polymer chemical process engineering applications. More recently Bég et al. [24] investigated (using the SparrowQuack-Boerner local non-similarity numerical method) the collective first order chemical reaction and Soret and Dufour diffusional influence on heat transfer from an inclined plate, with applications in solar energy systems. Bég et al. [25] have further studied numerically the effects of transverse magnetic field on magnetohydrodynamic convection from a stretching polymeric sheet embedded in a porous medium in the presence of Soret and Dufour effects. Bhargava et al. [26] extended the model in [25] to consider the oscillatory magneto-convection and species diffusion in a porous medium with Soret and Dufour effects with applications in magnetic materials processing. Rawat and Bhargava [27] studied viscous dissipation (Eckert number) effects in buoyancydriven micropolar convection heat transfer in a Darcian porous medium. Very recently Bég et al. [28] employed the implicit Keller box method to study Soret/Dufour effects on micropolar convection flow from a spherical body. Osalusi et al. [29] have used numerical shooting methods to study the rotating hydromagnetic disk flow with Soret/Dufour and Ohmic heating effects. In the context of combustion systems, Miller et al. [30] have described the effects of Soret and Dufour diffusion on laminar flame diffusion dynamics. Partha et al. [31] have analyzed the interaction of dispersion, buoyancy (Rayleigh number), mass transfer (Lewis number) an Soret/ Dufour gradients on magnetohydrodynamic convection in a porous regime. OKong'o and Bellan [32] have also studied Soret and Dufour effects in compressible gas flow with binary species mixing for mixtures of supercritical heptane and nitrogen and oxygen and hydrogen using fluctuation dissipation theory. They derived similarity solutions for velocity, temperature and mass fraction profiles, highlighting the strong influence of Soret and Dufour effects. The objective of the present work is to investigate the combined Soret and Dufour effects and also the influence of the inclination of the plate on the mixed convection heat and mass transfer over an inclined plate (solar energy collector) system using boundary layer theory. The effects of governing thermophysical parameters on heat and mass transfer characteristics are analyzed in detail. Numerical solutions are obtained using the Sparrow-Quack-Boerner local non-similarity numerical method and very efficient Runge-Kutta shooting quadrature., incorporating the applicable criteria that follow.

## 2. MATHEMATICAL MODEL

Consider the combined thermal convection and diffusion mass transfer in laminar boundary layer flow parallel to a flat plate which is inclined to the vertical with angle, $\gamma$, with free stream velocity, $u_\infty$, as depicted in figure 1. The temperature of the ambient medium is $T_\infty$ and wall temperature is $T_w$. The flow along the plate contains a species, A, slightly soluble in the fluid, B with the concentration at the plate surface being $C_w$ and the solubility of A in B far away from the plate is $C_\infty$. The streamwise coordinate, x, is measured from the leading edge of the plate, parallel to the plate and the transverse coordinate, y, normal to the plate in the outward direction, for flow regimes both above and beneath the plate. Following Sparrow et al. [15], we assume the Dufour effect may be described by a second order concentration derivative with respect to the transverse coordinate in the energy conservation equation, and the Soret effect by a second order temperature derivative in the concentration equation.

The forced flow, following Chen et al. [3] exists above the plate for g > 0 in the clockwise direction and beneath the plate for g < 0 (in the anti-clockwise direction). The fluid properties are assumed to be constant except the density variation in the buoyancy force term. Along with the Boussinesq approximation, the laminar boundary layer equations can be written with Soret and Dufour effects as follows:

$$\frac{\partial u}{\partial x}+\frac{\partial v}{\partial y}=0 \tag{1}$$

$$u\frac{\partial u}{\partial x}+v\frac{\partial v}{\partial y}$$

$$=\pm g\beta\sin\gamma\frac{\partial}{\partial x}\int_y^\infty (T-T_\infty)\mathrm{d}y\pm g\beta^*\sin\gamma\frac{\partial}{\partial x}\int_y^\infty (C-C_\infty)\mathrm{d}\,y$$

$$\pm g\beta(T-T_\infty)\cos\gamma\pm g\beta^*(C-C_\infty)\cos\gamma+v\frac{\partial^2 u}{\partial y^2} \tag{2}$$

$$u\frac{\partial T}{\partial x}+v\frac{\partial T}{\partial y}=\alpha\frac{\partial^2 T}{\partial y^2}+\frac{D_m K_T}{c_s c_p}\frac{\partial^2 C}{\partial y^2} \tag{3}$$

The first two terms on the RHS of equation (2) correspond to the streamwise pressure gradients induced by the combined buoyancy forces, with the plus and the minus signs representing, respectively, flows above and below the plate. The third and fourth terms correspond to the buoyancy forces generated by thermal and mass diffusion, with the plus and minus signs referring, respectively, to upward and downward forced flows. The final term in (2) on the right hand side is the viscous diffusion term. The initial and boundary conditions at the plate and in the free stream are:

$$u=u_\infty,\ T=T_\infty,\ C=C_\infty \quad \text{at } x=0$$

$$u=0,\ v=0,\ T=T_w,\ C=C_w \quad \text{at } y=0$$

$$u\to u_\infty,\ T=T_\infty,\ C=C_\infty \ \text{as}\ y\to\infty \tag{5}$$

where u and v denote the velocity components in the xand y-directions respectively, v is the kinematic viscosity, $\beta$ and $\beta^*$ are the coefficients of thermal expansion and concentration expansion, respectively, T and C are the temperature and concentration, respectively, $\rho$ is the density, $D_m$ is the mass diffusivity, $c_p$ is the specific heat capacity, $c_s$ is the concentration susceptibility, $\alpha$ is the thermal diffusivity, $T_m$ is the mean fluid temperature, $K_T$ is the thermal diffusion ratio. Also, as indicated by Chen et al. [3], equation (2) indicates the existence of both buoyancy induced streamwise pressure gradient terms and the buoyancy force terms for an inclined surface. The relative magnitude of these terms, however is controlled by the angle of inclination of the plate to the vertical, $\gamma$ . Chen et al. [3] have shown using an orderof-magnitude analysis that the buoyancy-induced streamwise pressure gradient terms can be neglected in comparison with the buoyancy force terms provided:

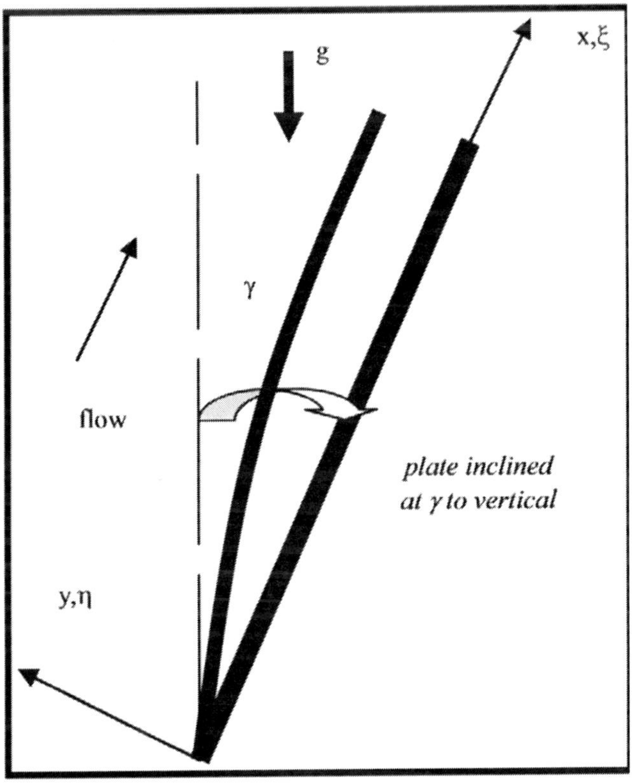


**Figure 1**. Physical model and coordinate system.

$$\frac{\delta}{x}\mathrm{Tan}\,\gamma \ll 1 \tag{6}$$

Chen et al. [3] have further shown that in terms of dimensionless boundary layer thickness $\eta_\delta$ (the $\eta$ value for which $y = \delta$), the condition (6) is also equivalent to the following condition:

$$\operatorname{Tan}\gamma \ll \frac{\mathrm{Re}_x^{1/2}}{\eta_\delta} \tag{7}$$

where denotes the local Reynolds number. Effectively the condition (6) or (7) is valid for Tan 3 ~ 30 i.e. angles of inclination, $\gamma \ll$ 72 ~ 88 degrees. In this situation, the buoyancy-induced streamwise pressure gradient terms are omitted in equation (2) which reduces to the much simpler form:

$$u\frac{\partial u}{\partial x}+v\frac{\partial v}{\partial y} = v\frac{\partial^2 u}{\partial y^2} \pm g\beta(T-T_\infty)\cos\gamma \pm g\beta^*(C-C_\infty)\cos\gamma \tag{8}$$

The governing boundary layer equations then comprise equations (1), (3), (4) and (8), with boundary conditions (6) subject to the condition given by (6) and (7). The special case of a vertical plate is retrieved from (8) for $\gamma = 0^\circ$ i.e. . Proceeding with the analysis we introduce the following dimensionless variables:

$$\begin{aligned}
&\eta = y(u_\infty/vx)^{1/2},\ \xi = \xi(x),\\
&f(\xi,\eta) = \psi(x,y)/(vu_\infty x)^{1/2}\\
&\theta(\xi,\eta) = (T-T_\infty)/(T_w-T_\infty),\\
&\phi(\xi,\eta) = (C-C_\infty)/(C_w-C_\infty)
\end{aligned} \tag{9}$$

where h is the pseudo-similarity variable, x is transformed x-coordinate which represents the thermal buoyancy effect, f is a reduced stream function, q is dimensionless temperature and f is the dimensionless concentration function. The stream function $\psi(x,y)$ satisfies the mass conservation equation (2) with

$$u = \frac{\partial \psi}{\partial y}, \quad v = -\frac{\partial \psi}{\partial x} \tag{10}$$

Implementing equations (9) and (10) into equations (8) and (3), (4), (5), we obtain:

$$f''' + \frac{1}{2} ff''' \pm \xi \cos\gamma (\theta + N\phi) = \xi \left[ f' \frac{\partial f'}{\partial \xi} - f'' \frac{\partial f}{\partial \xi} \right] \tag{11}$$

$$\frac{\theta''}{\Pr} + \frac{1}{2} f\theta' + Du\phi'' = \xi \left[ f' \frac{\partial \theta}{\partial \xi} - \theta' \frac{\partial f}{\partial \xi} \right] \tag{12}$$

$$\frac{\phi''}{\mathrm{Sc}} + \frac{1}{2} f\phi' + \mathrm{Sr}\theta'' = \xi \left[ f' \frac{\partial \phi}{\partial \xi} - \theta' \frac{\partial f}{\partial \xi} \right] \tag{13}$$

The transformed dimensionless boundary conditions are:

$$\eta = 0 : f' = 0, f = 0, \theta = 1, \phi = 1, \tag{14a}$$

$$\eta \to \infty : f' = 1, \theta = 0, \phi = 0 \tag{14b}$$

where [illegible] denotes differentiation with respect to $\eta$, Pr is the Prandtl number, Sc is the Schmidt number, the thermal buoyancy force parameter, $\xi = Gr_{x,t} / \mathrm{Re}_x^2$ (which is a measure of thermal buoyancy force effect on forced convection), [illegible] is a parameter quantifying the relative effects between buoyancy forces that arise from mass diffusion and thermal diffusion, [illegible] denotes local Grashof number for thermal diffusion,

$Gr_{x,c} = g\beta^* (C_w - C_\infty) x^3 / v^2$

denotes the local Grashof number for mass diffusion respectively,

$$Du = \frac{D_m K_T (C_w - C_\infty)}{c_s c_p \nu (T_w - T_\infty)}$$

is the Dufour number,

$$Sr = \frac{D_m K_T (T_w - T_\infty)}{\nu T_m (C_w - C_\infty)}$$

is the Soret number. The engineering design quantities of physical interest include the skin-friction coefficient (a function of local Reynolds number), Nusselt number function and Sherwood number function, which in dimensionless form are given by:

$$\frac{1}{2} C_f \, \mathrm{Re}_x^{1/2} = f''(\xi, 0) \tag{15}$$

$$Nu_x \, \mathrm{Re}_x^{-1/2} = -\theta'(\xi, 0) \tag{16}$$

$$Sh_x \, \mathrm{Re}_x^{-1/2} = -\phi'(\xi, 0) \tag{17}$$

## 3. NUMERICAL SOLUTIONS

We now obtain approximate solutions to Equations (11) - (13) based on the local similarity and local nonsimilarity methods developed by Sparrow et al. [33]. More recently this approach has been employed by Bég et al. [34] to simulate electromagnetic induction flows. Chang et al. [35] employed local nonsimilarity modeling to investigate unsteady rheological flow. Bég et al. [36] studied gravity-driven magneto-micropolar flow on a slope using both local nonsimilarity and network simulation methods. Details and convergence aspects are described in [34] to [36]. For the first level of truncation the $\xi$ derivatives in equations (10) - (13) can be neglected. Thus, the governing equations for the first level of the truncation are:

$$f''' + \frac{1}{2} ff'' + \xi \cos\gamma (\theta + N\phi) = 0 \tag{18a}$$

$$\frac{\theta''}{\Pr} + \frac{1}{2} f\theta' + Du\phi'' = 0 \tag{18b}$$

$$\frac{\phi''}{Sc} + \frac{1}{2} f\phi' + Sr\theta' = 0 \tag{18c}$$

subject to the boundary conditions:

$$\begin{aligned} &f(\xi,0) = 0,\ f'(\xi,0) = 0,\ \theta(\xi,0) = 1, \\ &\phi(\xi,0) = 1,\ f'(\xi,0) = 1,\ \theta(\xi,0) = 0, \\ &\phi(\xi,0) = 0 \end{aligned} \tag{19}$$

For the second level of truncation, we introduce,

$$F = \frac{\partial f}{\partial \xi}, \quad \Theta = \frac{\partial \theta}{\partial \xi}, \quad \Phi = \frac{\partial \phi}{\partial \xi} \tag{20}$$

and restore all of neglected terms in the first level of truncation. Thus, the governing equations are:

$$\begin{aligned} &f'' + \frac{1}{2} ff'' + \xi \cos\gamma (\theta + N\phi) = \xi [f'F' - f''F] \\ &\frac{\theta''}{\Pr} + \frac{1}{2} f\theta' + Du\phi'' = \xi [f'\Theta - \theta' F] \\ &\frac{\phi''}{\mathrm{Sc}} + \frac{1}{2} f\phi' + Sr\theta' = \xi [f'\Phi - \phi' F] \end{aligned} \tag{21}$$

subject to the boundary conditions:

$$\begin{aligned}
&f(\xi,0)=0,\ f'(\xi,0)=0,\ \theta(\xi,0)=1,\\
&\phi(\xi,0)=1,\ f'(\xi,0)=1,\ \theta(\xi,0)=0,\\
&\phi(\xi,0)=0
\end{aligned} \tag{22}$$

The introduction of the three new dependent variables [illegible] in the problem requires three additional equations with appropriate boundary conditions. This can be obtained by differentiating equations (21) with respect to $\xi$, and neglecting the terms $\partial F/\partial\xi$, $\partial\Theta/\partial\xi$, and $\partial\Phi/\partial\xi$, which leads to:

$$\begin{aligned}
&F''' + \frac{1}{2}(fF'' + Ff'') \pm \cos\gamma(\theta + N\phi) \pm \xi\cos\gamma(\Theta + N\Phi)\\
&= \xi\left[F'^2 - F''F\right] + \left[fF' - f''F\right]\frac{\Theta''}{\Pr} + \frac{1}{2}(f\Theta' + F\theta')\\
&\quad + Du\Phi''\\
&= \xi\left[F'\Theta - \Theta'F\right] + \left[f'\Theta - \theta'F\right]\frac{\Phi''}{Sc} + \frac{1}{2}(f\Phi' + F\phi')\\
&\quad + Sr\Theta''\\
&= \xi\left[F'\Phi - \Phi'F\right] + \left[f'\Phi - \phi'F\right]
\end{aligned} \tag{23}$$

$$\begin{aligned}
&F(\xi,0)=0,\ F'(\xi,0)=0,\ \Theta(\xi,0)=0,\\
&\Phi(\xi,0)=0,\ F'(\xi,0)=0,\ \Theta(\xi,0)=0,\\
&\Phi(\xi,0)=0
\end{aligned} \tag{24}$$

The coupled non-linear eqautions (18), (21) and (23) with the boundary conditions (19), (22) and (24) are solved using fourth order Rung-Kutta method with a shooting technique. This method has also been applied very recently to simulate viscoelastic flow in porous media by Bég and Makinde [37] and nanofluid thermal convection by Rashidi et al. [38], showing excellent efficiency compared with semi-analytical methods such as the differential transform technique with Padé approximants. The step size $\Delta\eta = 0.05$ is used to obtain the numerical solution with five-decimal place accuracy as the criterion of convergence.

# 4. RESULTS AND DISCUSSION

The values of Sr and Du have been selected to ensure that the product Sr Du is constant, assuming that the mean temperature is constant. The default values for the control parameters are selected as: Pr = 0.7 (air), Sr = 1, Du = 0.05 [i.e. Sr Du = 0.05], N = 1, Sc = 0.2 (hydrogen gas as the species diffusing in air). In all computations we present the variation of f', q and f versus h for the velocity, temperature and species diffusion boundary layers, and also

$$\frac{1}{2}C_f \, \mathrm{Re}_x^{1/2} = f''(\xi,0), \; Nu_x \, \mathrm{Re}_x^{-1/2} = -\theta'(\xi,0)$$

and [illegible] versus x as a simulation of skin friction function, Nusselt number function and Sherwood number, respectively. In figures 2 to 4 we have plotted the variation of dimensionless velocity, temperature and concentration function with the thermal buoyancy parameter, $\xi = Gr_{x,t}/\mathrm{Re}_x^2$.

For positive inclination angle of the plate (g = 70 degrees), we observe flow velocity (f') is strongly increased i.e. accelerated, with a rise in thermal buoyancy force parameter (x), in particular closer to the plate surface; further into the boundary layer, the profiles begin to converge for all values of x i.e. they tend towards unity as specified in the boundary conditions. A distinct velocity overshoot is identified for x values greater than or equal to unity; for x = 0 and 0.1, this overshoot vanishes. Strong thermal buoyancy therefore has a significant acceleration effect on the boundary layer flow. Temperature function (q) however, as illustrated in figure 2 is adversely affected with increasing thermal buoyancy parameter, x. Profiles all descend smoothly from the maximum value at the wall to zero in the free stream. The value N = 1 implies that the thermal and concentration (species diffusion) buoyancy forces are of the same order of magnitude. Dimensionless concentration function (f) is decreased also with increasing thermal buoyancy parameter, x, as shown in figure 4. The differences in values are however smaller over the same range of x values, compared with the temperature distribution (figure 3). In all cases the velocity, temperature and concentration are minimized with x = 0.

In figures 5 to 7 the effects of plate inclination on the dimensionless velocity, temperature and concentration function with coordinate transverse to the plate (h) are illustrated, again with both Soret and Dufour effects present, for N = 1. When g < 0 i.e. negative plate inclination, in figure 5, the velocity (f') is reduced at first i.e. flow is initially decelerated nearer the plate surface; however further away it is accelerated and infact overshoots. For the case of the vertical plate (g = 0 degrees) and for positive inclination, velocities are always monotonic distributions and no overshoot is present.

With g > 0 i.e. 30 degrees and 80 degrees, velocity is reduced i.e. flow is decelerated, largely owing to gravitational effects. Conversely in figure 6, with negative plate inclination (g < 0) the temperature (q) decreases slightly; for a vertical plate temperatures are greater than for the negatively inclined plate; temperatures are further increased marginally with positive inclination of the plate. In figure 7, a similar response is observed for the concentration distributions (f) which as with temperature are reduced with negative plate inclination, but enhanced with positive inclination and also for a vertical plate scenario (g = 0 degrees).

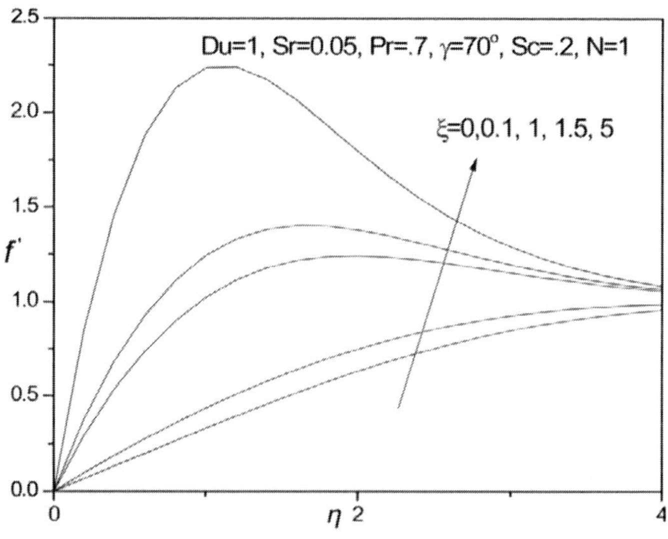


**Figure 2.** f' versus η for various ξ values.

**Figure 8** shows the response of temperature function through the boundary layer regime to Prandtl number, Pr. Larger Pr values imply a thinner thermal boundary layer thickness and more uniform temperature distributions across the boundary layer. Smaller Pr fluids have higher thermal conductivities so that heat can diffuse away from the vertical surface faster than for higher Pr fluids (thicker boundary layers). Physically Pr = 0.7 is representative for air or hydrogen and Pr = 1 for water. Pr = 10, 20 may correspond to oils and lubricants. Pr encapsulates the ratio of momentum diffusivity to thermal diffusivity. With a decrease in Pr temperatures as seen in figure 8 are substantially increased. The lowest temperatures correspond to the highest value of Prandtl number (Pr = 20). No temperature overshoot is observed.

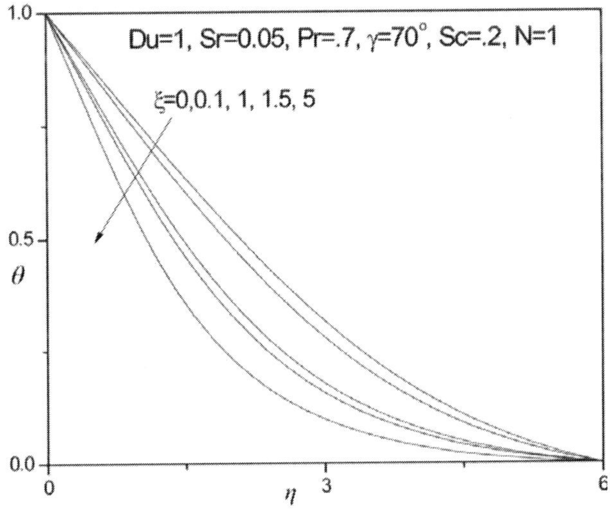


**Figure 3**. θ versus η for various ξ values.

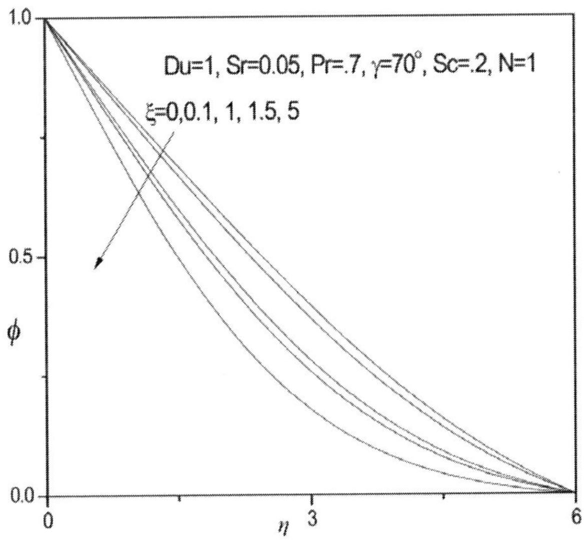


**Figure 4.** $\phi$ versus η for various ξ values.

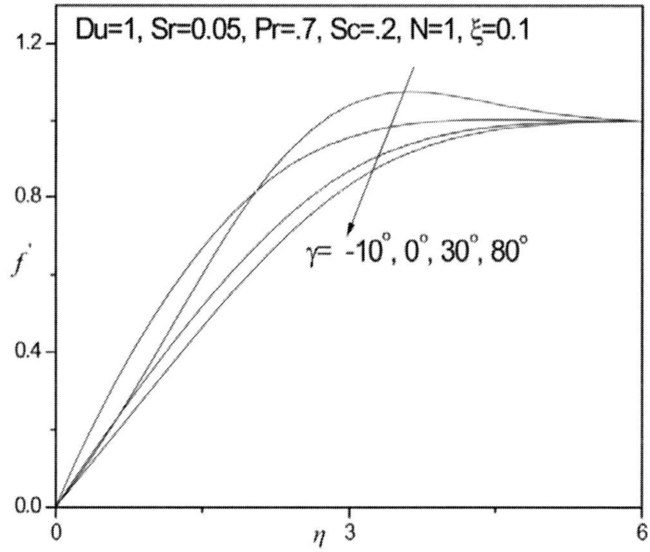


**Figure 5**. f' versus η for various γ values.

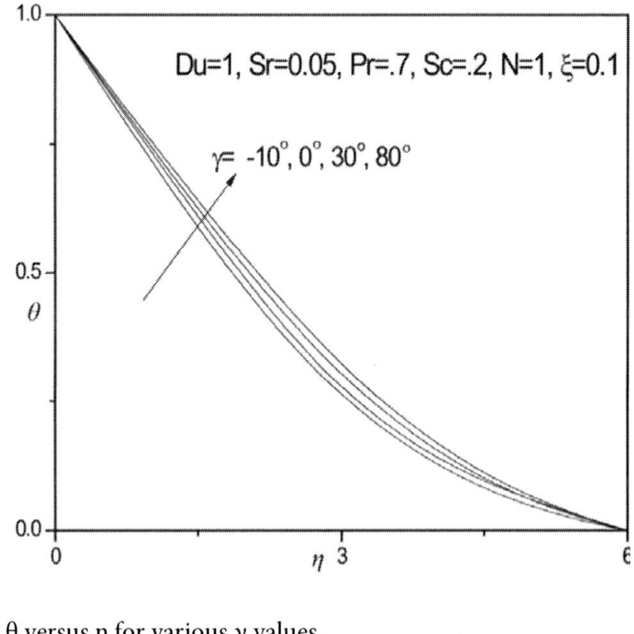


**Figure 6**. θ versus η for various γ values.

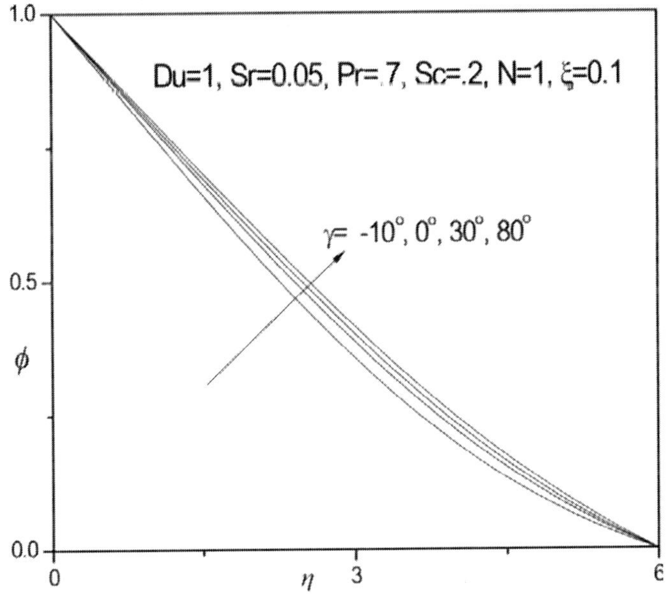


**Figure 7.** $\phi$ versus η for various γ values.

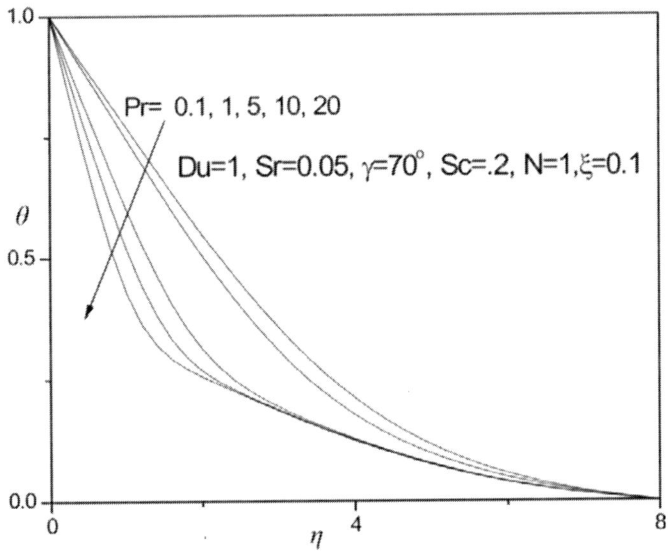


**Figure 8.** θ versus η for various Pr values.

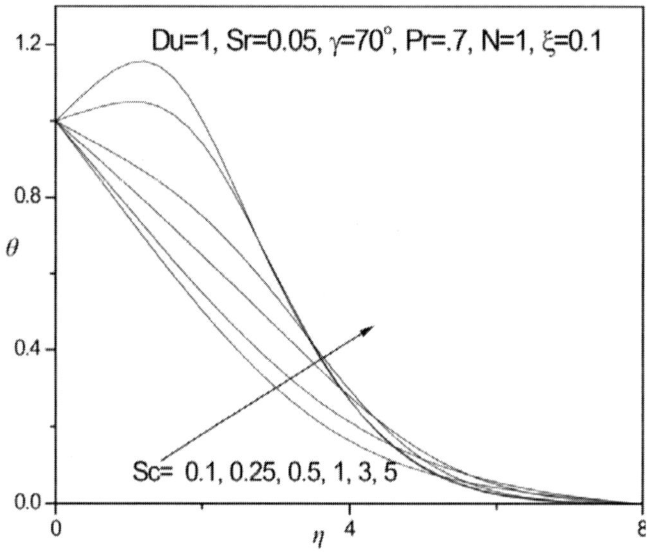


**Figure 9.** θ versus η for various Sc values.

In figure 9 the effect of Schmidt number, Sc on temperature profiles (q) is plotted, for low value of thermal buoyancy parameter (x = 0.1). Larger values of Sc correspond to Methanol diffusing in air (Sc = 1.0) and Ethylbenzene in air (Sc = 2.0), as indicated by Gebhart et al. [39]. When Pr and Sc differ, then the thermal and species diffusion regions are of different extents. Temperature is continuously decreased from the wall for Sc = 0.1 and Sc = 0.5. For Sc = 0.5 the profile is almost linear in decay. For Sc = 1, the concentration and velocity boundary layers will have approximately the same thickness; temperatures also decrease smoothly from the plate surface to the free stream. For Sc > 1 we observe a significant temperature overshoot a short distance from the plate surface. The highest temperature corresponds to Sc = 5 (maximum Schmidt number). For both Sc = 3 and 5, temperature profiles descend sharply after attaining the velocity overshoot. **Figure 10** indicates that concentration (f) is reduced continuously throughout the boundary layer, however, with increasing Schmidt number. Sc measures the relative effectiveness of momentum and mass transport by diffusion. Larger values of Sc are equivalent to reducing the chemical molecular diffusivity i.e. less diffusion therefore takes place by mass transport. As a result concentration is decreased in the boundary layer.

The influence of the concentration-to-thermal-buoyancy ratio parameter, N, on dimensionless velocity (f′), temperature (q) and concentration function (f) with coordinate transverse to the plate (h) are shown again with both Soret and

Dufour effects present, in figures 11 to 13. A positive rise in N clearly induces an increase in velocity as seen in figure 11; the flow is significantly accelerated for N – 20, where once again a velocity overshoot is computed at intermediate distance from the plate surface (h ~ 3). For N = 0.5, 1, 5, 10, no velocity overshoot is apparent although velocities are increased in the regime continuously. For N < 0 i.e. -1, the thermal and species buoyancy forces oppose each other and this causes a decrease in flow velocity compared with the scenario where both forces are absent (N = 0) or are aiding each other (N > 0). In figure 12 negative N causes a very slight rise in fluid temperature i.e. heats the fluid regime slightly; for N = 0, and for positive N (thermal and concentration buoyancy forces assisting each other), this trend is reversed with a decrease in temperature i.e. cooling of the boundary layer regime. In figure 13, a similar response for the concentration distribution is observed as in the case of the temperature distribution, with f values increasing fractionally with a negative N value and decreasing with positive N values. Opposing buoyancy forces (N < 0) therefore enhance species diffusion in the regime whereas aiding buoyancy forces (N > 0) inhibit species diffusion in the boundary layer. A similar response was reported by Rawat and Bhargava [27] and also Partha et al. [31].

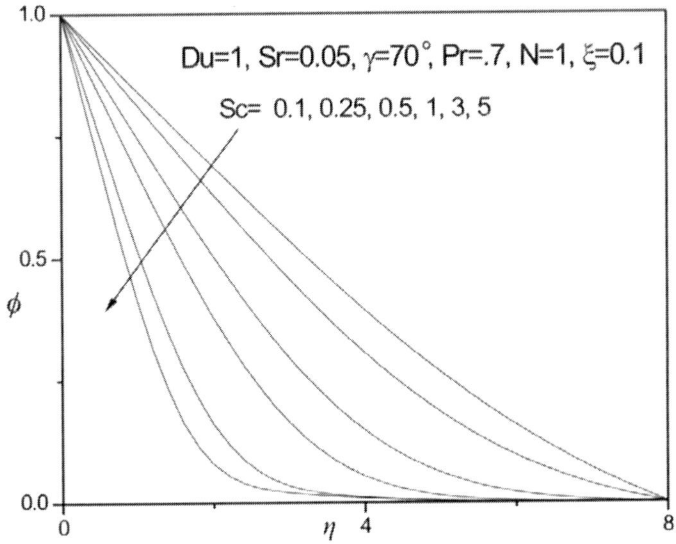


**Figure 10.** $\phi$ versus η for various Sc values.

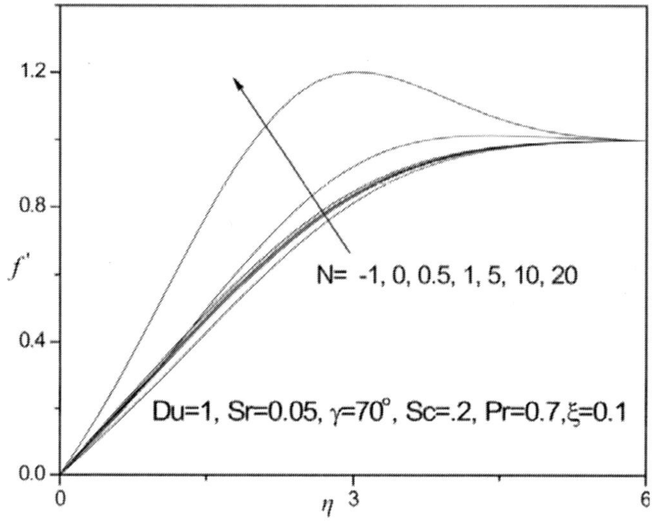


**Figure 11.** f' versus η for various N values.

In figures 14 and 15 the combined effect of Dufour number (Du) and Soret number (Sr) on the temperature and concentration distributions, respectively, in the regime are shown. We study the simultaneous increase (and decrease) of these parameters so that their product remains constant at 0.05, following Anghel et al. [21] and Postelnicu [22]. Increasing Dufour number causes a rise in temperature whereas an increase in Soret number cools the fluid i.e. reduces temperature, as observed in figure 14. Conversely in figure 15, we observe that a rise in Du decreases the concentration values in the boundary layer, whereas a rise in Soret number increases values.

In figures 16 and 17, we have presented the variation of the local Nusselt number, [illegible] and local Sherwood number, $Sh_x \, Re_x^{-1/2}$, respectively with the product of the thermal buoyancy force parameter, x, and Cosg i.e. [illegible] as originally computed by Chen et al. [3], with Soret and Dufour effects neglected. Our results agree almost exactly with the computations in [3]. We infer from figures 16 and 17 that in the absence of Soret and Dufour effects, the situations where N = 0 (buoyancy forces absent) and N > 0 (aiding buoyancy forces), generate a continuous increase in local Nusselt number, [illegible] with thermal buoyancy force parameter, x; for N = -1 (opposing buoyancy forces) the local Nusselt number remains unchanged for all values of thermal buoyancy parameter, x. Local Sherwood number, [illegible] as seen in figure 17, is boosted considerably with an increase in Schmidt number from 0.7 through 1.0 to 7.0, for N > 0; however it is reduced with N < 0 compared with N = 0 and N > 0

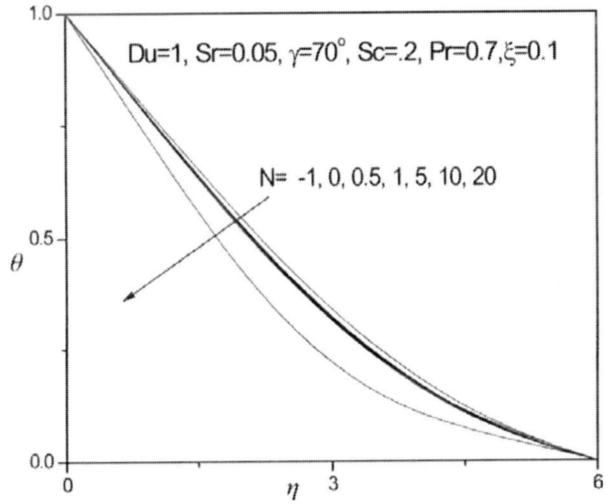


**Figure 12.** θ versus η for various N values.

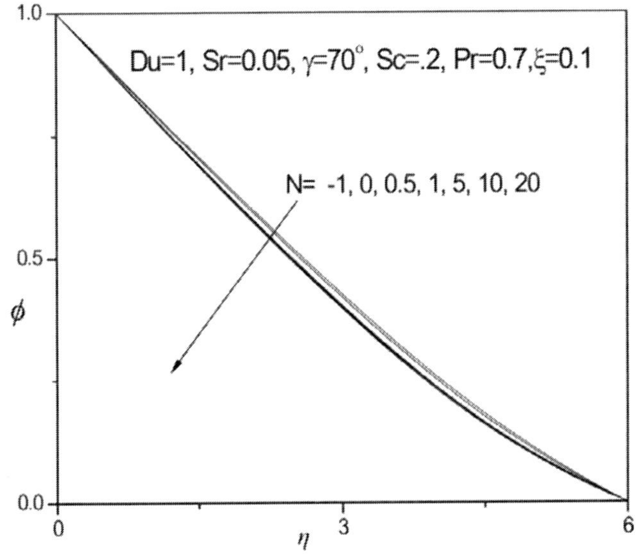


**Figure 13.** $\phi$ versus η for various N values.

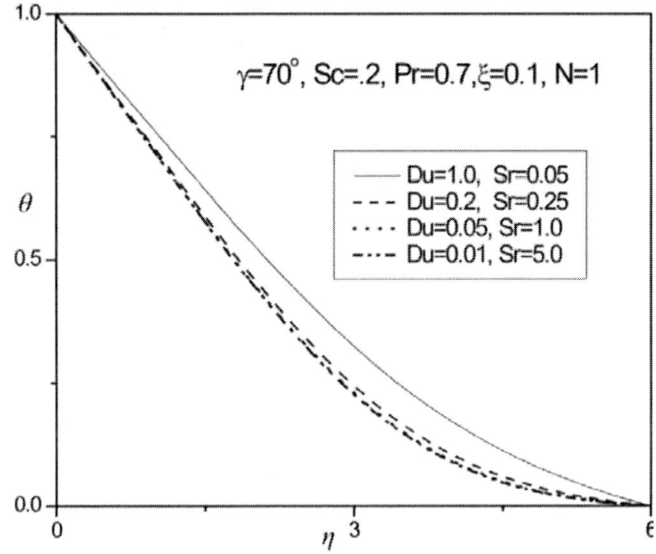


**Figure 14**. θ versus η for various Du & Sr values.

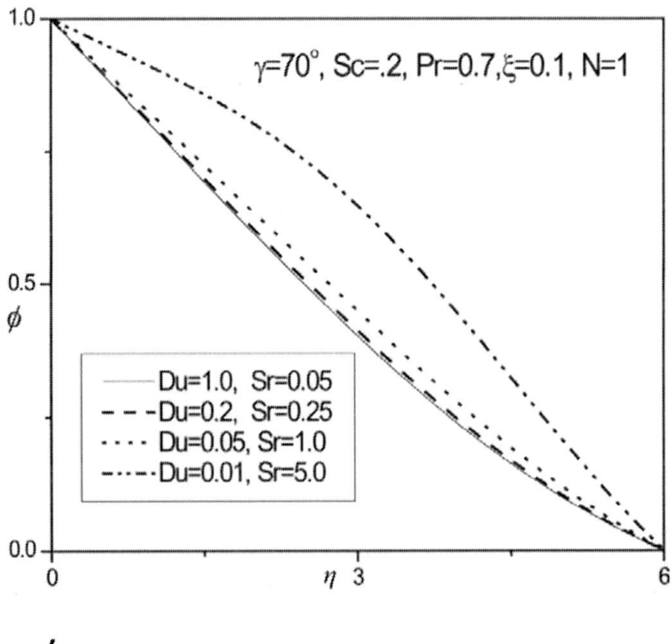


**Figure 15**. $\phi$ versus η for various Du & Sr values.

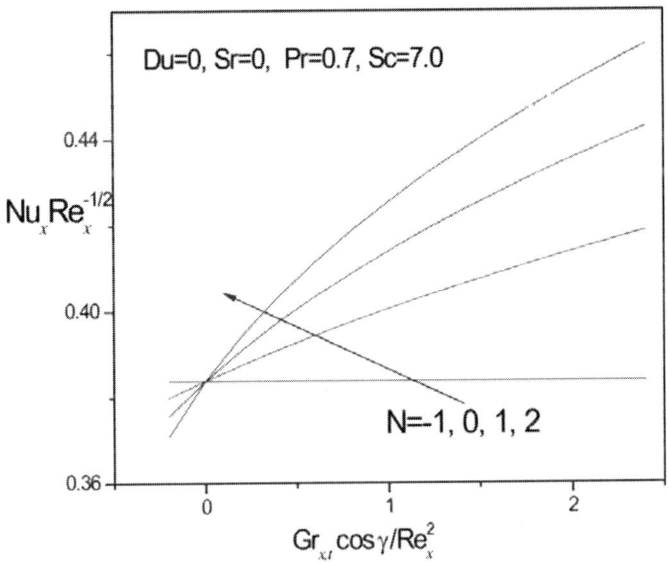


**Figure 16**. Local Nusselt number results.

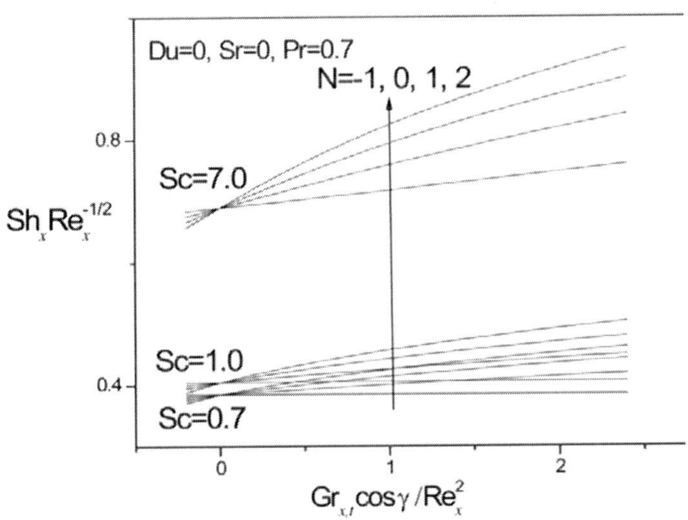


**Figure 17**. Local Sherwood number results.

We finally note that the local nonsimilarity method (LNM) has been extensively validated by the authors recently, as has the numerical shooting quadrature used in the present article, with other techniques including network simulation, finite elements and differential quadrature methods. Details of validation are provided in Bég et al. [24,25] and also in Bég et al. [34], Chang et al. [35] and Bég et al. [36]. There is great confidence therefore in the present numerical solutions and comparisons have been omitted herein for conservation of space.

## 5. CONCLUSIONS

Numerical solutions have been obtained for the mixed convection heat and mass transfer boundary layer flow along an inclined plate in the presence of Soret and Dufour thermo-diffusion effects, using the local nonsimilarity method and a Runge-Kutta shooting technique. For positive inclination of the plate (g > 0), a rise in Dufour number (Du) increases temperatures in the boundary layer but decreases the concentration; conversely an increase in Soret number (Sr) reduces temperature and elevates concentration values. For positive inclination angle of the plate (g = 70 degrees), flow is accelerated, with thermal buoyancy force parameter (x) whereas temperature (q) and concentration (f) are reduced with increasing thermal buoyancy parameter, x. For negative and positive plate inclinations (g < 0 and g > 0), the flow is accelerated and decelerated respectively; for these plate orientations however the temperature and concentration functions are decreased and increased respectively. Temperatures are enhanced and concentrations reduced in the regime with increasing Schmidt number. A positive increase in concentration-to-thermal-buoyancy ratio parameter, N, significantly increases flow velocity with the converse effect for negative N. A velocity overshoot is also identified for N = 20, at intermediate distance from the plate surface. With Soret and Dufour effects neglected, N > 0 induces a continuous rise in local Nusselt number, [illegible] and local Sherwood number, [illegible].

The present study is being extended to consider Soret/ Dufour effects in rotating disk flows [40] of interest in membrane oxygenators and also consider dusty (fluidparticle suspension) effects in photovoltaic collector systems [41]. It is also envisaged that homotopy methods will be employed in such studies [42]. The results of these investigations will be reported in the near future.

## REFERENCES


1. W. T. Kierkus, "An Analysis of Laminar Free Convection Flow and Heat Transfer about an Inclined Isothermal Plate," International Journal of Heat and Mass Transfer, Vol. 11, No. 2, 1968, pp. 241-252.

2. T. Fujii and H. Imura, "Natural-Convection Heat Transfer from a Plate with Arbitrary Inclination," International Journal of Heat and Mass Transfer, Vol. 15, No. 4, 1972, pp. 755-764.

3. T. S. Chen, C. F. Yuh and A. Moustsoglou, "Combined Heat and Mass Transfer in Mixed Convection along Vertical and Inclined Plates," International Journal of Heat and Mass Transfer, Vol. 23, No. 4, 1980, pp. 527- 537.

4. B. F. Armaly, T. S. Chen and N. Ramachandran, "Correlations for Laminar Mixed Convection on Vertical, Inclined and Horizontal Flat Plates with Uniform Surface Heat Flux," International Journal of Heat and Mass Transfer, Vol. 30, No. 2, 1987, pp. 405-408.

5. S. L. Lee and K. Hsu, "Interaction of Surface Suction/Blowing with Buoyancy Force on Mixed Convection Flow Adjacent to an Inclined Flat Plate," International Journal of Heat and Mass Transfer, Vol. 32, No. 10, 1989, pp. 1989-1991.

6. G. Wickern, "Mixed Convection from an arbitrarily Inclined Semi-Infinite Flat Plate—I. The Influence of the Inclination Angle," International Journal of Heat and Mass Transfer, Vol. 34, No. 8, 1991, pp. 1935-1945.

7. W. M. Yan, and C. Y. Soong, "Convective Heat and Mass Transfer along an Inclined Heated Plate with Film Evaporation," International Journal of Heat and Mass Transfer, Vol. 38, No. 7, 1995, pp. 1261-1269.

8. W. J. Sheu and M. C. Lin, "Thermal Ignition in Buoyancy-Driven Boundary Layer Flows along Inclined Hot Plates," International Journal of Heat and Mass Transfer, Vol. 39, No. 10, 1996, pp. 2187-2190.

9. C. Ludwig, "Diffusion Zwichen Unfleigh Erwarmten Orten Gleich Zusammengesetz Losungen," Sitzungbsber. Kaiser. Akad. Wiss. (Mathem.-Naturwiss. Cl.), Wien, Vol. 65, 1856, pp. 539.

10. C. Soret, "Sur l'état d'équilibre que prend, du point de vue de sa Concentration, une Dissolution Saline Primitivement Homogène, Dont Duex parties Sort Portèes à des Temperatures Différentes," C. R. Arch. Science Physical Nature, Genève, Vol. 2, 1879, pp. 48-61.

11. J. Platten and J. C. Legros, "Convection in Liquids," Springer, Berlin, 1984.

12. A. Mojtabi and M. C. Charrier-Mojtabi, "Double Diffusive Convection in Porous Media," Handbook of Porous Media, Marcel Dekker, New York, 2000, pp. 559-603.

13. O. E. Tewfik and J. W. Yang, "The Thermodynamic Coupling between Heat and Mass Transfer in Free Convection with Helium Injection," International Journal of Heat and Mass Transfer, Vol. 6, No. 10, 1963, pp. 915-923.

14. E. M. Sparrow, W. J. Minkowycz and E. R. G. Eckert, "Effects of Diffusion Thermo and Thermal Diffusion on Heat Transfer, Flow and Mass Transfer for the Helium-Air Boundary Layer in Stagnation Flow," ASME Journal of Heat Transfer, Vol. 86C, 1964, p. 508.

15. E. M. Sparrow, C. J. Scott, R. J. Frostron and W. A. Ebert, "Experiments on the Diffusion Thermo Effect in a Binary Boundary Layer with Injection of Various Gases," ASME Journal of Heat Transfer, Vol. 87, 1965, pp. 321-327.

16. C. R. A. Abreu, M. F. Alfradique and A. S. Telles, "Boundary Layer Flows with Dufour and Soret Effects. I: Forced and Natural Convection," Chemical Engineering Science, Vol. 61, No. 13, 2006, pp. 4282-4289.

17. W. Hort, S. J. Linz and M. Lucke, "Onset of Convection In Binary Gas Mixtures: Role of the Dufour Effect," Physical Review A, Vol. 45, No. 6, 1992, pp. 3737-3748.

18. R. M. L. Coelho and A. Silva-Telles, "Extended Graetz Problem Accompanied by Dufour and Soret Effects," International Journal of Heat and Mass Transfer, Vol. 45, No. 15, 2002, pp. 3101-3110.

19. A. Weaver and R. Viskanta, "Natural Convection Due to Horizontal Temperature and concentration Gradients: 2- Species Interdiffusion, Soret and Dufour Effects," International Journal of Heat and Mass Transfer, Vol. 34, No. 12, 1991, pp. 3121-3133.

20. N. G. Kafoussias and E. M. Williams, "Thermal-Diffusion and Diffusion-Thermo Effects on Mixed FreeForced Convective and Mass Transfer Boundary Layer Flow with Temperature Dependent Viscosity," International Journal of Engineering Science, 1995, pp. 1369- 1384.

21. M. Anghel, H. S. Takhar and I. Pop, "Dufour and Soret Effects on Free Convection Boundary Layer over a Vertical Surface Embedded in a Porous Medium," Studia University Babes-Bolyai Mathematica, Vol. XLV, No. 4, 2000, pp. 11-21.

22. A. Postelnicu, "Influence of Chemical Reaction on Heat and Mass Transfer by Natural Convection form Vertical Surfaces in Porous Media Considering Soret and Dufour Effects," Heat and Mass Transfer, Vol. 43, No. 6, 2007, pp. 595-602.

23. B. O. Anwar, R. Bhargava, S. Rawat and E. Kahya, "Numerical Study of Micropolar Convective Heat and Mass Transfer in a Non-darcian Porous Regime with Soret/Dufour Diffusion Effects," Emirates Journal of Engineering Research, Vol. 13, No. 2, 2008, pp. 51-66.

24. B. O. Anwar, A. B. Tasveer, A. Y. Bakier and V. Prasad, "Chemically-Reacting Mixed Convective Heat and Mass Transfer along Inclined and Vertical Plates with Soret and Dufour Effects: Numerical Solutions," International Journal of Applied Mathematics and Mechanics, Vol. 5, No. 2, 2009, pp. 39-57.

25. O. Anwar Bég, A. Y. Bakier and V. R. Prasad, "Numerical Study of Free Convection Magnetohydrodynamic Heat and Mass Transfer from a Stretching Surface to a Saturated Porous Medium with Soret and Dufour Effects," Computational Materials Science, Vol. 46, No. 1, 2009, pp. 57-65.

26. R. Bhargava, R. Sharma and O. Anwar Bég, "Oscillatory Chemically-Reacting MHD Free Convection Heat and Mass Transfer in a Porous Medium with Soret and Dufour Effects: Finite Element Modeling," International Journal of Applied Mathematics and Mechanics, Vol. 5, No. 6, 2009, pp. 15-37.

27. S. Rawat and R. Bhargava, "Finite Element Study of Natural Convective Heat and Mass Transfer in a Micropolar Fluid-Saturated Porous Regime with Soret/Dufour Effects," International Journal of Applied Mathematics and Mechanics, Vol. 5, No. 2, 2009, pp. 58-71.

28. O. Anwar Bég, V. R. Prasad, B. Vasu, N. Bhaskar Reddy, Q. Li and R. Bhargava, "Free Convection Heat and Mass Transfer from an Isothermal Sphere to a Micropolar Regime with Soret/Dufour Effects, International Journal of Heat and Mass Transfer, Vol. 54, 2011, pp. 9-18.

29. E. Osalusi, J. Side and R. Harris, "Thermal-Diffusion and Diffuse-Thermal Effects on Combined Heat and Mass Transfer of a Steady MHD Convective and Slip Flow Due to a Rotating Disk with Viscous Dissipation and Ohmic Heating," International Communications in Heat Mass and Transfer, Vol. 35, No. 8, 2008, pp. 908-918.

30. R. S. Miller, S. Palle and C. Nolan, "Effects of Soret and Dufour Diffusion on Laminar Diffusion Flames at Large Pressures," 56th Annual Meeting of Fluid Dynamics Division, American Physical Society, Meadowlands, New Jersey, November 23-25, 2003.

31. M. K. Partha, P. V. S. N. Murthy and G. P. Raja-Sekhar, "Soret and Dufour Effects in a Non-Darcy Porous Medium," ASME Journal of Heat Transfer, Vol. 128, No. 6, 2006, pp. 605-610.

32. N. OKong'o and J. Bellan, "Real-Gas Effects on Mean Flow and Temporal Stability of Binary Species Mixing Layers," AIAA Journal, Vol. 41, No. 12, 2003, pp. 2429-2443.

33. E. M. Sparrow, H. Quack and J. Boerner, "Local Nonsimilarity Boundary Layer Solutions," AIAA Journal, Vol. 8, No. 11, 1970, pp. 1936-1942.

34. O. A. Bég, A. Y. Bakier, V. R. Prasad and S. K. Ghosh, "Nonsimilar, Laminar, Steady, Electrically-Conducting Forced Convection Liquid Metal Boundary Layer Flow with Induced Magnetic Field Effects," International Journal of Thermal Sciences, Vol. 48, No. 8, 2009, pp. 1596-1606.

35. T-B. Chang, A. Mehmood, O. Anwar Bég, M. Narahari, M. N. Islam and F. Ameen, "Numerical Study of Transient Free Convective Mass Transfer in a Walters-B Viscoelastic Flow with Wall Suction," Communications in Nonlinear Science and Numerical Simulation, Vol. 16, No. 1, 2010, pp. 216-225.

36. O. A. Bég, J. Zueco and T. B. Chang, "Numerical Analysis of Hydromagnetic Gravity-Driven Thin Film Micropolar Flow along an Inclined Plane," Chemical Engineering Communications, Vol. 198, No. 3, 2011, pp. 312- 331.

37. O. Anwar Bég and O. D. Makinde, "Viscoelastic Flow and Species Transfer in a Darcian High-Permeability Channel," Journal of Petroleum Science and Engineering, Vol. 76, No. 3-4, 2011, pp. 93-99.

38. M. M. Rashidi and O. Anwar Bég, "DTM-Padé Modeling of Natural Convective Boundary Layer Flow of a Nanofluid Past a Vertical Surface," International Journal of Thermal and Environmental Engineering, Vol. 4, 2012, pp. 13-24.

39. B. Gebhart, et al., "Buoyancy-Induced Flows and Transport," Hemisphere, New York, 1988.

40. O. Anwar Bég, J. Zueco and L. M. López-Ochoa, "Network Numerical Analysis of Optically-Thick Hydromagnetic Slip Flow from a Porous Spinning Disk with Radiation Flux, Variable Thermophysical Properties and Surface Injection Effects," Chemical Engineering Communications, Vol. 198, No. 3, 2011, pp. 360-384.

41. J. Zueco, P. Eguía, E. Granada, J. Míguez and O. Anwar Bég, "An Electrical Network for the Numerical Solution of Transient MHD Couette Flow of a Dusty Fluid: Effects of Variable Properties and Hall Current," International Communications in Heat and Mass Transfer, Vol. 37, No. 10, 2010, pp. 1432-1439.

42. M. M. Rashidi, M. Rastegari and O. Anwar Bég, "A Study of Non-Newtonian Flow and Heat Transfer over a Non-Isother-Mal Wedge Using the Homotopy Analysis Method," Chemical Engineering Communications, 2011.

# CHAPTER 10

# Numerical Investigation on the Temperature Characteristics of the Voice Coil for a Woofer Using Thermal Equivalent Heat Conduction Models

Moo-Yeon Lee [1,*] and Hyung-Jin Kim [2]

[1] *Department of Mechanical Engineering, Dong-A University, Hadan 840, Saha-gu, Busan 604-714, Korea*
[2] *Graduate School of Mechanical Engineering, Dong-A University, Hadan 840, Saha-gu, Busan 604-714, Korea*

## ABSTRACT

The objective of this study is to numerically investigate the temperature and heat transfer characteristics of the voice coil for a woofer with and without bobbins using the thermal equivalent heat conduction models. The temperature and heat transfer characteristics of the main components of the woofer were analyzed with input powers ranging from 5 W to 60 W. The numerical results of the voice coil showed good agreement within ±1% of the data by Odenbach (2003). The temperatures of the voice coil and its units for the woofer without the bobbin were 6.1% and 5.0% on average, respectively; lower than those of the woofer with the bobbin. However, at an input power of 30 W for the voice coil, the temperatures of the main components of the woofer without the bobbin were 40.0% higher on average than those of the woofer obtained by Lee *et al.* (2013).

## KEYWORDS

heat transfer; temperature; voice coil; woofer

## 1. INTRODUCTION

Speakers are designed to reproduce the audio signals that they receive, project them at a higher level and provide a greater spread of sound above the normal acoustic level of the input powers [1]. Speakers are normally used with amplifiers to amplify the higher acoustic levels for providing a pleasing sound with greater spread. The speakers and equipment for applications, such as portable devices, vehicles and indoor and outdoor theaters, are mainly decided by the usage purposes and the available space, capacity and performance. The speakers are divided with regard to the types and structures of the units into tweeters, midrange and woofers [2]. Generally, the units of these speakers consist of a damper, bobbin, voice coil, top plate, yoke, magnet and bottom plate, as shown in Figure 1. These parts have important roles in deciding the performance of the speakers [3]. Figure 1 shows a cross-sectional view of the tested woofer. The voice coil is a key unit of the speakers that performs energy conversion across the electrical, mechanical and acoustic components [4]. The energy conversion efficiency of the voice coils is under 10%, due to thermal and mechanical losses of over 90%, as the input power supplied to the voice coil is transformed into audio sound signals [5]. When operating the speakers for a very long time with excessive input power, the voice coils could experience lower sound quality and harmonic distortion, due to a high temperature. In addition, in rare, but severe, cases, this could cause a fire [6]. Therefore, to improve the sound quality and the performance stability of the speakers, studies on the heat transfer characteristics of the voice coil and its units are necessary. Recently, Chang *et al.* (2013) experimentally and numerically investigated the total harmonic distortion (THD) in a moving-coil loudspeaker. They suggested a novel approach to reduce the THD of a moving-coil loudspeaker with non-linear parameters and supplied a 10% reduction of THD by displacing the initial position of the voice coil [7]. Hong *et al.*(2004) studied the performance improvement of a diaphragm for a micro-speaker using the Taguchi method in discrete design space. They reported that the second natural frequency of the diaphragm for the micro-speaker decreased by 37%, while maintaining a twisted mode shape [8]. Lee *et al.* (2013) analyzed the temperature characteristics of the voice coil for a woofer using a ferrofluid with the input acoustic signals. As a result, the temperature of the voice coil for the woofer with the ferrofluid was 51.0% lower than that without ferrofluid at an input power of 40 W [9].

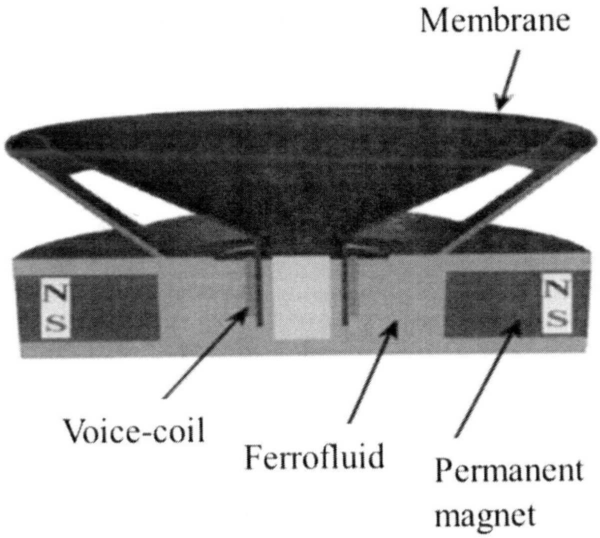


**Figure 1.** Schematic diagram of the tested woofer (cross-sectional view).

However, precise and concise studies on the temperature of heat transfer characteristics and on improvements for the performance of the woofer have been limited. Therefore, the objective of this study is to numerically investigate the characteristics and enhancement of the heat transfer of the voice coil for a woofer with and without bobbins using the thermal equivalent heat conduction models. In addition, due to the critical limitation of the maximum power of the woofers with the ohmic heat generated by the voice coils, the temperature characteristics of the voice coil for the tested woofer with and without a bobbin were observed while varying the input power.

# 2. NUMERICAL ANALYSIS

## 2.1. Numerical Method

A 200-W woofer with an 8-inch diaphragm consisting of a damper, bobbin, voice coil, top plate, yoke, magnet and bottom plate was investigated, as shown in Figure 1 [10]. The damper located over the voice coil could support the vertical motion of the voice coil and diaphragm. The bobbin was made of polyimide, which has heat-resistant properties. Like the human voice, the goal of the voice coil is to generate sound when input power is supplied to form a magnetic field. The voice coil was made of copper wound around the bobbin. Both the top and bottom plates were made of pure iron and were used for the magnetic field flow control and the magnetic field strength enhancement generated by the permanent magnet. The permanent magnets used in woofers are generally divided into ferrite, alnico and rare earth elements. The permanent magnet used in this study was of the ferrite type. The magnet generates a magnetic field to interact with the magnetic field generated by the voice coil.

A yoke consisting of pure iron was located at the center of the woofer for both the magnetic field flow control and the magnetic field strength enhancement generated by the voice coil [11]. Table 1 shows the specifications of the main components for the woofer. The notations used in Table 1 were D (Outer Diameter), d (Inner Diameter) and H (Height). 3D modeling of the woofer was done using SolidWorks.

**Table 1.** Specifications of the woofer.

| Components | Size (D × d × H, mm) | Specifications |
|---|---|---|
| Damper | 75 × 45 × 16 | Fiber paper |
| Bobbin | 36 × 35 × 35 | Polyimide |
| Voice coil(with bobbin) | 36.7 × 36.5 × 14 | Copper |
| Voice coil(without bobbin) | 36.7 × 35.2 × 27 | Copper |
| Top plate | 100 × 39 × 6 | Pure iron |
| Permanent magnet | 111 × 60 × 17 | Ferrite |
| Bottom plate | D: 100, H: 6 | Pure iron |
| Yoke | D: 35, H: 23 | Pure iron |

Figure 2 shows the configurations of the 3D modeling for the woofers with and without the bobbin. In order to analyze the effect of the bobbin, woofers with and without a bobbin were considered, as shown in Figure 2a,b respectively [12]. The heat transfer analysis was numerically performed using commercial numerical analysis software (ANSYS CFX, FLUID FLOW Tool) [13]. The mesh map of the tested physical model was made using tetrahedron elements with a maximum size of 2.0 mm and minimum size of 0.01 mm, comprising 406,210 elements and 74,372 nodes. Figure 3 shows the mesh map of the physical model. The heat transfer of the woofer was analyzed using the k-ε model [14]. Table 2 shows the numerical conditions used in the simulation. The temperature and pressure of the air inside the woofer system were set to 25 °C and 101.3 kPa. The gravitational force was 9.8 m/s$^2$. Additionally, the nominal impedance of the woofer was 8.0 Ω. In order to calculate the voice coil temperature, the input power supplied to the voice coil of the woofer was varied from 5 W, 15 W, 30 W, 45 W to 60 W. The working fluid was air. Table 3 shows the thermodynamic properties of the main components for the woofer. The thermal conductivities of the pure iron, copper, permanent magnet and air were 60.5, 401, 50 and 2.61 ×

$10^{-2}$ $Wm^{-1}$ $K^{-1}$, respectively. The specific heat capacities of the pure iron, copper, permanent magnet and air were $4.34 \times 10^2$, $3.85 \times 10^2$, $4.0 \times 10^2$ and $1.0044 \times 10^3$ J $kg^{-1}$ $K^{-1}$, respectively.

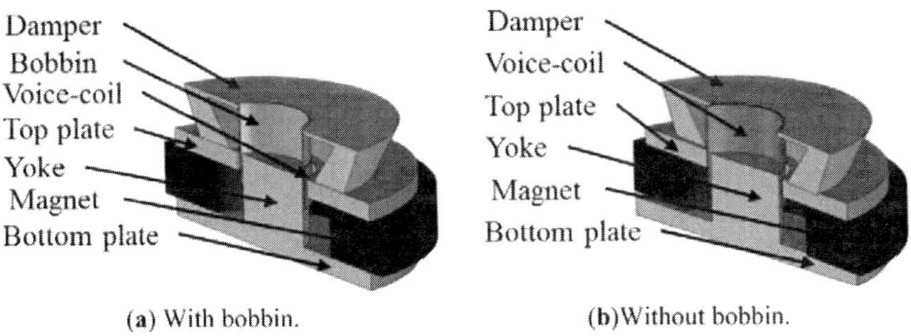


**Figure 2.** Configurations of the woofer with and without a bobbin.

**Figure 3.** Mesh map of the model.

**Table 2.** Numerical conditions.

| Components | Specifications |
|---|---|
| Woofer (allowable max. power, W) | 200 |
| Working fluid | Air |
| Pressure (kPa) | 101.3 |
| Input power (W) | 5, 15, 30, 45, 60 |
| Temperature (K) | 298 |
| Gravitational acceleration ($m/s^2$) | 9.8 |
| Nominal impedance (Ω) | 8.0 |

**Table 3.** Thermodynamic properties of the woofer.

| | Components | Specifications |
|---|---|---|
| Pure iron | Density ($kg\ m^{-3}$) | 7854 |
| | Specific heat capacity ($J\ kg^{-1}\ K^{-1}$) | $4.34 \times 10^2$ |
| | Thermal conductivity ($Wm^{-1}\ K^{-1}$) | 60.5 |
| | Thermal expansivity ($K^{-1}$) | $7.5 \times 10^{-4}$ |
| Copper | Density ($kg\ m^{-3}$) | 8933 |
| | Specific heat capacity ($J\ kg^{-1}\ K^{-1}$) | $3.85 \times 10^2$ |
| | Thermal conductivity ($Wm^{-1}\ K^{-1}$) | 401 |
| Ferrite magnet | Density ($Kg\ m^{-3}$) | 7300 |
| | Specific heat capacity ($J\ kg^{-1}\ K^{-1}$) | $4.0 \times 10^2$ |
| | Thermal conductivity ($W\ m^{-1}\ K^{-1}$) | 50 |
| Air (25 °C) | Density ($kg\ m^{-3}$) | 1.185 |
| | Specific heat capacity ($J\ kg^{-1}\ K^{-1}$) | $1.0044 \times 10^3$ |
| | Thermal expansivity ($K^{-1}$) | 0.003356 |
| | Dynamic viscosity ($kg\ m^{-1}\ K^{-1}$) | $1.831 \times 10^{-5}$ |
| | Thermal conductivity ($Wm^{-1}\ K^{-1}$) | $2.61 \times 10^{-2}$ |

## 2.2. Heat Transfer Models

Figure 4 shows the convergence criterion of the numerical analysis for the woofer with the bobbin at an input power of 15 W and an ambient temperature of 298 K. The temperature of the tested voice coil was increased with time. After 130 s, the temperature increased steadily and converged at 439 K. Then, the deviations of the voice coil temperature were within $10^{-3}$.Figure 5 shows the heat transfer models using the equivalent thermal circuit methods. The equivalent thermal circuit models with a combination of the serial and parallel configurations were developed for the prediction of the voice coil temperature of the tested woofers with and without the bobbins [15,16]. The temperatures of the tested woofers were predicted using the developed thermal models.

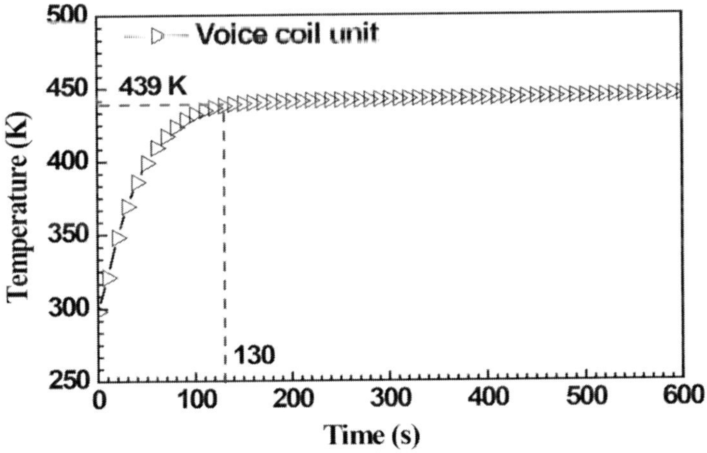


**Figure 4.** Convergence criterion of the numerical analysis.

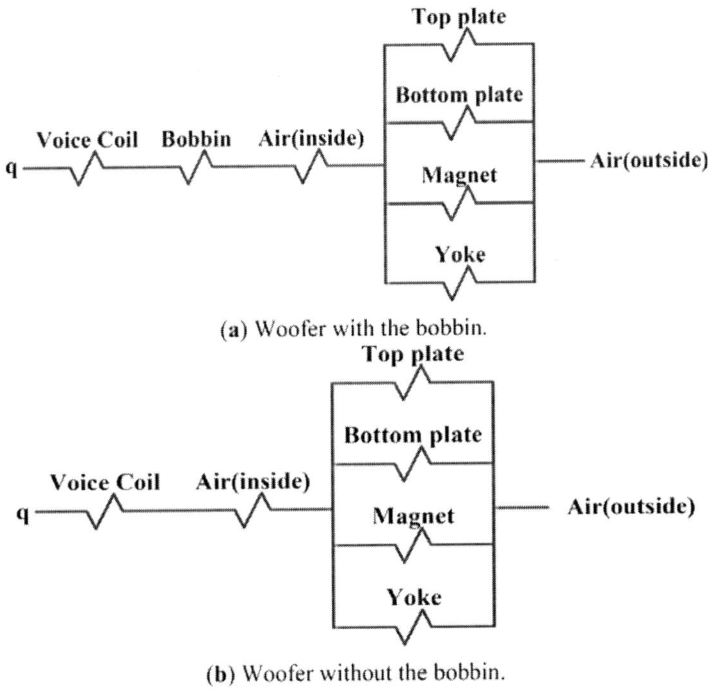


**Figure 5.** Heat transfer models using the thermal equivalent circuit technique for the tested woofers.

# 3. RESULTS AND DISCUSSION

## 3.1. Validation

Figure 6 shows the voice coil temperature of the tested woofer with the bobbin in comparison with the results by Odenbach (2003) [17]. Both results were compared with input powers ranging from 5 W to 60 W. The numerical results show good agreement within ±1% of the data by Odenbach (2003) [17]. In addition, the voice coil temperature of the woofer increased linearly with the input powers. The voice coil temperature could be theoretically calculated using Equation (1). The temperature increase of the voice coil could be calculated by the resistance increase with the rising of the input power, as mentioned in Ciprian (2005) [18].

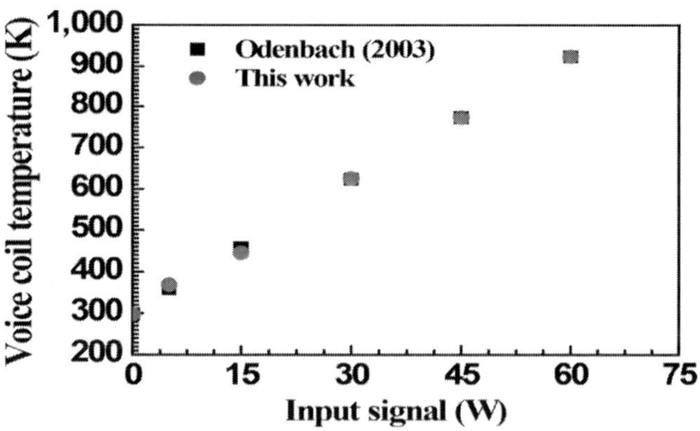


**Figure 6.** Voice coil temperature of the tested woofer with the bobbin in comparison with previous data.

$$R_{th} = T_{coil}/P_e \tag{1}$$

where $P_e$ is the applied electric power.

## 3.2. Heat Transfer Characteristics

Table 4 shows the heat transfer characteristics of the woofers with and without the bobbins and input powers ranging from 5 W to 60 W. Because the heat of the voice coils is generated proportionally to the increase of the input powers, the heat generated at the voice coils by the input powers was transferred toward the center direction of the other components, as shown in Table 4. The heat of the voice coil was transferred to the top plate and then to the permanent magnet. In addition, the heat transfer from the voice coil to other components of the woofer without the bobbin was superior to that with the bobbin. This is because the heat transfer area of the voice coil of the woofer without the bobbin was increased with the decrease of the thermal resistance.

**Table 4.** Heat transfer characteristics for the main components of the woofers with and without the bobbin varying with the input power ranging from 5 W to 60 W.

| Input / Units | 5 W | 15 W | 30 W | 45 W | 60 W |
|---|---|---|---|---|---|
| Speaker with the bobbin | 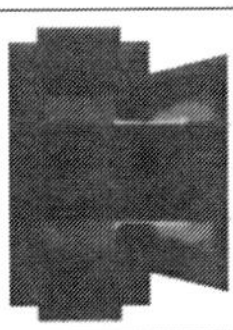 |  |  |  |  |
| Speaker without the bobbin | 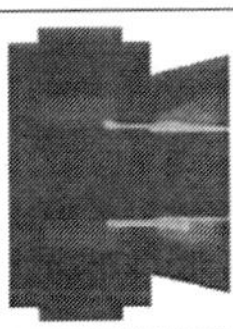 | 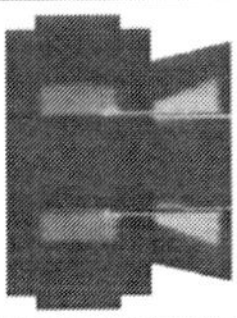 |  |  |  |

Figure 7 shows the effects of the bobbin on the voice coil temperature for the woofers with the input powers. The voice coil temperature of the woofer without the bobbin was 6.1% lower on average than that with the bobbin within the tested input power ranges, because the heat generated at the woofer without the bobbin with the input powers was dissipated effectively with less thermal resistance [19]. Furthermore, the effects of the bobbin were increased with the input powers. The voice coil temperature of the woofer without the bobbin at the input power of 60 W was decreased by 104 °C compared with the woofer with the bobbin. This is because the heat transfer rate of the voice coil of the woofer without the bobbin was faster than that of the woofer with the bobbin, as shown in Table 4.

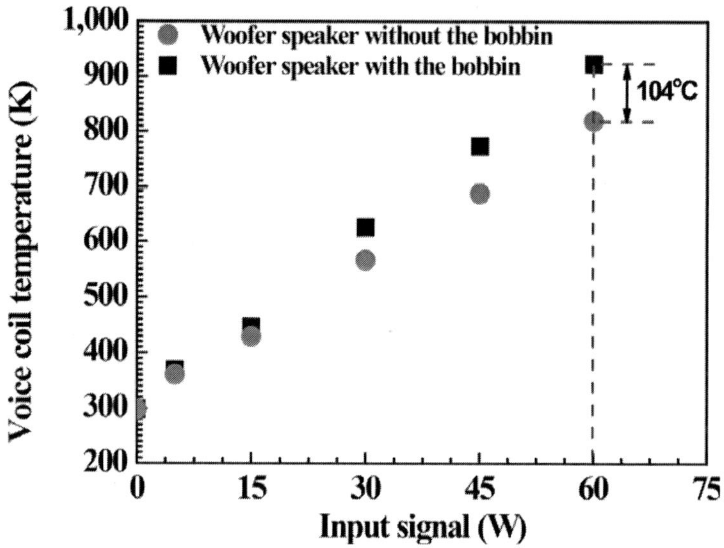


**Figure 7.** The effects of the bobbin on the voice coil temperature for the woofers with the input powers.

Figure 8 shows the temperature and heat transfer characteristics of the main components, except for the voice coil, for the woofers with and without the bobbin. The temperatures of the main components of the woofer without the bobbin were 5.0% lower on average than those of the woofer with the bobbin. The temperatures of the top plate for woofers with and without the bobbins increased by 160.0% and 128.9%, respectively, with the increase of the input powers from 5 W to 60 W. However, the increasing rate of the temperature of the main components for the woofer without the bobbin was lower than that of the woofer with the bobbin. In addition, the temperature of the top plate for the woofer without the bobbin was 12.0% lower than that for the woofer speaker with the bobbin at an input power of 60 W. Based on the results of Figures 7and 8, the bobbin was a key parameter related to the heat transfer of the main components for the woofers. Its effect was greater with the input power increasing.

Figure 9 shows the temperature characteristics of the components of the ferrofluid and normal woofers obtained from Lee*et al.* (2013) [9]. At the input powers of 10 W and 50 W of the voice coil, the main components of the woofer without the bobbin tested in this study showed lower temperatures than those of the woofer with the bobbin obtained from Lee *et al.*(2013) [9], although they showed higher temperatures than the ferrofluid woofer. Generally, the ferrofluid woofer shows more effective heat transfer characteristics than the normal woofers [20]. At the input power of 30 W of the voice coil, the temperatures of the main components of the woofer without the bobbin were 9.8% lower on average than those of the ferrofluid woofer obtained from Lee *et al.* (2013) [9]. However, they were 40.0% higher on average than those of the main components of the ferrofluid woofer obtained from Lee *et al.* (2013) [9].

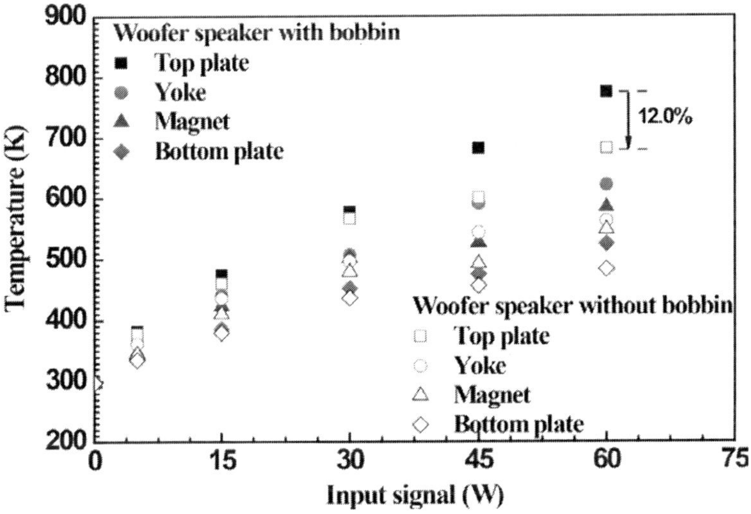


**Figure 8.** Temperature and heat transfer characteristics of the main components for the woofers with and without the bobbins.

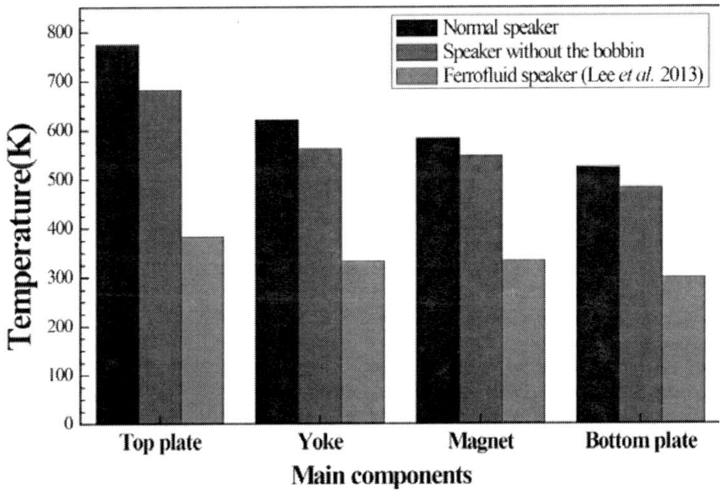


**Figure 9.** Temperature characteristics of the main components of the ferrofluid and normal woofers at an input power of 30 W.

Temperatures computed in this study are theoretically obtained by the numerical model applying the simple thermal equivalent heat conduction methods, but practical values would limit the temperatures to lower values. In addition, for improving the reality of the simulation model, we will consider and develop the thermal modeling of the woofer considering the transient phenomena reflecting the frequency and input power effects in the next article.

# 4. CONCLUSIONS

The temperature and heat transfer characteristics of the voice coil and its components for the woofers with and without bobbins using the thermal equivalent heat conduction models were numerically investigated with the input powers. The effects of the bobbin on the temperature and heat transfer of the voice coil and the main components for the woofers were also considered. The numerical results of the voice coil with the input currents showed good agreement within ±1% of the data by Odenbach (2003) [17], and the voice coil temperature of the woofer increased linearly with the input powers. The presence of the bobbin was a very important parameter related to the temperature and heat transfer characteristics of the main components of the woofers. The heat transfer from the voice coil to other components of the woofer without the bobbin was superior to that with the bobbin, due to the enhanced heat transfer rate and reduced thermal resistance at the voice coil of the woofer without the bobbin. The voice coil temperature of the woofer without the bobbin was 6.1% lower on average than that with the bobbin with all input powers. The temperatures of the main components of the woofer without the bobbin were 5.0% lower on average than those of the woofer with the bobbin. The temperatures of the top plate for the woofers with and without the bobbins increased by 160.0% and 128.9%, respectively, with the increase of the input powers from 5 W to 60 W. At the input power of 30 W of the voice coil, the temperatures of the main components of the woofer without the bobbin were 9.8% lower on average than those of the woofer obtained from Lee *et al.* (2013) [9]. However, they were 40.0% higher on average than those of the main components for the ferrofluid woofer obtained from Lee *et al.* (2013) [9].

# ACKNOWLEDGMENTS

This work was supported by the Dong-A University research fund.

## AUTHOR CONTRIBUTIONS

Moo-Yeon Lee is the first and the corresponding author. He designed the research and wrote the paper. All results and discussions are performed by Moo-Yeon Lee. Hyung-Jin Kim is a co-author. He performed the CFD analysis and drew figures in the paper. Both authors have read and approved the final manuscript.

## REFERENCES


1. Lemarquand, G.; Ravaud, R.; Lemarquand, V.; Depollier, C. Mechanical properties of ferrofluids in loudspeakers.*Audio Eng. Soc. Conv* **2008**, *125*, 1–5.
2. Lee, T.K.; Kim, B.S. The analysis of the non-linear parameter for the structure of an automotive woofer speaker. *J. Korean Soc. Mech. Tech* **2012**, *14*, 19–24. (In Korean).
3. Kim, S.K. Recent technology trends in home speaker. *J. Korean Inst. Electr. Electron. Mater. Eng. (KIEEME)* **2006**, *19*, 33–43. (In Korean).
4. Choe, C.S. The cause of the fire investigation practices—Karaoke speaker for the cause of the fire analysis (II). *Korea Fire Prot. Assoc* **2006**, *116*, 40–43. (In Korean).
5. Awrejcewics, J.; Koruba, Z. *Classical Mechanics: Applied Mechanics and Mechatronics*; Springer: New York, NY, USA, 2012.
6. Kim, S.J. Overview of speaker industry trends and film speaker technology. *J. Korean Inst. Electr. Electron. Mater. Eng. (KIEEME)* **2006**, *19*, 13–23. (In Korean).
7. Chang, C.; Wang, C.C.; Shiah, Y.C.; Huang, J.H. Numerical and experimental analysis of harmonic distortion in a moving-coil loudspeaker. *Commun. Nonlinear Sci. Numer. Simul* **2013**, *18*, 1902–1915.
8. Hong, D.K.; Woo, B.C.; Ahn, C.W. A study on performance improvement of daiphgram for micro speaker using table of orthogonal array. Proceedings of the KSPE Autumn Conference, Busan, South Korea, 21–22 October 2004; pp. 298–301.
9. Lee, M.Y.; Kim, H.J.; Lee, W.Y. Numerical analysis on temperature characteristics of the voice coil for woofer speaker using ferrofluid. *J. Korean Magn. Soc* **2013**, *23*, 166–172. (In Korean).

10. Kim, H.J.; Kim, D.W.; Lee, M.Y. Experimental study on the heat transfer characteristics of woofer speaker unit. *J. Korea Acad. Ind. Coop. Soc* **2014,** *15*, 2623–2627. (In Korean).

11. Lee, M.Y.; Kim, H.J. Heat transfer characteristics of the speaker using nano-sized ferrofluid. *Entropy* **2014.** submitted for publication.

12. Oh, S.J. *Fundamentals of Loudspeaker Engineering*; SeokHakDang: Seoul, South Korea, 2006; pp. 108–113.

13. Menter, F.; Ferreira, J.C.; Esch, T.; Konno, B.; Germany, A.C. The SST turbulence model with improved wall treatment for heat transfer predictions in gas turbines. Proceedings of the International Gas Turbine Congress, Tokyo, Japan, 2–7 November, 2003. IGTC2003-TS-059.

14. Speziale, C.G.; Thangam, S. Analysis of an RNG based turbulence model for separated flows. *Int. J. Eng. Sci* **1992,** *30*, 1379–1388.

15. Koh, S.G.; Lee, K.J.; Kang, J.H.; Sung, K.H.; Kim, C.J. Development of a temperature prediction tool for voice coils in loudspeakers using CFD. In Proceedings of the KSME Spring Conference, Gangwon-do, South Korea, 23–25 April 2008; pp. 41–44.

16. Incropera, F.P.; Lavine, A.S.; Bergman, T.L.; DeWitt, D.P. *Principles of Heat and Mass Transfer*; Wiley: NewYork, NY, USA, 2013; pp. 112–117.

17. Odenbach, S. Ferrofluids magnetically controlled suspensions. *Colloids Surf. A Physicochem. Eng. Asp* **2003,** *217*, 171–178.

18. Ionescu, C.; Codreanu, N.D.; Golumbeanu, V.; Svasta, P. Thermal simulation of a high power loudspeaker. Proceedings of the IEEE Spring Seminar on Electronics Technology, Wiener Neustadt, Austria, 19–22, May, 2005; pp. 134–139.

19. Borwick, J. *Loudspeaker and Headphone Handbook*, 3rd ed; FocalPress: Waltham, MA, USA, 2001; pp. 76–81.

20. Mayer, D. Future of electrotechnics: Ferrofluids. *Adv. Electr. Electron. Eng* **2011,** *7*, 9–14.

# CHAPTER 11

# Natural and Artificial Methods for Regeneration of Heat Resources for Borehole Heat Exchangers to Enhance the Sustainability of Underground Thermal Storages: A Review

Tomasz Sliwa [1,*] and Marc A. Rosen [2]

[1] *Drilling and Geoengineering Department, AGH University of Science and Technology, al. Mickiewicza 30, 30-059 Cracow, Poland*

[2] *Faculty of Engineering and Applied Science, University of Ontario Institute of Technology, Oshawa, ON L1H 7K4, Canada*

## ABSTRACT

The concept of borehole heat exchanger (BHE) field exploitation is described, along with problems regarding the sustainability of heat resources in rock masses. A BHE field sometimes has problems with the stability of the heat carrier temperature during long-term exploitation. The main reason for this is an insufficient heat stream with which to transfer heat by conduction in rock. Possibilities for the regeneration of heat in rock masses, based on experiences at the Geoenergetics Laboratory (Drilling, Oil and Gas Faculty, AGH University of Science and Technology), are described.

## KEYWORDS

borehole heat exchanger; geoenergetics; heat pump; underground thermal energy storage keyword

## 1. INTRODUCTION

Borehole thermal energy storages (BTESs) are increasing in interest and implementation, especially with systems for heating and/or cooling. The use of traditional fossil fuels alone to keep buildings at approximately 20 °C is wasteful, as there is a significant mismatch in the qualities of energy supply and demand. That is, fossil fuels are too valuable for use in heating/cooling. They can be used in more sophisticated systems and to meet higher quality energy needs. Heat pump technology is more efficient for heating and cooling, in part because it matches the quality of energy demand and supply better. Geothermal heat can also be extracted from non-geothermal waters, but for such processes it is necessary to locate appropriate geological layers and to solve problems regarding mineralization (Tomaszewska and Pająk 2012) [1].

One technology useful for helping to meet building heating and cooling demands is the borehole heat exchanger (BHE). This technology is often advantageous in most locations, and is independent of geological conditions. Shallow holes (to depths of about 200 m) are used for heat/cool storage and extraction. Deeper BHEs are generally only used for heating (Sapinska-Sliwa *et al.*, 2015) [2], also with use old oil and gas wells (Sliwa and Gonet, 2004) [3]. Typical constructions for BHEs are presented in Figure 1. The most popular construction is the U-tube design (a), while the least common is the coaxial type (c). Shown in Figure 1 are several key parts of the BHEs, including the borehole wall (1) and U-tube wall (4), the grout (2) and interior of the U-tube (3) and the annulus space (5), as well as the external column (8), the inner column (6) and the interior of inner column (7).

Generally the best heat transfer condition is observed for the coaxial system, but this depends on many factors, including good installation execution (workmanship).

Borehole heat exchanger fields are in operation for heat extraction and cold extraction (heat injection); some reversing systems exist that provide heat and cool extraction. When the number of BHEs is low (one or few) there usually is no problem with heat carrier temperature during the long period of exploitation (often years or decades). When number of BHE is high there is the possibility for a systematic yearly increase or decrease in heat carrier temperature (Signorelli *et al.*, 2005) [5]. Such a continuous increase (or decrease) in BHE temperature is very much dependent on the main heat load type (heating or cooling), and not only on the number of BHEs. The effect of the number of boreholes is described below with examples of fields.

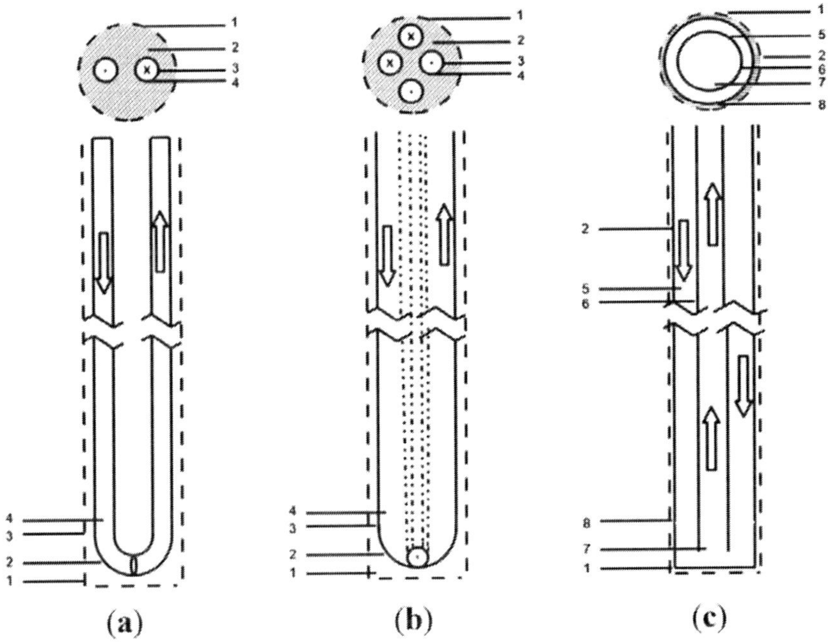


**Figure 1.** Borehole heat exchanger (BHE) constructions: (**a**) single U-tube; (**b**) double U-tube; (**c**) coaxial (Gonet *et al.* 2011) [4]. Legend: 1—borehole wall, 2—grout, 3—interior of U-tube, 4—U-tube wall, 5—annulus space, 6—inner column (pipe), 7—interior of inner column (pipe), 8—external column.

An example of the operating temperatures of a BHE field in which more heat is extracted than heat is input during the summer is shown in Figure 2. The calculations for that field accounts for 65 BHEs. The annual heat extraction is 534 MWh, and heat input is only 210 MWh. So an annual energy deficit of 324 MWh exists, which accounts for the systematic decrease in the temperature of the heat carrier. In Figure 2 are shown (1) the mean maximum temperatures during cooling mode (heat injection into the rock mass); and (2) the mean minimum temperatures during heating mode (heat extraction from the rock mass). The legend for this figure is the same as for Figure 3 and Figure 4.

The variability of the maximum and minimum temperatures during 25 years of operation for a BHE field with 50 wells is shown in Figure 3. The system serves a poultry hatchery. Only 585 MWh of heat is extracted from rock mass, while 830

MWh is injected. The ground temperatures thus increase every year. The calculations for example BHE fields have been made with EED3.16 software (Blomberg *et al.*, 2008) [7].

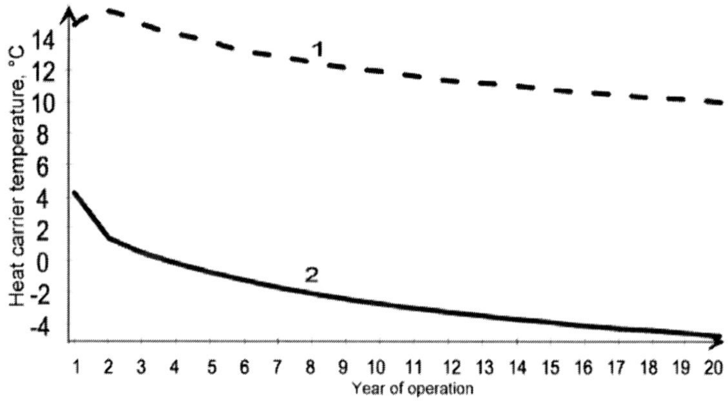


**Figure 2.** Variability of maximum and minimum heat carrier temperatures for a BHE field with 50 wells during 20 years of operation (Jaszczur and Sliwa, 2013) [6]. Legend: 1—maximum temperature during cooling mode (heat injected into rock mass), 2—minimum temperature during heating mode (heat extracted from rock mass).

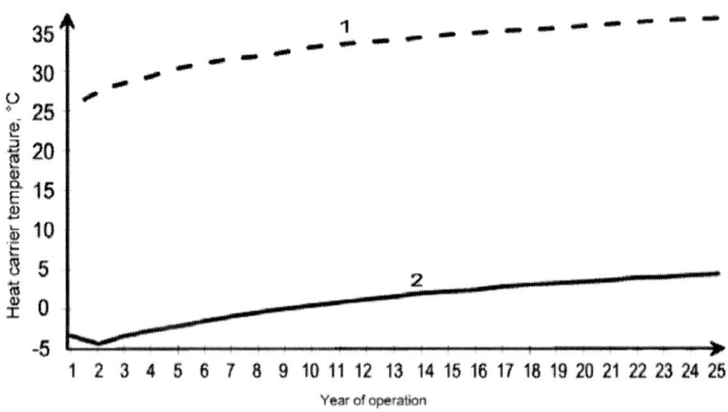


**Figure 3.** Variation of maximum and minimum temperatures of a heat carrier with load time for 25 years of system operation. Legend: 1—maximum temperature during cooling mode (heat injected into rock mass); 2—minimum temperature during heating mode (heat extracted from rock mass).

Figure 4 shows the behavior of the temperature of the heat carrier when approximately balancing the amount of heat produced and heat input to the rock mass. The maximum and minimum temperatures of the heat carrier seem to be stable during the long period of exploitation.

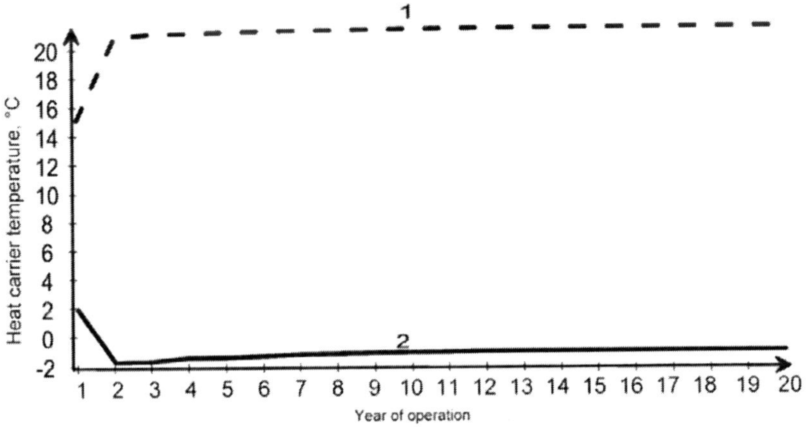


**Figure 4.** Minimum and maximum temperatures of a heat carrier for the operation of a near-balanced BHE field. Legend: 1—maximum temperature during cooling mode (heat injected into rock mass), 2—minimum temperature during heating mode (heat extracted from rock mass).

Research on BHE fields and heat storage in rock masses are in progress in many universities. The oldest work was carried out in Sweden. Luleå University of Technology, Luleå, Sweden had the first large-scale borehole heat store, constructed in 1982–1983 (Nordell, 1994) [8].

In addition to the work at AGH University of Science and Technology, studies are being conducted at the Technical University of Ostrava (VSB), Czech Republic among others. The BHE arrangement on this university campus is described by Bujok *et al.* (2012) [9]. The system at the University of Ontario Institute of Technology, Oshawa, Canada is described by Koohi-Fayegh and Rosen (2012) [10]. Systems with heat pumps and BHEs exist among others in Karlsruhe Institute of Technology, Polytechnic University of Turin, University of Western Ontario and, with a deep BHE, RWTH Aachen University.

Some of the most important experience and scientific output in the area of geothermal heat pumps has been done with BHEs in Europe (Sweden (Gehlin *et al.*, 2015) [11], Switzerland (Rybach and Eugster, 2010) [12] and Germany

(Weber *et al.*, 2015) [13]); in the Americas (Canada (Raymond *et al.*, 2015) [14] and USA (Boyd *et al.*, 2015) [15]); and in Japan in Asia (Sasada, 2012) [16].

The largest installation of BHEs currently under construction is for Ball State University, Indiana, U.S., where 4100 vertical loops are being installed (Boyd *et al.*, 2015) [15]. That installation has BHE depths ranging from 122 to 152 m (Lund *et al.*, 2010) [17]. A general overview of that borehole thermal energy storage (BTES), along with discussions of the general procedures for design and construction, are to be found in Lee (2012) [18].

The main aim of the present article is to review and critically discuss the various geological and meteorological factors affecting the design, number, depth and distribution of BHEs. A BHE field project needs to take into consideration various technical options for heat exchange. Appropriate analysis and design can reduce the number of BHEs, and enable the construction of adequate BHE fields even when the field surface area is insufficient.

## 2. APPROACH AND METHODOLOGY

The present research utilizes experimental data on borehole heat exchangers, so as to provide realistic results and interpretations.

### 2.1. Geoenergetics Laboratory

At the end of the 1990s the idea of converting existing, old boreholes to borehole heat exchangers (BHEs) was proposed at the Faculty of Drilling, Oil and Gas, AGH University of Science and Technology, Poland. After the research of Sliwa (2002) [19], work began on building a BHE research facility in the area of the university.

The Laboratory of Geoenergetics was inaugurated in 2007, when the first commercial Thermal Response Test (TRT) was performed in Poland (Gonet and Sliwa, 2008) [20]. The first equipment of the laboratory was a TRT set. The drilling of borehole heat exchangers began at AGH-UST in 2008. The Laboratory is based on five borehole heat exchangers of various designs. Additionally the types of grout and additive (Stryczek *et al.*, 2013) [21] influence the properties of cement. Research on the dependence of energy efficiency with construction and technology is important, and was therefore the first research task. The construction of the BHEs is described in Table 1. A schematic of the Laboratory of Geoenergetics installation is shown in Figure 5. The devices shown in that figure are designated as follows: 1—coaxial borehole heat exchanger, 2—single U-tube borehole heat exchanger filled with ordinary cement stone, 3—single U-tube borehole heat exchanger filled with cement stone of high thermal conductivity, 4—single U-tube borehole heat exchanger filled with gravel and sealed by clay in the interval of high permeability (*i.e.*, the interval 3–15 m deep), 5—double U-tube borehole heat exchanger, 6 - buffer tank for the heat transfer medium on the cold side (cold reservoir), 7—buffer tank for the heat transfer medium on the warm side (heat accumulator), 8—heat pump, 9—fan coil for heating or cooling the Laboratory room, 10—water heater/cooler (to dissipate excess electrical power), 11—water supply network to fill the system, 12—diaphragm expansion vessels, 13—plate heat exchanger for shifting heat between cold and heat storage tanks, 14—set of valves to change the direction of fluid flow in a coaxial borehole heat exchanger, 15—ventilation unit of the auditorium of the Faculty of Drilling, Oil and Gas at the Academy of Science and Technology which complements the heating or air conditioning equipment, 16—additional heat exchanger for heat exchange between BHEs and solar collectors directly (with no tanks), for direct heat regeneration of the rock mass, 17—additional heat exchanger—supporting heat exchanger in tanks—additional heat exchange surface to increase the efficiency, 18—stationary flat-plate solar collector, 19—stationary vacuum solar collector, 20—stationary collector "heat pipe", 21—solartrack (mobile) flat collector, 22—solartrack (mobile) vacuum solar collector, 23—heating pipes installed in the parking lot for snow melting, 24—intake for extracting air from auditorium, 25—fresh air supply to auditorium, 26—discharge of exhaust air to the atmosphere, 27—intake for fresh atmospheric air, 28—recuperator, *i.e.*, a rotary heat exchanger (heat recovery ventilator), 29—peak heat source—heat from district heating.

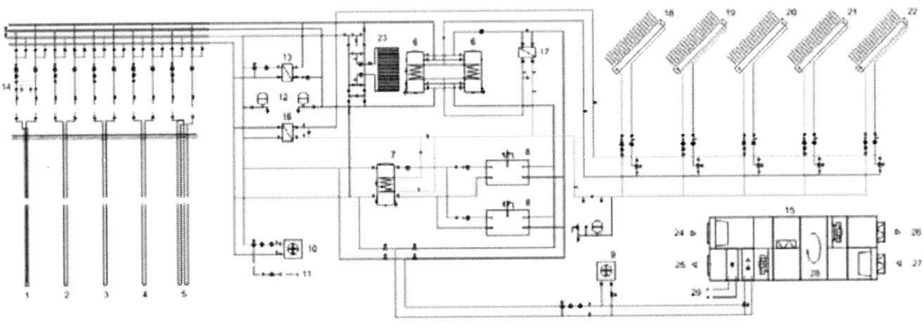

**Figure 5.** Schematic of Laboratory of Geoenergetics installation (Sliwa *et al.* 2015) [22]. Legend: 1—coaxial borehole heat exchanger, 2—single U-tube borehole heat exchanger filled with ordinary cement stone, 3—single U-tube borehole heat exchanger filled with cement stone of high thermal conductivity, 4—single U-tube borehole heat exchanger filled with gravel and sealed by clay in the interval of high permeability (*i.e.*, the interval 3–15 m deep), 5—double U-tube borehole heat exchanger, 6—buffer tank for the heat transfer medium on

the cold side (cold reservoir), 7—buffer tank for the heat transfer medium on the warm side (heat accumulator), 8—heat pump, 9—fan coil for heating or cooling the Laboratory room, 10—water heater/cooler (to dissipate excess electrical power), 11—water supply network to fill the system, 12—diaphragm expansion vessels, 13—plate heat exchanger for shifting heat between cold and heat storage tanks, 14—set of valves to change the direction of fluid flow in a coaxial borehole heat exchanger, 15—ventilation unit of the auditorium of the Faculty of Drilling, Oil and Gas at the Academy of Science and Technology which complements the heating or air conditioning equipment, 16—additional heat exchanger for heat exchange between BHEs and solar collectors directly (with no tanks), for direct heat regeneration of the rock mass, 17—additional heat exchanger—supporting heat exchanger in tanks—additional heat exchange surface to increase the efficiency, 18—stationary flat-plate solar collector, 19—stationary vacuum solar collector, 20—stationary collector "heat pipe", 21—solartrack (mobile) flat collector, 22—solartrack (mobile) vacuum solar collector, 23—heating pipes installed in the parking lot for snow melting, 24—intake for extracting air from auditorium, 25—fresh air supply to auditorium, 26—discharge of exhaust air to the atmosphere, 27—intake for fresh atmospheric air, 28—recuperator, *i.e.*, a rotary heat exchanger (heat recovery ventilator), 29—peak heat source—heat from district heating.

### 2.2. Ground Characteristics at Test Site

The lithological-stratigraphic profile under the Lab is presented in Table 2. The thermal conductivity of rocks and the volumetric specific heat are included. At a depth of 3–4 m, a highly permeable layer begins. Its bottom is at a depth of 15 m. In this layer, underground water flows (Figure 6).

### 2.3. Experimental Capability and Measurable Quantities

The Laboratory of Geoenergetics, after drilling and constructing the BHEs, was built on the surface. The installation is equipped with (Sliwa *et al.*, 2015) [22]:

1. A heating-cooling system for the Faculty Auditorium, which has a seating capacity of 160
2. Solar collectors for regeneration of heat in the rock mass, with individual measurement sensors

**Table 1.** Detailed parameters of BHEs of Laboratory of Geoenergetics (Gonet *et al.*, 2011) [4].

| No. | Parameter | Pictorial Configuration | Value | | | | |
|---|---|---|---|---|---|---|---|
| | | | LG 1 (BHE 1) | LG 2 (BHE 2) | LG 3 (BHE 3) | LG 4 (BHE 4) | LG 5 (BHE 5) |
| 1 | Depth of borehole | - | 78 m | 82 m | 78 m | 78 m | 78 m |
| 2 | Depth at which exchanger`s tubes are seated | | 78 m | | | | |
| 3 | Design of borehole heat exchanger | - | 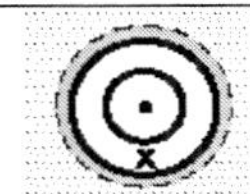 coaxial | | 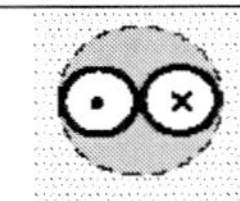 single U-tube | | 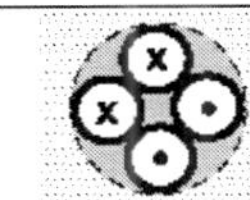 double U-tube |
| 4 | Borehole diameter $D_o$ (drill tool diameter) | 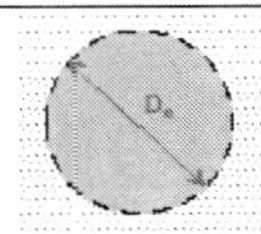 | 143 mm | | | | |
| 5 | Distance between tubes in heat exchanger, k |  | n/a | | 80 mm | | n/a |
| 5` | Distance between tubes in heat exchanger, k | 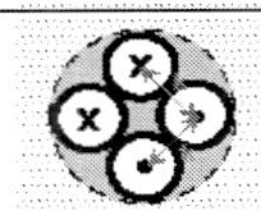 | n/a | | n/a | | 70 mm |

**Table 1.** *Cont.*

| No. | Parameter | Pictorial Configuration | Value | | | | |
|---|---|---|---|---|---|---|---|
| | | | LG 1 (BHE 1) | LG 2 (BHE 2) | LG 3 (BHE 3) | LG 4 (BHE 4) | LG 5 (BHE 5) |
| 6 | Material used for sealing heat exchanger (grout) | 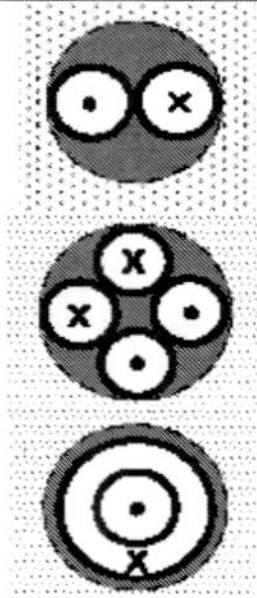 | cement, $\lambda = 1.2\ Wm^{-1}{\cdot}K^{-1}$ | cement, $\lambda = 1.2\ Wm^{-1}{\cdot}K^{-1}$ | Thermocem, increased thermal conductivity cement, $\lambda = 2.0\ Wm^{-1}{\cdot}K^{-1}$ | Gravel 8–16 mm of grain and clayey plugs, $\lambda = 1.8\ Wm^{-1}{\cdot}K^{-1}$ | cement, $\lambda = 1.2\ Wm^{-1}{\cdot}K^{-1}$ |
| 7 | Outer diameter,of tubes in heat exchanger | 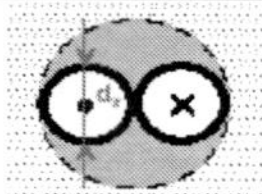 | n/a | | 40 mm | | 32 mm |
| 7` | Outer diameter of outer tubes in heat exchanger | 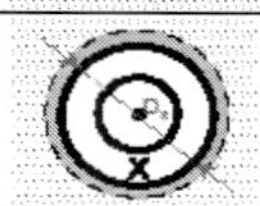 | 90 mm | | | n/a | |

**Table 1.** *Cont.*

| No. | Parameter | Pictorial Configuration | Value | | | | |
|---|---|---|---|---|---|---|---|
| | | | LG 1 (BHE 1) | LG 2 (BHE 2) | LG 3 (BHE 3) | LG 4 (BHE 4) | LG 5 (BHE 5) |
| 7'' | Outer diameter of inner tubes in heat exchanger | 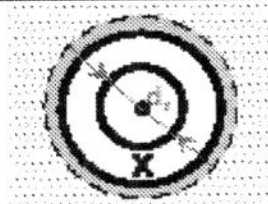 | 40 mm | n/a | | | |
| 8 | Thickness of tube wall in heat exchanger | 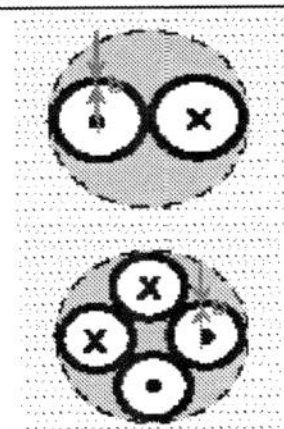 | n/a | | 2.4 mm | | 2.4 mm |
| 8' | Thickness of outer tube wall in heat exchanger | 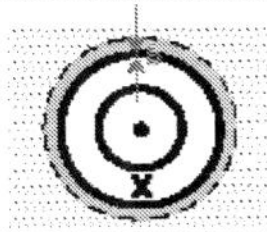 | 5.4 mm | n/a | | | |
| 8'' | Thickness of inner tube wall in heat exchanger | 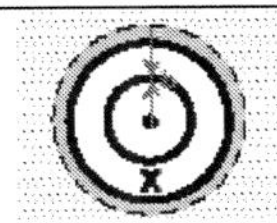 | 2.4 mm | n/a | | | |

**Table 1.** *Cont.*

| No. | Parameter | Pictorial Configuration | Value | | | | |
|---|---|---|---|---|---|---|---|
| | | | LG 1 (BHE 1) | LG 2 (BHE 2) | LG 3 (BHE 3) | LG 4 (BHE 4) | LG 5 (BHE 5) |
| 9 | Material of tubes in heat exchanger | 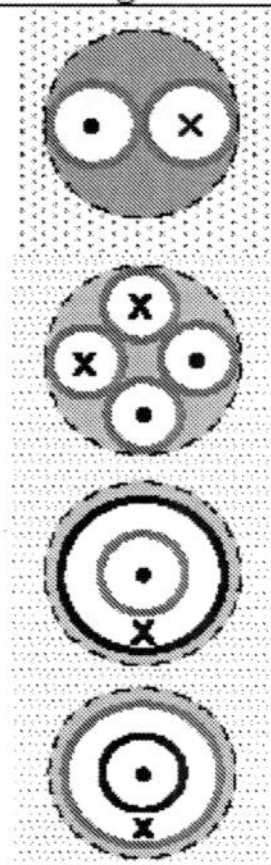 | Polyethylene, $\lambda = 0.42\ Wm^{-1} \cdot K^{-1}$ | | | | |

3. A system for snow melting of the parking lot in front of the Laboratory
4. A fan for heating and cooling glycol using atmospheric air

**Table 2.** Lithological and stratigraphic profile of BHE 2 (AGH LG-2 well) with thermal parameters of rocks (Gonet *et al.*, 2011) [4].

| No. | Top, m | Bottom, m | Thickness, m | Lithology | Stratigraphy | Thermal Conductivity, $\lambda$, $W \cdot m^{-1} \cdot K^{-1}$ | Volumetric Specific Heat, $c_v$, $MJ \cdot m^{-3} \cdot K^{-1}$ |
|---|---|---|---|---|---|---|---|
| 1 | 0.0 | 2.2 | 2.2 | Anthropogenic ground (dark grey fill with debris) | Quaternary (Pleistocene, Holocene) | 1.600 | 2.000 |
| 2 | 2.2 | 2.6 | 0.4 | Aggregate mud (grey ground) | | 1.600 | 2.200 |
| 3 | 2.6 | 4.0 | 1.4 | Fine, dusty and slightly clayey sand | | 1.000 | 2.000 |
| 4 | 4.0 | 6.0 | 2.0 | Fine sand | | 1.200 | 2.500 |
| 5 | 6.0 | 15.0 | 9.0 | Sand and slag mix, slag | | 1.800 | 2.400 |
| 6 | 15.0 | 30.0 | 15.0 | Grey clay | Tertiary (Miocene) | 2.200 | 2.300 |
| 7 | 30.0 | 78.0 | 48.0 | Gray clay slate | | 2.100 | 2.300 |
| | | | Weighted average | | | **2.039** | **2.309** |

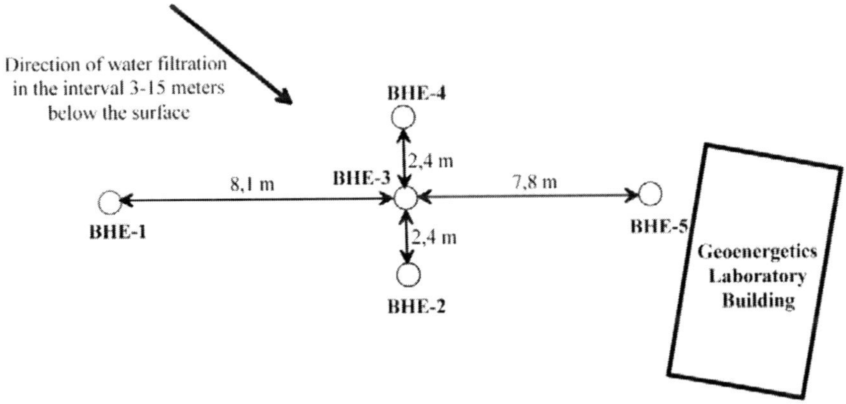


**Figure 6.** Location of borehole exchangers of Laboratory of Geoenergetics and direction of underground water flow.

Heat transfers in the Laboratory are listed in Table 3 and shown in Figure 7, where arrows indicate the possible direction of heat flow. The installation has many possibilities for heat regeneration in the rock mass, or heat rejection (when necessary).

**Table 3.** Natural and artificial heat sources for regeneration of energy resources in BHE field of Geoenergetics Laboratory.

| No. | Natural Heat Sources for Regeneration BHE of Field | Artificial Heat Sources for Regeneration BHE of Field |
|---|---|---|
| 1 | Underground water flow | Heating-air-conditioning system of Auditorium |
| 2 | Sun operation on surface (also wind and frost) | Solar collectors |
| 3 | Heat transfer from the sides of the BHE space | Snow melting installation |
| 4 | Heat transfer from the bottom of the BHE space | Fan (heater/cooler) |

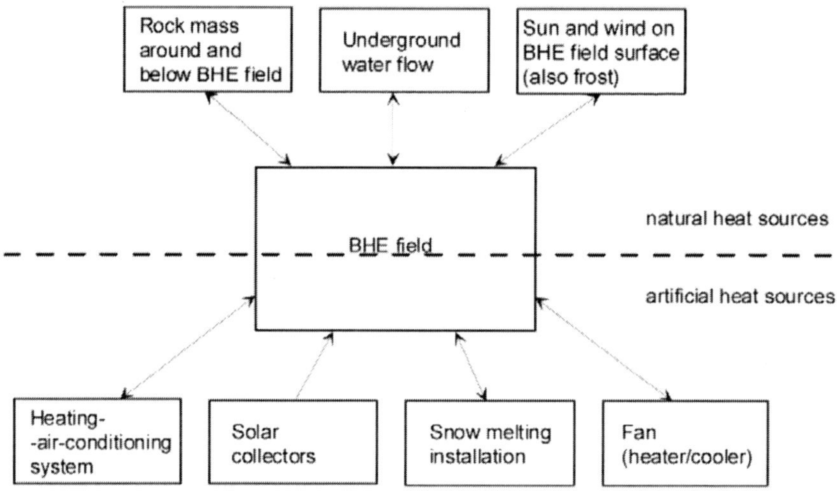


**Figure 7.** Schematic of directions of heat flows for various devices of the Geoenergetics Laboratory and by natural heat transfer. Note that only the solar collectors have one-way heat transfer.

# 3. NATURAL HEAT RESOURCES REGENERATION

Solar radiation on the ground surface and warm ambient air/wind are two significant sources of heat for BHE fields. These sources are especially important when the depth of wells is not great. Temperatures profiles for the ground down to a 30 m depth are shown in Figure 8 for a borehole not using a BHE. This well was made especially for temperature measurement. In a typical direct evaporation BHE field, the distance of this well from active BHEs (*i.e.*, exploited ones) is about 5 m. Notable changes in temperature are observed to a depth of about 15 m. No inflow of heat from exploited BHEs is observed in the field.

In deep BHEs, heat regeneration from solar radiation and ambient air is not as significant, and the importance declines as well depth increases. For BHEs in former oil wells (Sliwa *et al.*, 2014) [23] or other deep BHEs (Sapinska-Sliwa *et al.*, 2015) [2], including directional wells (Knez, 2014) [24], it can be assumed for modeling that ground surface temperature is constant at the average annual ambient air temperature. The relation between heat extraction rate and temperature of the heat carrier fluid, under various conditions, is described by Eskilson (1987) [25]. Eskilson's research was used as the basis for the software EED, which is in use around the world by designers.

Hellström (1991) [26] describes formulas for the ground thermal resistance for several types of borehole heat exchangers. The resistance has a big impact on the heat transfer between the heat carrier and the rock mass. The impact of temperature distribution and geothermal gradient is described by Kurevija *et al.* (2011) [27].

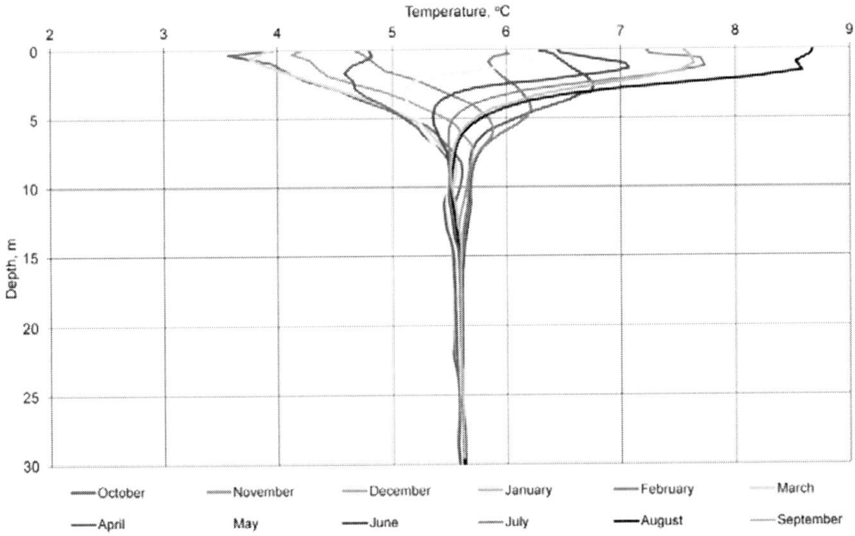


**Figure 8.** Mean ground temperatures in a well not using an active BHE near Krakow (in the village of Pałecznica, 50 km northeast) broken down by month for the period October 2013 to September 2014.

The second group of heat sources for resource regeneration is geothermic heat from around and under the BHE field. When a geothermal source heat pump (GSHP) system is the only regeneration heat source, heat conduction from rocks alone may not be sufficient, and the temperature of the BHE field may drop every year, as shown in Figure 2. The renewability of heat in such BHEs fields is not complete (Cataldi, 2001, Signorelli *et al.*, 2004) [28,29]. After a few years of temperature drop, the temperature takes on a low value at which point heat from surrounding rocks can conduct adequately to the field. Then the temperature of the BHE field reaches a value at which the extracted heat has the same value as heat conducted from surrounding rocks for regeneration.

The third group of natural heat sources for regeneration is convection. The sensitivity of the loop outlet temperatures and heat exchange rates to hydrogeological, system and meteorological factors was analyzed over 6-month and 25-year operation periods by Dehkordi and Schincariol (2014) [30]. Long-term exploitation of BTES (10 years) is also evaluated by Chiasson *et al.* (2000) [31]. Other researchers have analyzed the effect of waterflow in geological layers on heat transfer efficiency and heat loss (Diao *et al.*, 2004; Fujii *et al.*, 2005; Hecht-Méndez *et al.*, 2013; Molina-Giraldo *et al.*, 2011; Fan *et al.*, 2007) [32,33,34,35,36].

In the case where underground water flow is present, heat transfer can accompany the mass flow. In Figure 9 the results are shown of a numerical simulation of a deep BHE with a permeability layer (Sliwa and Gonet, 2005) [37]. At depths of 300–400 m, the isotherm curves are clearly disturbed.

In Figure 10, the temperature distribution is shown around the five BHEs of the Geoenergetics Laboratory. In the interval 3–15 m deep is a moist layer with water flow (as shown in Figure 6). A cross-section at a depth of 9 m (inside the underground water layer) is shown in Figure 11. The increased heat convection is visible according to the direction of the flow of water.

## 4. ARTIFICIAL HEAT RESOURCES REGENERATION

The only sources of heat existing in the Geoenergetics Laboratory are described here. In real projects, it is necessary to take into consideration all possible options.

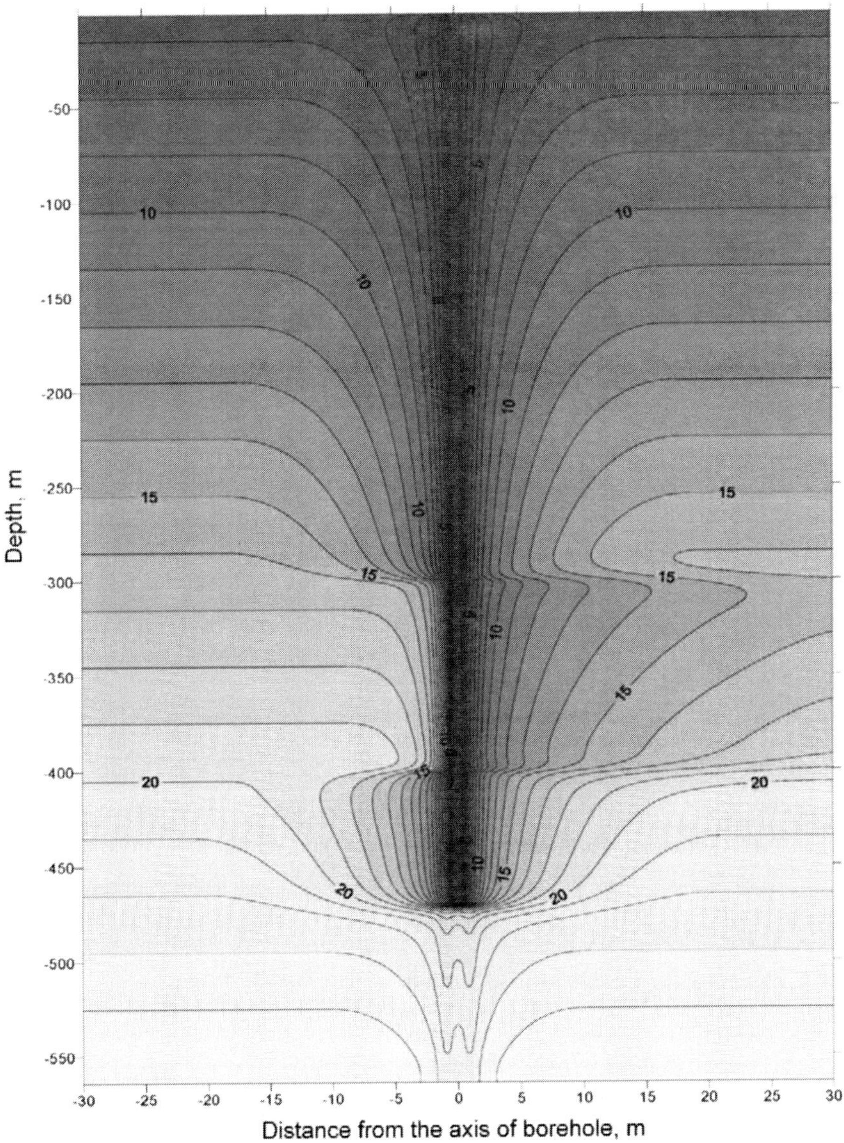


**Figure 9.** Isotherms (in °C) for a cross-section through a deep BHE (Sliwa, 2002) [19]; example of underground water flow impact to rock mass temperature distribution.

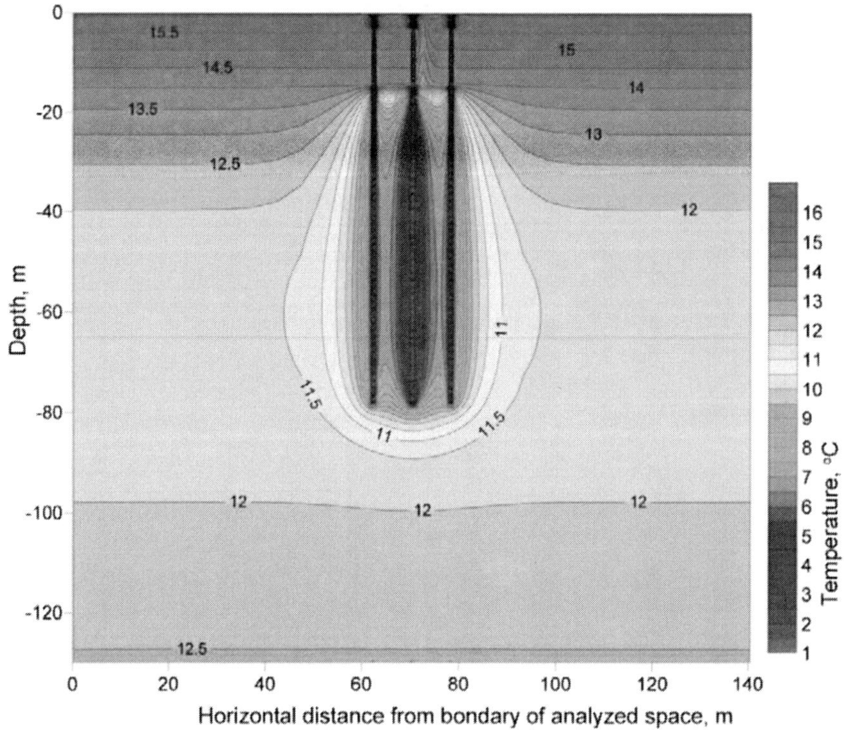


**Figure 10.** Vertical cross-section through the BHEs of Laboratory of Geoenergetics (wells No. 1, 3 and 5, as shown in Figure 6) showing isotherms (in °C) from numerical simulation (Gonet *et al.*, 2011) [4]; interval between −3 and −15 m with water filtration gives just natural temperature distribution.

## 4.1. Heating/Air-Conditioning System of the Auditorium

BHE fields typically work by injecting heat into the surrounding rock mass. Hotels, office buildings, commercial facilities, university buildings, and private residences can use BHEs for heating and air-conditioning. The simple payback time for investment costs of such installations can be fewer than 10 years. Air-conditioning processes of the auditorium in AGH University of Science and Technology. Drilling, Oil and Gas Faculty correspond to heat injection into BHEs. The origin of the heat is the sun in addition to electrical energy for the compressor. So, cooling of building interiors with use of heat pumps and BHEs fields is indirectly a process of solar heat storage. If the amount of the total heat from air-conditioning is lower than the total energy extracted from rocks in winter time, what has occurred can be termed a regeneration of heat resources.

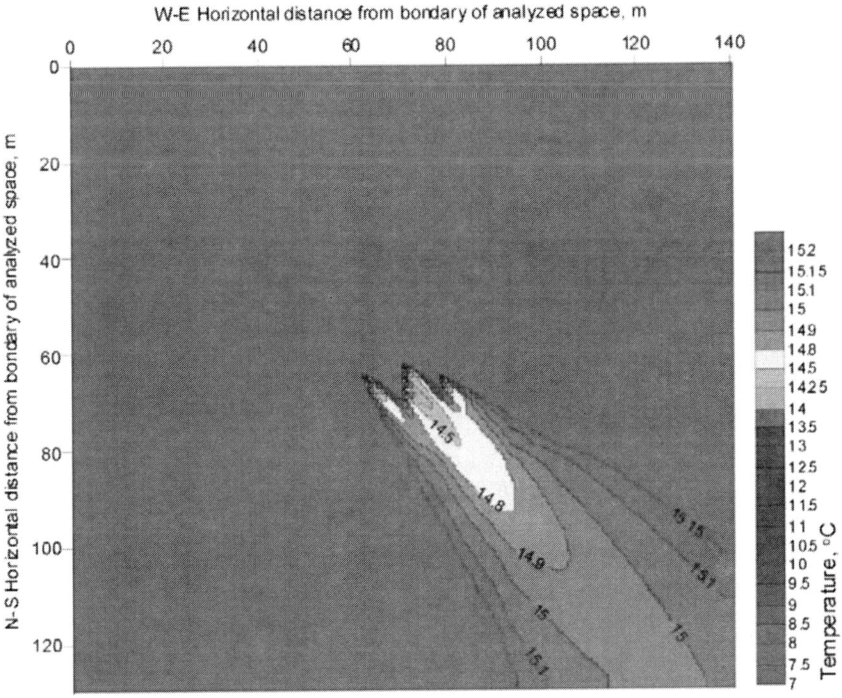


**Figure 11.** Horizontal cross-section at a level 9 m below the ground surface showing isotherms (in °C) based on simulation for exploited BHEs at the Geoenergetics Laboratory (Gonet *et al.*, 2011) [4].

Natural ways of heat regeneration can result in full or only partial temperature regeneration. When full temperature renewability is not achieved due to supplemental heat resources, it is necessary to use artificial sources of heat.

Using heat pumps can avoid the need for separate heating and cooling equipment. Also operational costs of heating and cooling are often lower than traditional methods. In countries with temperate climates, the amount of energy from air-conditioning injected into rock masses in summer is usually lower than the heat extracted from ground in winter time.

## 4.2. Solar Collectors

Five solar collectors have been investigated in terms of heat recovery performance. The best results regarding efficiency is achieved for a collector "heat pipe," while the lowest efficiency is observed for a stationary flat-plate collector. Much better efficiency is observed when the temperature of the working fluid of the heat pipe, glycol, is low. In typical heating systems, for domestic hot water heating for example, the efficiency of solar collectors is about

60%–70%. In case that involve circulating the heat carrier from BHEs at low temperatures, the efficiency can be higher than 80%. In the Laboratory of Geoenergetics, the heating stream was also observed at night. That observation occurred in summer, when the ambient air temperature was high. About 0.5 kW heating power can be attained with five solar collectors, using low-temperature glycol to transfer heat into rock masses.

### 4.3. Snow Melting Installation

In the laboratory testing facility, a heating rate of over 33 kW was possible. Over a minimum duration of 32 min, it was possible to attain 58.5 MJ of heat transfer from the parking area, which was 1880 $m^2$. The system's working fluid is a glycol solution. The time to raise the temperature of 1.50 $m^3$ glycol from 0 to 10 °C was observed.

The snow melting installation works like a solar collector in summer. It receives heat from solar radiation and warm air on the surface. The unit heating rate observed is 17.5 $W/m^2$ when the maximum solar energy (900–1000 $W/m^2$) is received from the collectors. The unit heat extracted from the parking area is 31.1 $kJ/m^2$ over a period of 32 min.

Research was also undertaken by partially shading the parking area. It is necessary to prepare long term observations (at least one year) to determine seasonal effects and outputs.

The installation for snow melting can be used in summer not only for heat extraction, but also for cooling the surface of road to protect against ruts (Heliasz and Ostaficzuk, 2002) [38].

### 4.4. Fan (Heater/Cooler)

In the Geoenergetics Laboratory, research was undertaken involving the heating of 1.0 $m^3$ of glycol solution. For the snow melting installation, the initial temperature of the heat carrier was 0 °C, and the final temperature was 10 °C. Poniedziałek and Sliwa (2013) [39] demonstrate that it is possible to obtain examples of research. Various air temperatures with time of heating have been measured. For the average air temperature (18.1 °C), the average heating time was 233 min, leading to an average heating power of 4.18 kW.

Using the fan is probably the most inexpensive way for heating, considering investments and operational costs. The fan can heat the working medium when the air temperature is higher than the temperature of the glycol solution, and also in warmer winter days.

Many examples of use of waste heat for regeneration of underground resources can be cited. Some examples are described by Hellström and Gehlin (1997), Andersson *et al.* (2009) and, for waste cold, Andersson *et al.* (2003) [10,41,42].

### 4.5. Further Discussion

The paper addresses the medium/long- term sustainability of BHE fields, which is important since an unbalanced load between winter and summer operations can lead to a failure of the ground-coupled heat pump system. This result is consistent with other studies, but the results are more quantified for the Laboratory of Geoenergetics in Poland.

Details relating to the results demonstrated with regard to the Laboratory of Geoenergetics are presented in brief for this paper to allow it to focus on the general concepts raised. Specific details like measurement equipment used, error analysis and numerical models are presented in many of the references already cited, and work is ongoing to prepare a detailed article on these engineering and scientific details.

## 5. CONCLUSIONS

The following conclusions are drawn from the review and analysis, based on the Laboratory of Geoenergetics:

1) Natural ways of heat regeneration can permit full or partial temperature restoration. For supplemented heat resources, when natural full renewability is not possible but necessary (or advantageous), artificial sources of heat can be used.

2) In real projects, it is necessary to take into consideration all possible options for obtaining heat. Sometimes it is possible to obtain a low-temperature heat flux from other sources, such as waste heat.

3) Nowadays designers should analyze all energetic aspects of buildings and other structures as well as the neighborhood of designed projects.

4) BHE fields can be effectively used for energy storage. This is particularly important for renewable energy sources, which are often characterized by intermittency (like wind energy for example). Energy storage also enables greater use of distributed energy sources.

5) Natural heat regeneration options for BHEs storages are sun, wind, surrounding rocks and underground water. Artificial options that can be used include solar heat using, for example, air-conditioning installations,

solar collectors, fans, snow melting installations working in reverse mode and waste heat from industry.

Natural and artificial heat sources for regeneration of the energy resources in a BHE field, based findings from the Laboratory of Geoenergetics, are summarized in Table 4.

**Table 4.** Natural and artificial heat sources for regeneration of the energy resources in a BHE field.

| No. | Heat Source | Origin | Comments |
|---|---|---|---|
| 1 | Underground water flow | Natural | Positive in the case of heating and cooling systems with deficit of heat or cool. Negative when storing heat or cool. |
| 2 | Solar energy incidence on surface (wind and frost also) | Natural | Heat gain positive when heat deficit/storage. Heat loss positive when cold deficit/storage. |
| 3 | Heat transfer from the sides of the BHE space | Natural | Positive when heat deficit. Negative when cold storage. |
| 4 | Heat transfer from the bottom of the BHE space | Natural | Positive when heat deficit. Negative when cold storage. |
| 5 | Heating-air-conditioning system of Auditorium | Artificial, need extra installation | Should be used in both directions of heat transfer when possible. |
| 6 | Solar collectors | Artificial, need extra installation | Can be used only for heat recovery when heat deficit in annual balance or for heat storage. |
| 7 | Snow melting installation | Artificial, need extra installation | In reverse mode can be used as a heat source when there is a deficit of heat. |
| 8 | Fan (heater/cooler) | Artificial, need extra installation | Can be used in the case of deficit of heat or cool in annual balance. Can be used for heat or cool storage. |

## ACKNOWLEDGMENTS

The paper was founded in Statutory Research carried out at the Drilling, Oil and Gas Faculty, AGH University of Science and Technology, funded by the Polish Ministry of Science and Higher Education (No. 11.11.190.555). The authors thank Zbigniew Jezuit and Anna Bieda from AGH-UST and Agnieszka Pycia and Tomasz Nowosiad from the Student Scientific Association GEOWIERT AGH-UST DOGF for help in preparing of some pictures.

## AUTHOR CONTRIBUTIONS

Tomasz Sliwa and Marc A. Rosen equally contributed to the paper. Tomasz Sliwa designed the sites of in situ measurements and did the modeling, and performed the experiments and analyzed the data. Marc A. Rosen formulated and depicted the innovative proposals within the paper. He also collected the research resources and systematized the content, analyzed the accuracy and reliability of the method, and participated in the editing of the paper. All authors read and approved the final manuscript.

## REFERENCES


1. Tomaszewska, B.; Pająk, L. Geothermal water resources management-economic aspects of their treatment. *Miner. Resour. Manag. (Gospodarka Surowcami Mineralnymi)* **2012**, *28*, 59–70. Available online: https://www.min-pan.krakow.pl/Wydawnictwa/GSM284/tomaszewska-pajak.pdf (accessed on 24 August 2015).
2. Sapinska-Sliwa, A.; Rosen, M.A.; Gonet, A.; Sliwa, T. Deep Borehole Heat Exchangers: A Conceptual Review. In Proceedings of the World Geothermal Congress 2015, Melbourne, Australia, 19–25 April 2015; pp. 1–11. Available online: http://www.geothermal-energy.org/pdf/ IGAstandard/WGC/ 2015/ 28045.pdf? (accessed on 24 August 2015).
3. Sliwa, T.; Gonet, A. The closing wells as heat source. *Acta Montan. Slovaca* **2004**, *9*, 300–302. Available online: http://actamont.tuke.sk/pdf/2004/n3/37sliwa.pdf (accessed on 24 August 2015).
4. Gonet, A.; Sliwa, T.; Stryczek, S.; Sapinska-Sliwa, A.; Jaszczur, M.; Pająk, L.; Złotkowski, A. *Metodyka Identyfikacji Potencjału Cieplnego Górotworu wraz z Technologią Wykonywania i Eksploatacji Otworowych Wymienników Ciepła (Methodology for the Identification of Potential Heat of the Rock Mass Along with Technology Implementation and Operation of Borehole Heat Exchangers)*; AGH University of Science and Technology Press: Krakow, Poland, 2011; p. 439. (In Polish)
5. Signorelli, S.; Kohl, T.; Rybach, L. Sustainability of Production from Borehole Heat Exchanger Fields. In Proceedings of the World Geothermal Congress 2005, Antalya, Turkey, 26–28 January 2005; pp. 1–6. Available online: http://www.geothermal-energy.org/pdf/IGAstandard/WGC/ 2005/ 1450.pdf? (accessed on 8 September 2015).
6. Jaszczur, M.; Śliwa, T. The analysis of long-term borehole heat exchanger system exploitation. *CAMES* **2013**, *20*, 227–235. Available online: http://cames.ippt.gov.pl/pdf/CAMES_20_3_6.pdf (accessed on 24 August 2015).
7. Blomberg, T.; Claesson, J.; Eskilson, P.; Hellström, G.; Sanner, B. Earth Energy Designer (EED 3.16), Software within Building Physics and Ground Heat Storage. Available online: http://www.buildingphysics.com/manuals/EED3.pdf(accessed on 24 August 2015).

8. Nordell, B. Borehole Heat Store Design Optimization. Ph.D. Thesis, Division of Water Resources Engineering, Luleå University of Technology, Luleå, Sweden, 1994; p. 217. Available online: http://www.ltu.se/cms_fs/1.5013!/file/FINAL_NORDELL_PHD_red.pdf (accessed on 24 August 2015).
9. Bujok, P.; Klempa, M.; Koziorek, J.; Rado, R.; Porzer, M. Ocena Wpływu Warunków Klimatycznych na Bilans Energetyczny Górotworu na Obszarze Poligonu Badawczego VSB—TU Ostrava. Available online: http://journals.bg.agh.edu.pl/DRILLING/2012-01/W_2012_1_07.pdf (accessed on 22 September 2015). (In Polish).
10. Koohi-Fayegh, S.; Rosen, M.A. Examination of thermal interaction of multiple vertical ground heat exchangers. *Appl. Energy* **2012**, *97*, 962–969.
11. Gehlin, S.; Andersson, O.; Bjelm, L.; Alm, P.G.; Rosberg, J.E. Country Update for Sweden. In Proceedings of the World Geothermal Congress 2015, Melbourne, Australia, 19–25 April 2015; Available online: https://pangea.stanford.edu/ERE/db/WGC/papers/WGC/2015/01021.pdf (accessed on 8 September 2015).
12. Rybach, L.; Eugster, W.J. Sustainability aspects of geothermal heat pump operation, with experience from Switzerland. *Geothermics* **2010**, *39*, 365–369.
13. Weber, J.; Ganz, B.; Schellschmidt, R.; Sanner, B.; Schulz, R. Geothermal Energy Use in Germany. In Proceedings of the World Geothermal Congress 2015, Melbourne, Australia, 19–25 April 2015; Available online: http://www.geothermal-energy.org/pdf/IGAstandard/WGC/2015/01045.pdf? (accessed on 24 August 2015).
14. Raymond, J.; Malo, M.; Tanguay, D.; Grasby, S.; Bakhteyar, F. Direct Utilization of Geothermal Energy from Coast to Coast: A Review of Current Applications and Research in Canada. In Proceedings of the World Geothermal Congress 2015, Melbourne, Australia, 19–25 April 2015; Available online: http://www.geothermal-energy.org/pdf/IGAstandard/WGC/2015/28041.pdf? (accessed on 24 August 2015).
15. Boyd, T.L.; Sifford, A.; Lund, J.W. The United States of America Country Update 2015. In Proceedings of the World Geothermal Congress 2015, Melbourne, Australia, 19–25 April 2015; pp. 1–12. Available online: http://www.geothermal-energy.org/pdf/IGAstandard/WGC/ 2015/01009.pdf? (accessed on 24 August 2015).
16. Sasada, M. Geothermal Heat Pumps in Japan. Japan International Geothermal Symposium, Geothermal Research Society of Japan, Japan,

2012. Available online: http://www.geothermal-energy.org/pdf/IGAstandard/Japan/2012/p17-Masakatsu_Sasada(Eng-Jpn).pdf? (accessed on 8 September 2015).

17. Lund, J.W.; Gawell, K.; Boyd, T.L.; Jennejohn, D. The United States of America Country Update 2010. In Proceedings of the World Geothermal Congress 2010, Bali, Indonesia, 25–29 April 2010; pp. 1–18. Available online: http://www.geothermal-energy.org/pdf/IGAstandard/ WGC/2010/0102.pdf? (accessed on 24 August 2015).

18. Lee, K.S. *Underground Thermal Energy Storage*; Springer Science & Business Media: Heidelberg, Germany, 2012; p. 152.

19. Sliwa, T. Techniczno-ekonomiczne problemy adaptacji wykorzystanych odwiertów na otworowe wymienniki ciepła (Technical and Economic Problems of Adaptation of Used Wells into Borehole Heat Exchangers). Ph.D. Thesis, AGH University of Science and Technology in Krakow, Krakow, Poland, 2002; p. 128.

20. Gonet, A.; Sliwa, T. Thermal Response Test on the Example of Borehole Heat Exchangers in Ecological Park Of Education and Amusement "OSSA". *Transport. Logist.* **2008**, *6*, 42–47.

21. Stryczek, S.; Wiśniowski, R.; Gonet, A.; Złotkowski, A.; Ziaja, J. Influence of polycarboxylate superplasticizers on rheological properties of cement slurries used in drilling technologies. *Arch. Min. Sci.* **2013**, *58*, 719–728. Available online: http://mining.archives.pl/index.php/index.php?option=com_remository&Itemid=0&func=fileinfo&id=645&lang=pl (accessed on 24 August 2015).

22. Sliwa, T.; Knez, D.; Gonet, A.; Sapinska-Sliwa, A.; Szłapa, B. Research and Teaching Capacities of the Geoenergetics Laboratory at Drilling, Oil and Gas Faculty AGH University of Science and Technology in Kraków (Poland). In Proceedings of the World Geothermal Congress 2015, Melbourne, Australia, 19–25 April 2015; pp. 1–14. Available online: http://www.geothermal-energy.org/pdf/IGAstandard/WGC/2015/09012.pdf? (accessed on 24 August 2015).

23. Sliwa, T.; Rosen, M.A.; Jezuit, Z. Use of oil boreholes in the Carpathians in geoenergetics systems historical and conceptual review. *Res. J. Environ. Sci.* **2014**, *8*, 231–242. Available online: http://scialert.net/qredirect.php?doi=rjes.2014.231.242&linkid=pdf (accessed on 24 August 2015).

24. Knez, D. Stress state analysis in aspect of wellbore drilling direction. *Arch. Min. Sci.* **2014**, *59*, 71–76.

25. Eskilson, P. Thermal Analyses of Heat Extraction Boreholes. Ph.D. Thesis, Department of Mathematical Physics, Lund Institute of Technology, Lund, Sweden, 1987. Available online: http://www.buildingphysics.com/Eskilson1987.pdf(accessed on 8 September 2015).

26. Hellström, G. Ground Heat Storage, Thermal Analyses of Duct Storage Systems. Ph.D. Thesis, Department of Mathematical Physics, Lund Institute of Technology, Lund, Sweden, 24 May 1991.

27. Kurevija, T.; Vulin, D.; Krapec, V. Influence of Undisturbed Ground Temperature and Geothermal Gradient on the Sizing of Borehole Heat Exchangers. In Proceedings of the World Renewable Energy Congress 2011, Linköping, Sweden, 8–13 May 2011; pp. 1360–1367. Available online: http://www.ep.liu.se/ecp/057/vol5/017/ecp57vol5_017.pdf(accessed on 8 September 2015).

28. Cataldi, R. Sustainability and Renewability of Geothermal Energy. In Proceedings of the International Scientific Conference on Geothermal Energy in Underground Mines, Ustron, Poland, 21–23 November 2001; Available online: http://www.geothermal-energy.org/pdf/IGAstandard/Poland/2001/a4.pdf? (accessed on 24 August 2015).

29. Signorelli, S.; Kohl, T.; Rybach, L. Sustainability of Production from Borehole Heat Exchanger Fields. In Proceedings of the Twenty-Ninth Workshop on Geothermal Reservoir Engineering, Stanford University, Stanford, CA, USA, 2004; pp. 1–6. Available online: http://www.geothermal-energy.org/pdf/IGAstandard/SGW/2004/Signorelli.pdf? (accessed on 24 August 2015).

30. Dehkordi, E.; Schincariol, R.A. Effect of thermal-hydrogeological and borehole heat exchanger properties on performance and impact of vertical closed-loop geothermal heat pump systems. *Hydrogeol. J.* **2014**, *22*, 189–203.

31. Chiasson, A.C.; Rees, S.J.; Spitler, J.D. A preliminary assessment of the effects of ground-water flow on closed-loop ground-source heat pump systems. *ASHRAE Trans.* **2000**, *106*, 380–393.

32. Diao, N.; Li, Q.; Fang, Z. Heat transfer in ground heat exchangers with groundwater advection. *Int. J. Therm. Sci.* **2004**,*43*, 1203–1211.

33. Fujii, H.; Itoi, R.; Fujii, J.; Uchida, Y. Optimizing the design of large-scale ground-coupled heat pump systems using groundwater and heat transport modelling. *Geothermics* **2005**, *34*, 347–364.

34. Hecht-Méndez, J.; de Paly, M.; Beck, M.; Bayer, P. Optimization of energy extraction for vertical closed-loop geothermal systems considering groundwater flow. *Energy Convers. Manag.* **2013**, *66*, 1–10.

35. Molina-Giraldo, N.; Blum, P.; Zhu, K.; Bayer, P.; Fang, Z. A moving finite line source model to simulate borehole heat exchangers with groundwater advection. *Int. J. Therm. Sci.* **2011**, *50*, 2506–2513.

36. Fan, R.; Jiang, Y.Q.; Yao, Y.; Shiming, D.; Ma, Z.L. A study on the performance of a geothermal heat exchanger under coupled heat conduction and groundwater advection. *Energy* **2007**, *32*, 2199–2209.

37. Sliwa, T.; Gonet, A. Theoretical model of borehole heat exchanger. *J. Energy Res. Technol.* **2005**, *127*, 142–148.

38. Heliasz, Z.; Ostaficzuk, S. Możliwości i sposoby wykorzystania w Polsce ciepła odpadowego i energii geotermalnej do odśnieżania i odladzania—koncepcje i problemy (Possibilities and ways to use in Poland the waste heat and geothermal energy for snow melting and de-icing—Concepts and problems). *Prace Wydz. Nauk Ziemi Uniw. Śląskiego***2002**, *17*, 243–256. (In Polish).

39. Poniedziałek, M.; Sliwa, T. The Use of a Heater and Borehole Heat Exchangers for Regeneration of Heat Resources in the Rock Mass on the Example of the Geoenergetics Laboratory. *AGH Drill. Oil Gas* **2013**, *30*, 353–359. Available online: http://journals.bg.agh.edu.pl/DRILLING/2013.30.2/drill.2013.30.2.353.pdf (accessed on 24 August 2015).

40. Hellström, G.; Gehlin, S. Direct Cooling of Telephone Switching Stations Using a Borehole Heat Exchanger. In Proceedings of the 7th International Conference on Thermal Energy Storage, Megastock'97, 1997, Sapporo, Japan, 18–21 June; pp. 509–514.

41. Andersson, O.; Rydell, L.; Algotsson, T. Efficient usage of waste heat by applying a seasonal energy storage (BTES) at ITT Water & Wastewater AB, Emmaboda, Sweden. Available online: http://intraweb.stockton.edu/eyos/energy_studies/content/docs/effstock09/Session_11_1_Case%20studies_Overviews/101.pdf (accessed on 8 September 2015).

42. Andersson, O.; Hellström, G.; Nordell, B. Heating and Cooling with UTES in Sweden—Current Situation and Potential Market Development. Available online: https://pure.ltu.se/portal/files/350717/artikel.pdf (accessed on 8 September 2015).

# CHAPTER 12

# Calculation Method for Forced-Air Convection Cooling Heat Transfer Coefficient of Multiple Rows of Memory Cards

Takayuki Atarashi[1,2*], Tetsuya Tanaka[1], Shigeyasu Tsubaki[3], Shigeki Hirasawa[2]

[1]*Hitachi, Ltd., Infrastructure Systems Company, Toshima, Tokyo, Japan*
[2]*Department of Mechanical Engineering, Kobe University, Kobe, Japan*
[3]*Hitachi, Ltd., Information and Telecommunication Systems Company, Hatano, Japan*

## ABSTRACT

Forced-air convection cooling of high-power electronic devices is widely used, but it has a problem that a rise in temperature of the air used to cool the upstream devices decreases the cooling capability for the downstream devices. In this study we made an experimental apparatus including a memory card array and measured the effect of the rise in temperature of the air on the heat transfer coefficient of the memory cards that were downstream in the air flow. Using these measurements, we devised a simple calculation model, called the thermal diffusion layer model, to calculate the heat transfer coefficient of multiple rows of memory cards. The rise in temperature of downstream memory cards due to higher temperature air can be evaluated with a parameter representing the delay of thermal mixing for air. The heat transfer coefficient calculated with the thermal diffusion layer model agreed with our experimental results.

## KEYWORDS

Forced Convection, Heat Transfer Performance, LSI Cooling, Memory Array

## 1. INTRODUCTION

The heat density of high-load electronics, such as high-performance servers and the inverters used in vehicles, has been increasing as a result of higher processing speeds and higher power outputs. Focusing on the heat density of high-performance server-CPU (Central Processing Unit) modules, the trend in development since 2000 has been away from bi-polar devices and toward low-power-consumption CMOS (Complementary Metal Oxide Semiconductor) devices. However, continued demand for higher performance has reversed the trend, and the heat density of CPU modules has begun to rise again. The method used to cool electronic devices depends on the heat generation level and the operating environment, whereas the available methods include natural convection cooling [1] , forced-air convection cooling, heat pipe cooling [2] , and liquid cooling. Forced-air convection cooling is the most often used because of its practical advantages in cost, installation, and maintenance. For these reasons, we have been developing a forced-air convection cooling structure to cool multiple LSI (Large Scale Integration) packages and memory arrays that are mounted on a flat wiring board in electronic equipment. A parallel flow method is used in which the LSI packages and arrays are cooled sequentially by air that flows parallel to the wiring board. A problem with this method is that the heat from the upstream devices raises the temperature of the cooling air so that its cooling capability is decreased remarkably for the downstream devices. Yanagida [3] analyzed the temperature distribution of air after heating devices and proposed a method that used the thermal diffusion layer model to calculate temperature distribution of LSI packages on a printed wiring board. The authors developed highly efficient cooling methods using inclined strip fins and air-jet cooling methods that were not affected by an increase in the temperature of the cooling air [4] [5] . Kheirandish et al. [6] analyzed forced convective laminar flow in a channel with a heated obstacle attached to one wall. Recently, Nakayama [7] reviewed various cooling methods of electronic devices. In this work we study forced-air convection cooling of multiple rows of memory cards placed on a printed wiring board. Each memory card has many DRAM-LSIs (Dynamic Random Access Memory LSIs) mounted on both sides (Figure 1). Some studies [8] on this topic reported on the cooling performance of a single memory card, but very few have dealt with rows of memory cards.

In this study, we made an experimental memory card apparatus with four rows in the flow direction and eight rows in the parallel direction. We measured the effect of a rise in the temperature of the air on the heat transfer coefficient of the downstream memory cards. Using these measurements, we then devised a simple model for calculating the heat transfer coefficient of multiple rows of memory cards. We compared the calculation and experimental results and found their values to be in good agreement.

## 2. EXPERIMENTAL APPARATUS

Figure 2 shows the experimental apparatus. The test section is placed in an acryl duct. The outside of the duct is thermally insulated with 50-mm-thick of glass wool. Cooling air from a blower flows into the test section after flow rate G and temperature $T_{ain}$ are measured in chambers 1 and 2.Figure 3 shows the test section. Memory cards are four rows in the flow direction and eight rows in the parallel direction in the duct. Each memory card has five model-LSIs mounted on both sides. In the heat transfer experiment, any one of the four rows in the flow direction is replaced with heating model-LSIs and others are non-heating model-LSIs. The heating model-LSI is aluminum block with a stainless steel heater with heating rate Q of each model-LSI is 0.2 W. The surface temperature $T_s$of the heating model-LSI was measured with a T-type thermocouple, and the heat transfer coefficient h was obtained using Equation (1).

$$h = \frac{Q}{A_S (T_S - T_{ain})} \tag{1}$$

Here, $T_{ain}$ is the inlet air temperature, and $A_s$ is the surface area. Also, the surface temperatures of non-heating model-LSIs were measured with thermocouples. In this paper, the number of rows of memory cards in the flow direction is expressed as $N_{card}$ = 1 - 4 and number of rows of the memory LSIs in each card is expressed as $N_{lsi}$ = 1 - 5.

DRAM-LSI

**Figure 1.** Memory card.

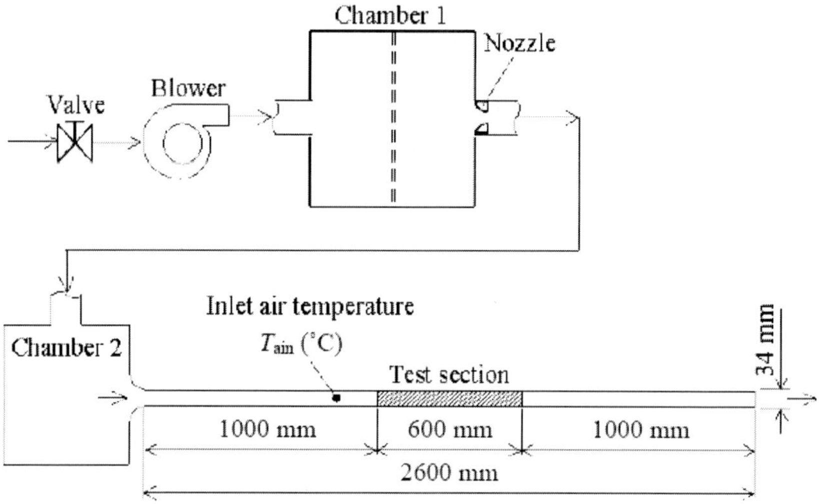


**Figure 2**. Experimental apparatus.

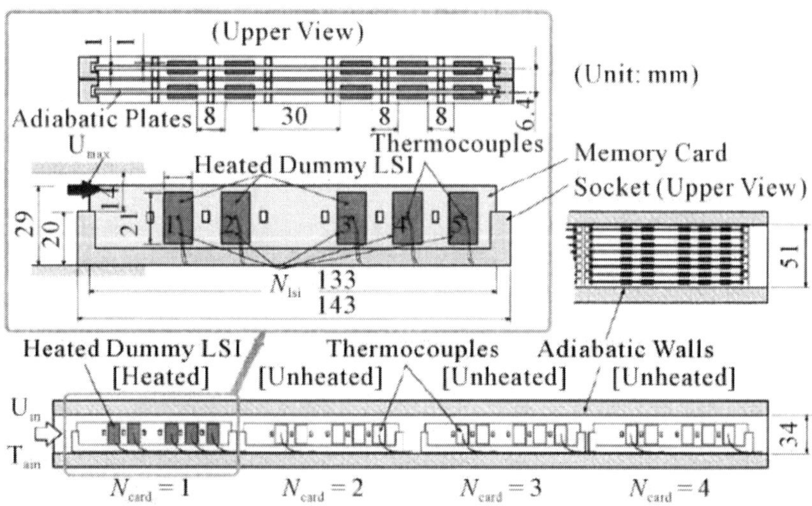


**Figure 3**. Test section of memory card array.

# 3. EXPERIMENTAL RESULTS OF TEMPERATURE RISE OF LSIS AND THERMAL DIFFUSION LAYER MODEL

First, we measured the temperature rise of non-heating LSIs when they were heated by the air flow from the upstream memory card. Figure 4 shows the experimental results of the temperature rise $\Delta T_a$ of the non-heating LSIs at the $N_{lsi}$ = 1 or 5 position on the downstream memory cards at the $N_{card}$ = 2 - 4 positions when the heating memory card was placed at the $N_{card}$ = 1 - 3 positions.Figure 5 shows the parameter $\alpha_t = \Delta T_a/\Delta T_m$ which means the delay of thermal mixing in air, $\Delta T_a$ is the measured temperature rise of the non-heating LSI, and $\Delta T_m$ is the calculated temperature rise of completely mixed air. The temperature rise $\Delta T_m$ was calculated with the air flow rate and heating rate of the LSI. When the air is not mixed completely, $\alpha_t$ is larger than 1. The parameter $\alpha_t$decreases downstream and comes close to 1 at all heating memory card positions, $N_{card}$.

Figure 6 shows the relation between the delay of thermal mixing in air $\alpha_t$ and the distance from the heat source X for various heating positions and flow velocities. We can obtain the relationship between the parameter $\alpha_t$ and the distance from the heat source X, similar to the thermal diffusion layer model proposed by Yanagida [3] .

Figure 7 shows the thermal diffusion layer model for our experimental apparatus. The dimensions of the heating LSI were as follows: height a = 1 mm, width b = 21 mm, and length L = 10 mm. The area of the thermal diffusion layer A is wider than the front area of the LSI (a ´ b) by the thermal diffusion thickness $\delta_t$. Hence, we can write A = (a ´ b) + (2a + b) ´ $\delta_t$. The delay in thermal mixing in the air flow $\alpha_t = \Delta T_a/\Delta T_m$ can be calculated with $A_0/A$, where $A_0$ is the cross-sectional area of the air duct. The thermal diffusion thickness $\delta_t$ is proportional to $\sqrt{X}$ , and the proportionality constant is obtained using our experimental data shown in Figure 6, as follows:

$$\delta_t = 0.0032\sqrt{X} \tag{2}$$

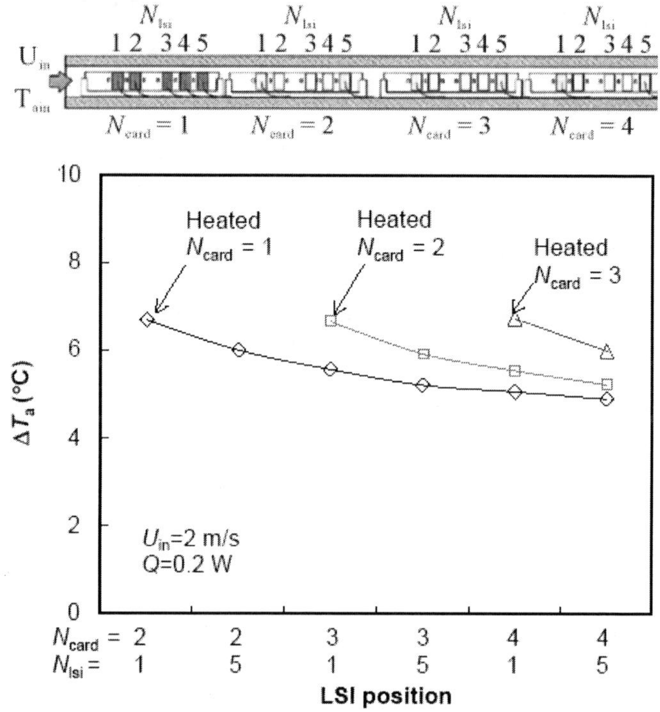


**Figure 4**. Temperature rise of LSIs.

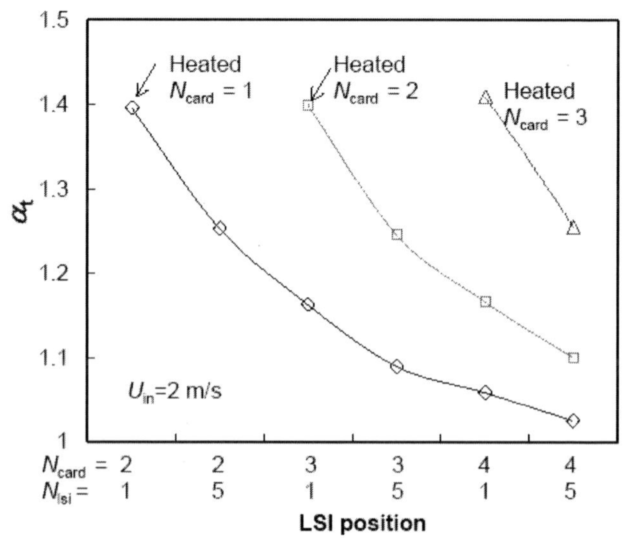


**Figure 5**. Delay of thermal mixing in air.

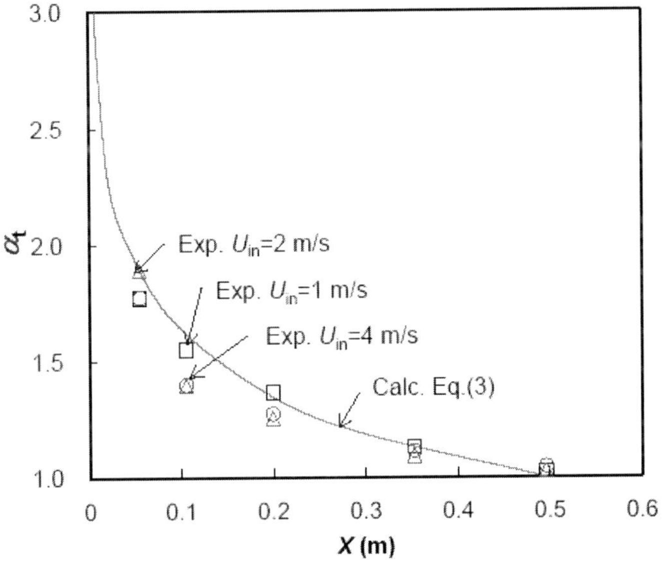


**Figure 6**. Relation between delay of thermal mixing in air and distance from heat source.

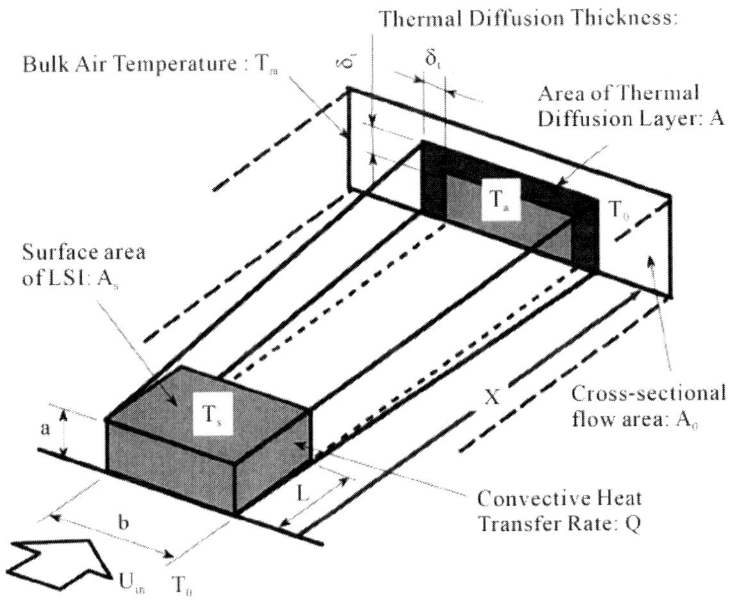


**Figure 7**. Thermal diffusion layer model.

The parameter $\alpha_t$ and the distance from the heat source X have following relationship using Equation (2):

$$\alpha_t = \frac{A_0}{A} = \frac{A_0}{ab + 0.0032(2a+b)\sqrt{X}} \tag{3}$$

Calculation results with Equation (3) agree with the experimental results in Figure 6.

# 4. EXPERIMENTAL RESULTS OF HEAT TRANSFER COEFFICIENT AND COMPARISON WITH CALCULATION USING THERMAL DIFFUSION LAYER MODEL

Figure 8 shows the experimental results of the heat transfer coefficient h of LSIs at the $N_{lsi}$ = 1 - 5 positions on the memory card at the $N_{card}$ = 1 position with a change in the inlet air velocity $u_{in}$when the heating memory card was placed at the $N_{card}$ = 1 position. We see that the heat transfer coefficient h of the downstream LSIs is lower than that of the upstream LSIs, because of the temperature rise in the air flow.

The heat transfer coefficient $h_S$ on a flat surface of a single LSI can be calculated with the following equations [9] by using the Reynolds number, $\mathrm{Re} = \frac{u_{max} L}{\nu_a}$ :

$$h_S = 0.664\,\mathrm{Re}^{0.5}\,\mathrm{Pr}_a^{0.33}\,\frac{\lambda_a}{L} \tag{4}$$

Here, $u_{max}$ is the air velocity above the LSI's surface, L is the length of LSI, $n_a$ is the dynamic viscosity of air, $Pr_a$ is the Prandtl number of air, and $l_a$ is thermal conductivity. The broken line inFigure 8 shows the result of calculation $h_S$ using Equation (4), and it agrees with the experimental results of the $N_{lsi}$ = 1 data.

Next, we calculated the heat transfer coefficient $h_N$ coupled with the temperature rise in air using $\alpha_t$. The heat transfer coefficient $h_N$ can be calculated with the following equation:

$$h_N = \frac{1}{\frac{A_S}{GC_p}\sum_{i-1}^{N-1}\alpha_{1,i} + \frac{1}{h_S}} \quad (5)$$

Here G is the air flow rate, and $C_p$ is the specific heat of air. The solid lines in Figure 8 show the results of the calculation $h_N$ with Equation (5), and they agree with the experimental results of the $N_{lsi}$ = 2 - 5 data.

Figure 9 and Figure 10 show the experimental results of the heat transfer coefficient h of the LSIs at the $N_{lsi}$ = 1 - 5 positions on the memory cards at the $N_{card}$ = 2 and 4 positions with a change in the inlet air velocity $u_{in}$ when the heating memory card was placed at the $N_{card}$ = 2 and 4 positions. The heat transfer coefficient h in Figure 9 and Figure 10 is almost same as that in Figure 8, and the calculations $h_S$, $h_N$ using Equations (4) and (5) agree with the experimental results.

Figure 11 shows the experimental results of the overall heat transfer coefficient $h_{overall}$ of the LSIs at the $N_{lsi}$ = 1 and 5 positions on the memory cards at the $N_{card}$ = 2 and 4 positions when all memory cards at the $N_{card}$ = 1 - 4 positions are heating. The results of the calculation with Equation (5) agree with the experimental results. Therefore, the thermal diffusion layer model of Equation (5) is a precise calculation of heat transfer coefficient of multiple rows of memory cards, despite that it is very simple. Our proposal of the thermal diffusion layer model of Equation (5) is thus very practical.

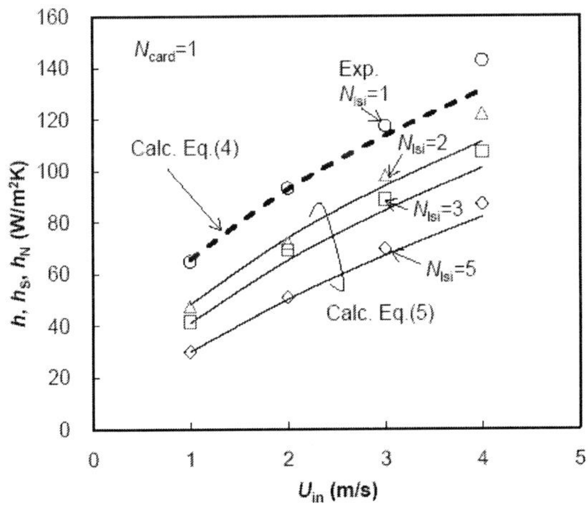


**Figure 8.** Heat transfer coefficient of LSIs $N_{lsi}$ = 1 - 5 on memory card $N_{card}$ = 1.

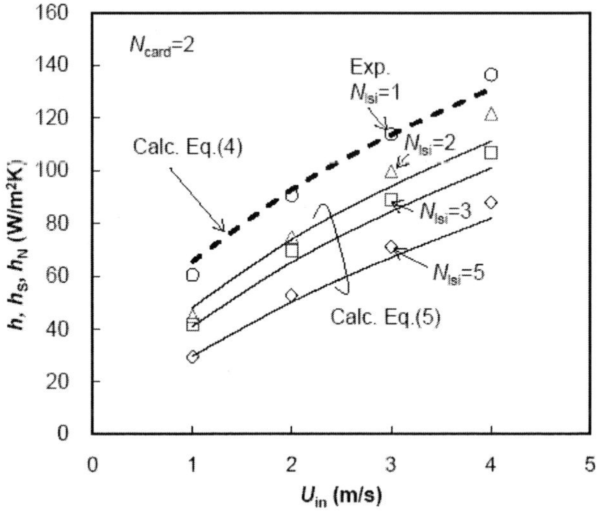


**Figure 9**. Heat transfer coefficient of LSIs $N_{lsi}$ = 1 - 5 on memory card $N_{card}$ = 2.

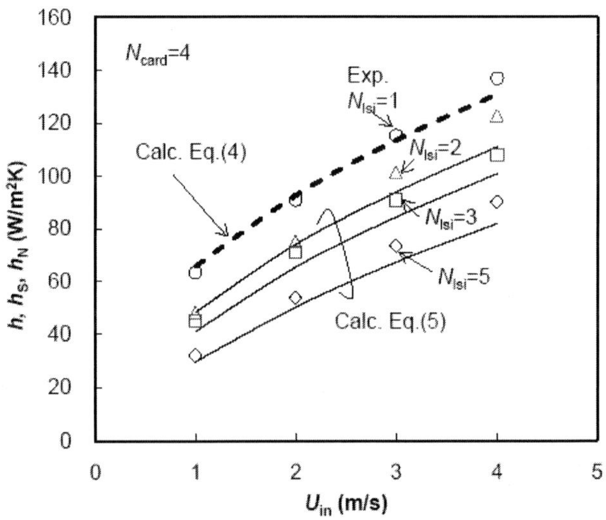


**Figure 10**. Heat transfer coefficient of LSIs $N_{lsi}$ = 1 - 5 on memory card $N_{card}$ = 4.

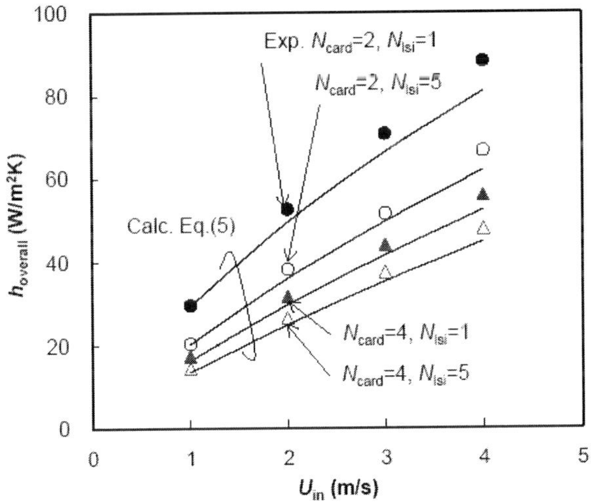


**Figure 11**. Overall heat transfer coefficient.

## 5. CONCLUSIONS

We made an experimental apparatus including a memory card array and measured the effect of a rise in the temperature of air on the heat transfer coefficient of memory cards at downstream positions in the air flow. Using these measurements, we proposed a simple calculation model of the heat transfer coefficient of multiple rows of memory cards. The following results were obtained.

1) The rise in temperature of the downstream LSI resulting from a temperature rise in the air heated by an upstream heating source can be evaluated with a parameter called the delay of thermal mixing in air, $\alpha_t$. The parameter $\alpha_t$ decreases downstream and comes close to 1 at a large distance from the heat source X and $\alpha_t$ does not depend on the heating position or flow velocity.

2) The heat transfer coefficient coupled with the rise in temperature of the air through the parameter $\alpha_t$ can be calculated with Equation (5) and the thermal diffusion layer model. The calculation agreed with the experimental results. There results indicate that the thermal diffusion layer model is a simple and precise method of calculating heat transfer coefficient of multiple rows of memory cards.

# REFERENCES


1. Nakazato, N., Hirasawa, S. and Mato, T. (1998) Natural Convection Cooling in Vertical Finned Plates in a Cabinet for Communication Equipment. IEICE Transactions on Elcctronics, 3, 421-426.

2. Toyoda, H. and Kondo, Y. (2013) Effect of Non-Condensable Gas Leakage on Long Term Cooling Performance of Loop Thermosyphon. Journal of Electronics Cooling and Thermal Control, 3, 131-135. http://dx.doi.org/10.4236/jectc.2013.34014

3. Yanagida, T. (1988) A Calculation Method for Temperature Distribution of IC Packages on a Printed Wiring Board. Transactions of JSME, Series B, 54, 1730-1735.

4. Atarashi, T., Hatada, T. and Daikoku, T. (1994) Forced Air Cooling Radiators for Electronic Equipment. Transactions of JSME, Series B, 60, 620-626.

5. Atarashi, T., Tanaka, T. and Tsubaki, S. (2012) Optimization of Heat Sinks with Plate Fins in Air-Jet Cooling. Proceedings of 3rd International Forum on Heat Transfer (IFHT2012), Paper No. 150.

6. Kheirandish, Z., Nassab, S.A.G. and Vakilian, M. (2013) Second Law Analysis of Forced Convective Cooling in a Channel with a Heated Wall Mounted Obstacle. Journal of Electronics Cooling and Thermal Control, 3, 101-110.http://dx.doi.org/10.4236/jectc.2013.33012

7. Nakayama, W. (2014) Max Jakob Award Paper—Heat in Computers: Applied Heat Transfer in Information Technology. Transaction of ASME, Journal of Heat Transfer, 136, Article ID: 013001.

8. Yoneda, N., Arita, J. and Hosokawa, K. (1999) Thermal Analysis of Memory Module. Proceedings of JSME Thermal Engineering Conference, 31-32.

9. Holman, J.P. (1981) Heat Transfer (International Student Edition). McGraw Hill International Book Co.

# CHAPTER 13

# Influence of Orientation and Radiative Heat Transfer on Aluminum Foams in Buoyancy-Induced Convection

Marijn Billiet *, Sven De Schampheleire, Henk Huisseune and Michel De Paepe

*Department of Flow, Heat and Combustion Mechanics, Ghent University, Sint-Pietersnieuwstraat 41, Ghent 9000, Belgium*

## ABSTRACT

Two differently-produced open-cell aluminum foams were compared to a commercially available finned heat sink. Further, an aluminum plate and block were tested as a reference. All heat sinks have the same base plate dimensions of four by six inches. The first foam was made by investment casting of a polyurethane preform and has a porosity of 0.946 and a pore density of 10 pores per linear inch. The second foam is manufactured by casting over a solvable core and has a porosity of 0.85 and a pore density of 2.5 pores per linear inch. The effects of orientation and radiative heat transfer are experimentally investigated. The heat sinks are tested in a vertical and horizontal orientation. The effect of radiative heat transfer is investigated by comparing a painted/anodized heat sink with an untreated one. The heat flux through the heat sink for a certain temperature difference between the environment and the heat sink's base plate is used as the performance indicator. For temperature differences larger than 30 ∘C, the finned heat sink outperforms the in-house-made aluminum foam heat sink on average by 17%. Furthermore, the in-house-made aluminum foam dissipates on average 12% less heat than the other aluminum foam for a temperature difference larger than 40 ∘C. By painting/anodizing the heat sinks, the heat transfer rate increased on average by 10% to 50%. Finally, the thermal performance of the horizontal in-house-made aluminum foam heat sink is up to 18% larger than the one of the vertical aluminum foam heat sink.

## KEYWORDS

experimental; heat sink; natural convection; metal foam; finned heat sink; radiation; heat transfer

**PACS:** J0101

## 1. INTRODUCTION

Electronic components are omnipresent in daily life products, such as cars, computers and power converters. Hence, it is important to reduce the chance of premature electronics failure. The main causes of electronics failure are over-voltage, elevated operating temperature, moisture and electrostatic discharge (ESD) [1]. To avoid failures caused by elevated operating temperatures, electronic components are cooled. Besides the increased lifetime, some components, such as LEDs, are also more efficient at lower operating temperatures [2,3,4].

Due to the miniaturization of electronic devices and the resulting rise of heat fluxes [5,6], electronic devices need to be cooled by heat sinks. Two types of heat sinks can be distinguished: passive and active heat sinks. The passive heat sink dissipates heat by buoyancy-induced convection. The active heat sink is equipped with a fan to induce an air flow through it and thereby dissipates heat by forced convection. From an economic and ecological point of view, passive heat sinks are favorable over active heat sinks, because they are less expensive, more reliable, require less maintenance and do not use energy.

Current heat sinks use fins to increase their surface-to-volume ratio. These fin shapes can become quite complex, e.g., three-dimensional surface extensions [7,8,9]. One promising material that can be used as three-dimensional surface extensions is high-porous open-cell metal foam.

Open-cell metal foam is a porous media that has a very high volumetric porosity $\phi$ which is defined as:

$$\phi = \frac{V_{\text{total}} - V_{\text{solid}}}{V_{\text{total}}} \tag{1}$$

The volumetric porosity is typically larger than 0.85. A high porosity will be beneficial for mobile applications due to the reduced weight of the heat sink.

The structure of high-porous open-cell metal foam can be found in Figure 1. The metal foam consists of cells that are connected through pores. Each cell is made of several struts, which connect in the nodes. The cell size of metal foam is commonly given as the pores per linear inch (PPI)-value. The porosity $\phi$ and the PPI-value are typically used to characterize open-cell metal foam in the literature. Nevertheless, other methods for characterizing metal foam exist [10].

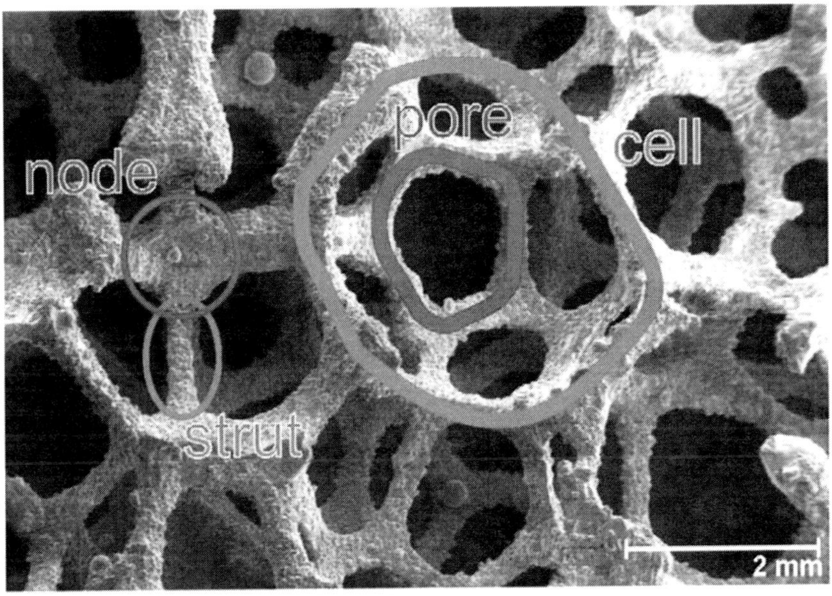


**Figure 1.** The structure of open-cell metal foam.

The work of Ashby [11], Banhart [12] and Dukhan [13] describes the different production methods of open-cell metal foam. Two methods that are relevant to this work are investment casting of a polymer foam precursor and leachable bed casting. A product example of each production method is shown in Figure 2. In the first method, an open-cell polymer foam is made by reticulation of a polyurethane foam [14]. Then, the investment casting technique is used to make an exact metallic copy of this open-cell polymer foam. In the second method, metal is cast over a stacked bed of solvable spheres, e.g., salt spheres or sand with a polymer bonding agent [15]. After solidification, these spheres are removed.

**Figure 2.** An example of a metal foam produced by (**a**) investment casting of a polymer foam precursor and (**b**) leachable bed casting.

To date, metal foam used in forced convection is extensively described in the literature [16,17,18]. However, open literature considering buoyancy-induced convection is limited.

Aluminum foam in buoyancy-induced convection was first investigated by Bhattacharya *et al.* [19]. The authors tested different aluminum foam heat sinks in a horizontal and vertical orientation. Furthermore, their finned aluminum foam heat sink was compared to a commercially available finned heat sink. Qu *et al.* [20] developed a heat transfer correlation for copper foam in buoyancy-induced convection at an arbitrary orientation. Besides the orientation, the correlation also included the effects of the foam structure and the height of the heat sink. Recently, De Schampheleire *et al.* [21,22] investigated the influences of porosity, pore size, foam height and radiative heat transfer on the thermal performance of aluminum foam heat sinks in buoyancy-induced convection, both numerically and experimentally. Based on these results published in the open literature, the following conclusions can be drawn:

- The heat transfer of highly-porous metal foam in buoyancy-induced convection increases with decreasing porosity [19,20,21];
- The heat transfer of highly porous metal foam in buoyancy-induced convection decreases with increasing pore density [19,20,21];

- The heat dissipated by the metal foam heat sink increases with increasing heat sink height until it reaches an asymptotic value. The asymptotic value is reached when the effect of the increasing viscous drag and the temperature drop along the heat sink's height equals the effect of increasing surface area [20,21];
- De Schampheleire *et al.* [22] indicated the importance of the radiative heat transfer of metal foam heat sinks in buoyancy-driven convection. The authors found that up to 30% of the total heat transfer of aluminum foam heat sinks in buoyancy-induced convection is due to thermal radiation;
- The experimental data of both Bhattacharya *et al.* [19] and Qu *et al.* [20] suggest a non-significant effect of the orientation on the thermal performance of metal foam heat sinks.

The first aim of this study is to compare the thermal performance of an aluminum foam heat sink made by investment casting of a polymer precursor, an aluminum foam heat sink made by leachable bed casting and a finned heat sink in buoyancy-induced convection. To the best of our knowledge, an experimental comparison between a finned heat sink and an aluminum foam heat sink has not yet been done. Bhattacharya *et al.* [19] only compared a finned metal foam heat sink with the correlation of an optimal finned heat sink and concluded that the finned metal foam heat sink performed better. Further, the influence of thermal radiation on buoyancy-induced convection is investigated. In contrast to the work that has already been done, the influence of the foam production method and the height of the foam was investigated. Finally, this paper investigated the influence of the heat sink's orientation on its thermal performance.

# 2. METHODOLOGY

## 2.1. Experimental Setup

No standardized methodology exits for testing heat sinks in buoyancy-induced convection. Hence, the design of the experimental setup was based on experimental setups found in the literature [19,20,21,23,24]. The design of the experimental setup aimed to achieve a one-dimensional heat flux through the heat sink.

The experimental setup, depicted in Figure 3, consists of a frame, a rotation mechanism and an insulated box. The rotation mechanism enables the setup to test the heat sinks in different orientations.

Figure 4 and Figure 5 are a top view and a cross-section of the insulated box, respectively. To test different heat sink sizes in future work, four heater elements with different dimensions are placed in an array. The heat sink that will be tested is placed on top of one or multiple heater elements. In the current study, only the 4″ × 6″ heater element was used.

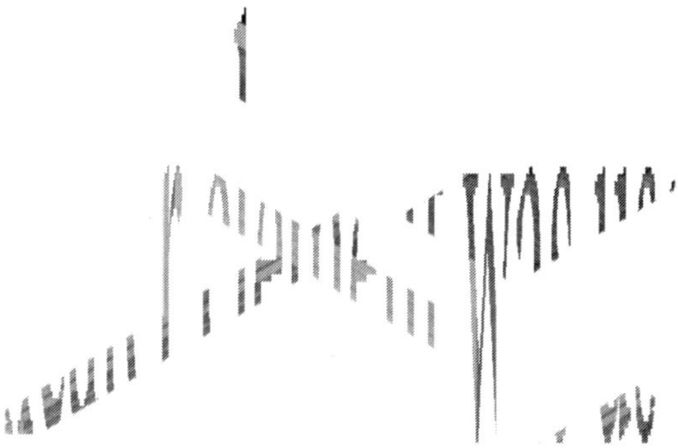

**Figure 3.** Overview of the experimental setup.

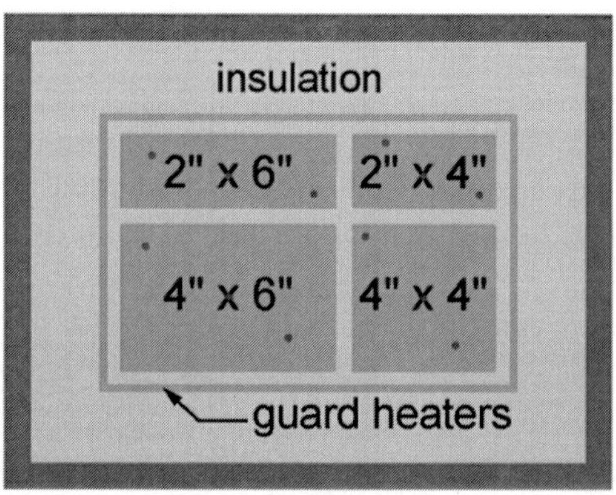


**Figure 4.** Top view of the insulated box. The blue dots indicate the locations of the thermocouples.

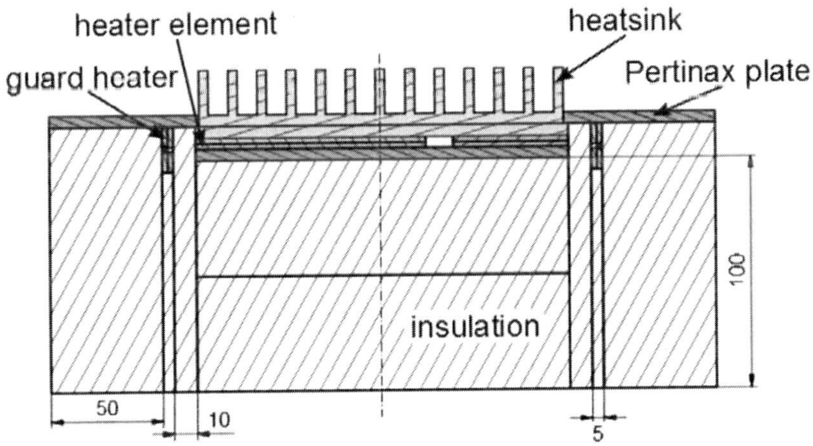


**Figure 5.** A cross-section of the insulated box.

Figure 6 represents the cross-section of one of the four heater elements. This heater element consists of a stack of metal plates and an electrical film heater (Omega® KH series, OMEGA Engineering, Inc., Stamford, CT, USA). The two copper plates ensure a uniform temperature distribution at the base plate of the heat sink, resulting in a uniform heat flux through the base plate. The uniform temperature distribution was verified using an IR-camera. The temperature uniformity was within the uncertainty of the camera. The copper plates are also used to fix the thermocouples in between them. The aluminum plate is added to compensate for different base plate thicknesses. Further, layers of thermal paste ($k = 0.8$ W/(m.K)) and thermal pads ($k = 6$ W/(m.K)) are added in between to avoid insulating air gaps between the metal plates. The two largest and the two smallest electrical film heaters are each respectively powered by an Elektro Automatik® PS 8160-04 2U (Helmholtzstrasse, Germany) and a TTi® PLH120-P (Fort Worth, TX, USA).

As depicted in Figure 5, guard heaters are only used at the sides, while the bottom surfaces of the heater elements are well insulated. The insulation material used is Microtherm ($k = 0.0221$ W/(m.K)). Below the main heaters, there is 100 mm of insulation. The thickness of insulation next to the guard heaters is 50 mm and in between the guard and main heater there is 10 mm of insulation.

The side guard heaters consist of 2 copper plates and an electrical film heater (Omega® KHLV series). These side guard heaters are powered by a TTi® PL303QMD-P. When only part of the main heaters is used, which is the case in this study, the other heater elements serve as extra guard heaters.

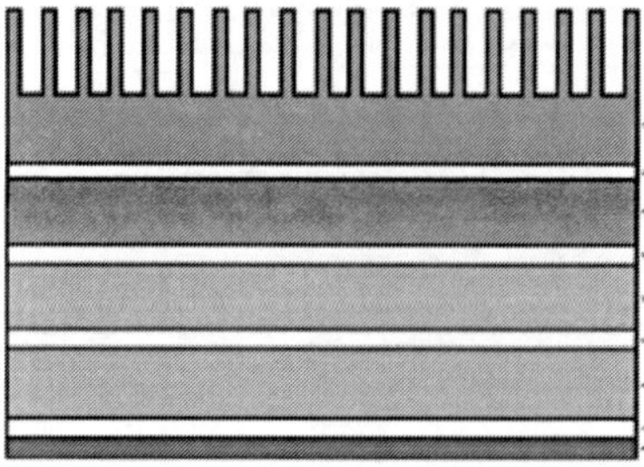


**Figure 6.** Cross-section of a heater element.

The heat sink is firmly pressed against the heaters by a 5 mm-thick plate of Pertinax® ($k$ = 0.2 W/(m.K)), as shown inFigure 5. In addition, this plate forms one smooth surface with the base plate of the heat sink to avoid any disturbance of the air flow.

The temperature of each heater element is measured by 3 K-type thermocouples (Dj = 0.5 mm) and the temperature of each side guard heater is measured by 2 K-type thermocouples (Dj = 0.5 mm). In both cases, the thermocouples are placed between the two copper plates (Figure 5 and Figure 6) into a machined slot of 0.5 mm, which is not visible on the figure. The ambient temperature is measured by 4 K-type thermocouples (Dj = 0.75 mm). These 4 thermocouples are placed at the four corners of the experimental setup at a height of 15 cm above the Pertinax® plate (Fibox, Espoo, Finland). All thermocouples are read out by a data acquisition system (Keithley® 2700, Keithley Instruments Inc., Cleveland, OH, USA).

## 2.2. Test Samples

Table 1 lists the tested heat sinks. All tested heat sinks have the same base plate dimensions (4″ × 6″). First, a flat aluminum plate (= bare base plate) and two solid aluminum blocks were tested. Further, two different aluminum foams were tested. The first aluminum foam (reticulated22 and reticulated40) was made in-house by investment casting of a polymer foam precursor. This aluminum foam has a very high porosity (94.6%) and a moderate PPI-value (10 PPI). Both values are given by the manufacturer. The average cell diameters measured in two perpendicular directions are respectively 3.4 mm and 4.8 mm. The measured pore density is 6 PPI in one direction and 8 PPI in the perpendicular direction. This value is lower than specified by the manufacturer. The second aluminum foam tested (spheres40) was bought from Alveotec® (Venissieux, France) [25]. This foam was produced using leachable bed casting. This foam has a relatively low porosity (85%) and PPI-value (2.5 PPI). Both values are given by the manufacturer. The average cell diameters measured in two perpendicular directions are respectively 8.1 mm and 8.3 mm. The measured pore density is 2.5 PPI in one direction and 2.6 PPI in the perpendicular direction. Although the Alveotec heat sink consists only of 4 pores in the height direction, the heat sink is representative for the Alveotec foam properties. Brun [26] demonstrated that 4 pores in every direction are enough to construct a representative elementary volume for thermo-hydraulic calculations. For constructing the aluminum foam heat sinks, the aluminum foam was glued onto the base plate using thermal conductive epoxy (Loctite® ESP110, Düsseldorf, Germany). Finally, a finned heat sink (Type S120), which was produced by Aavid® (Aavid Thermalloy, Laconia, NH, USA) [27], was also put to the test. After the first experiments, all heat sinks, except the finned heat sink, were painted with Kontakt Chemie® (Jena, Germany) Graphit 33. For the finned heat sink, an anodized version was bought and tested.

**Table 1.** Overview of the test samples.

| Name | Width (mm) | Length (mm) | Height (mm) | # Fins | Porosity (-) | PPI | $\upsilon$ (m$^2$/m$^3$) | Material |
|---|---|---|---|---|---|---|---|---|
| plate | 150.3 ± 0.05 | 101.4 ± 0.05 | 0 | - | - | - | - | Al 1050 |
| block22 | 150.0 ± 0.05 | 99.0 ± 0.05 | 22.0 ± 0.05 | - | 0 | - | 79 | Al 1050 |
| block40 | 150.0 ± 0.05 | 99.0 ± 0.05 | 40.0 ± 0.05 | - | 0 | - | 59 | Al 1050 |
| reticulated22 | 150.0 ± 0.05 | 100.9 ± 0.05 | 22.2 ± 0.05 | - | 0.946 | 10 | 486 | Al 1050 |
| reticulated40 | 149.3 ± 0.05 | 101.0 ± 0.05 | 40.0 ± 0.05 | - | 0.946 | 10 | 486 | Al 1050 |
| spheres40 [25] | 149.8 ± 0.05 | 97.5 ± 0.05 | 40.0 ± 0.05 | - | 0.85 | 2.5 | 360 | Al 1050 |
| finned22 [27] | 150.0 ± 0.05 | 101.0 ± 0.05 | 22.0 ± 0.05 | 25 | 0.65 | - | 352 | Al 6060 |

## 2.3. Measuring Procedure

First, the temperature difference between the heat sink's base plate and the ambient air was set between 10 ∘C and 70 ∘C. Then, the side guard heaters' temperature was controlled to the average temperature of the heat sink's base plate using a PID-controller. Finally, when the steady state was reached, the power dissipated by the heat sink was measured as a function of the temperature difference between the base plate of the heat sink and the ambient air. Steady state was attained when the standard deviation of 100 temperature measurements of the heat sink's base plate at a sampling frequency of 0.17 Hz was below 0.02 K. Some of the experiments were repeated to confirm the consistency of the experimental results.

The results are reported as the heat dissipated by the heat sink as a function of the temperature difference between the heat sink's base plate and the ambient air. This is used as a measure for the thermal performance of the heat sink. For the Alveotec® foam heat sink, the heat dissipated by the heat sink per base plate surface area was also calculated to compensate for the small differences in base plate dimensions.

### 2.4. Uncertainty Analysis

The uncertainty analysis used in this work follows the textbook of Taylor [28].

The dimensions of all components are measured using a caliper with an accuracy of 0.05 mm.

All thermocouples are calibrated using a dry-block calibrator (Druck® DBC 150, Rho, Italy) and a reference thermometer (Fluke® 1523, Everett, WA, USA). After calibration, the average uncertainty of the thermocouples is ±0.11 ∘C., and the maximum uncertainty is ±0.2 ∘C.

$$C_{\dots} \qquad (2)$$

Equation (2) gives the power dissipated by the heat sink. The power supplied by the electrical heaters (Pe) is measured internally by the power supplies. The relative uncertainty of the power measurement depends on the actual power, which is measured and ranges between 4.7% (at 5 W) and 1.5% (at 50 W).

The heat loss through the insulation (Qloss) is estimated using a 3-dimensional finite elements conduction model. The geometry used in the model resembles the actual experimental setup, shown in Figure 5. Two parts were simplified in the model's geometry. First, the heater elements were replaced by a solid block of copper at a constant temperature. Further, the tested heat sink was modeled as a flat plate with an equivalent convection coefficient. At the other boundaries, a fixed convection coefficient of 5 W/K was assumed. The free air temperature was set to 20 ∘C, which was comparable to the real testing conditions. To verify the estimation of the heat loss, the results of the aluminum plate experiments were compared to correlations from the literature [29,30] (see Figure 7).

## 3. RESULTS AND DISCUSSION

First, the thermal performance of the in-house-made aluminum foam heat sink was compared to that of the finned heat sink and of the solid aluminum block. Later, a comparison between the in-house-made aluminum foam and the Alveotec® foam was made. Then, the influence of painting was investigated. Finally, the effect of orientation on the thermal performance was studied.

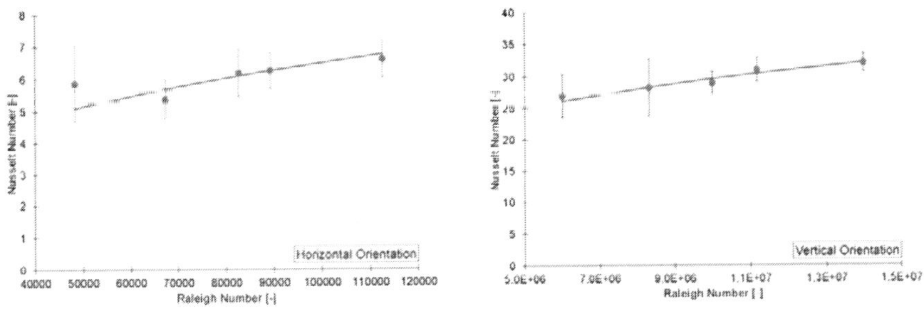


**Figure 7.** A comparison between a painted flat plate in horizontal and vertical orientation with the correlation found in the literature [29,30].

### 3.1. Comparison to Commercial Finned Heat Sink

The in-house-made aluminum foam heat sink (reticulated22) was compared to the finned heat sink (Aavid® S120) and the solid aluminum block in buoyancy-induced convection. All three tested heat sinks have the same height of 22 mm. Figure 8shows the heat dissipated by the heat sink as a function of the temperature difference between the heat sink's base plate and the ambient air for all three heat sinks.

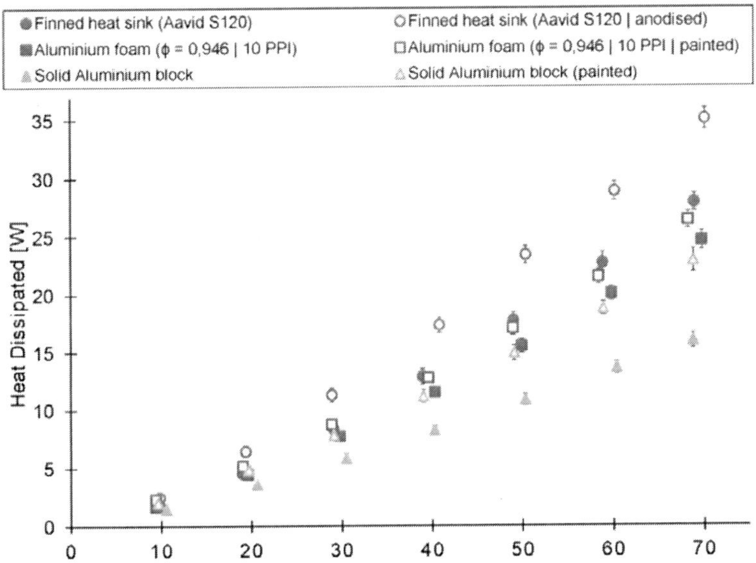


**Figure 8.** The heat dissipated by the untreated and the painted/anodized heat sink as a function of the temperature difference between the ambient air and the heat sink's base plate (heat sink's height = 22 mm).

If the temperature difference is below 40 ∘C, no significant difference exists between the heat dissipated by the finned and the aluminum foam heat sink. However, the heat dissipated by the finned heat sink is on average 17% larger for a temperature difference higher than 40 ∘C. Both the finned and the aluminum foam heat sink clearly outperform the solid aluminum block, except for the lowest temperature difference measured.

The possible performance gain due to the higher surface-to-volume ratio (Table 1) of the aluminum foam heat sink compared to the finned heat sink is not found in the experimental results due to the higher hydraulic resistance of the aluminum foam heat sink.

The difference in performance between the finned and foamed heat sink (reticulated22) is explained by considering the smoke visualizations in Figure 9. The same technique was also used by Taji *et al.* [31] to visualize the air flow through a heat sink. In Figure 9, smoke generated by burning incense is used to visualize the air flow through the finned and aluminum foam heat sink. The smoke enters both heat sinks at the same location, *i.e.*, at the side near the base plate. However, the exit location differs. In the finned heat sink, the smoke exits at the center of the heat sink. In the case of the aluminum foam heat sink, the smoke exits at the top of the heat sink before it reaches the heat sink's center. After exiting the aluminum foam heat sink, the smoke deflects to the center to rise due to a chimney effect. Due to this deflection, cold air cannot enter the heat sink at the top; thus, no multi-chimney pattern [8] is formed. Hence, only part of the heat exchanging surface area of the aluminum foam adds to the convective heat transfer. Due to the higher hydraulic resistance of metal foam compared to fins, the heat exchanging surface area of foams is used less efficiently than in finned heat sinks. This explains why the metal foam, even though having a larger surface-to-volume ratio (see Table 1), has a smaller thermal performance than the finned heat sink. This shallow penetration by the air in the metal foam was also reported by De Schampheleire *et al.* [22], who performed numerical simulations. Moreover, a higher temperature difference induces a larger buoyancy force, but the air flow rate through the metal foam will increase less than the one of the finned heat sink, due to its higher hydraulic resistance. This will lower the local convection coefficient.

### 3.2. Comparison between the Alveotec® and the In-House-Made Aluminum Foam

Figure 10 compares the thermal performance of the in-house-made foam (reticulated40) to that of the Alveotec® foam (spheres40) [25]. The figure shows the heat dissipated by the heat sink as a function of the temperature difference between the heat sink's base plate and the ambient air.

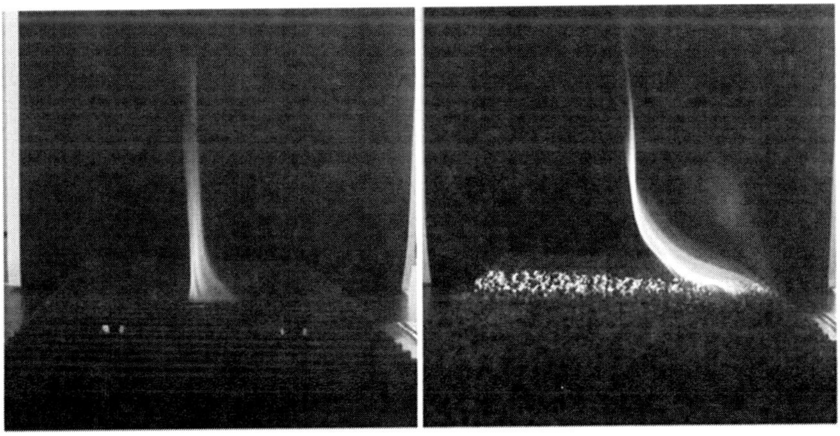

(a) (b)

**Figure 9.** Visualization of the air flow through (**a**) the finned heat sink (Aavid® S120) and (**b**) the in-house-made aluminum foam heat sink (reticulated22) using smoke (ΔT= 70 ∘C, horizontal orientation).

At temperature differences below 50 ∘C, no significant difference in thermal performance exists between the Alveotec®and the in-house-made foam. Nevertheless, at larger temperature differences, the Alveotec® foam heat sink dissipates up to 12% more heat than the in-house-made aluminum foam heat sink. These results were expected based on the conclusions of previous studies [19,20,21]. These previous studies concluded that the thermal performance of metal foam in buoyancy-induced convection increased with decreasing porosity and PPI-value. A lower PPI-value results in larger pores and, thus, a lower hydraulic resistance, but also a lower surface-to-volume ratio. However, this decrease is partly compensated by the decrease of porosity, which increases the surface-to-volume ratio. Moreover, in buoyancy-induced convection, the hydraulic resistance has to be overcome by a small buoyancy force. Hence, a low hydraulic resistance has probably the most important influence on the thermal performance. Finally, it has to be noted that the in-house-made metal foam was bonded to the base plate using thermal conductive epoxy. This can lead to a higher contact resistance, thereby decreasing the thermal performance [32] according to De Schampheleire *et al.* [21], the influence of contact resistance is relatively low (<5%) in buoyancy-induced convection.

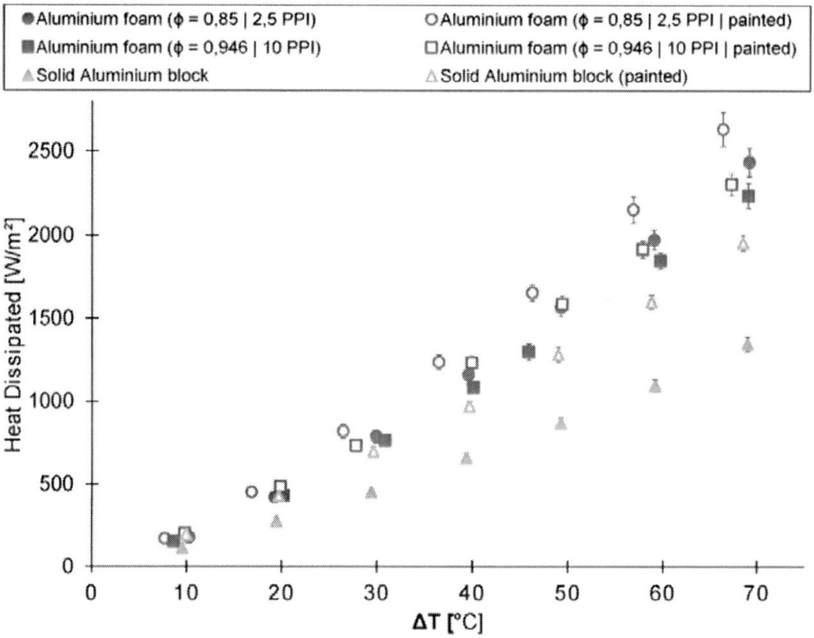


**Figure 10.** The heat dissipated by the unpainted and painted heat sink as a function of the temperature difference between the ambient air and the heat sink's base plate (heat sink's height = 40 mm).

### 3.3. Radiative Heat Transfer

To test the influence of thermal radiation on the total heat transfer in buoyancy-induced convection, an untreated and painted/anodized version of each heat sink was tested. Both painting and anodizing increase the emissivity of the heat sink and, consequently, the heat transfer by thermal radiation. The emissivity of the in-house-made aluminum foam was experimentally determined for both the unpainted and painted version using a thermal emissivity measuring instrument (INGLAS® TIR 100-2, Friedrichshafen, Germany). The emissivity of the unpainted and painted in-house-made aluminum foam is, respectively, 0.540 ± 0.005 and 0.894 ± 0.005. The emissivity of the other heat sinks was not tested, but will be of the same order, except for the emissivity of the untreated finned heat sink and the untreated solid aluminum block. The emissivity of those will be much lower than the one of the in-house-made foam, because their surfaces are very smooth. This emissivity lies between 0.05 and 0.25 [33].

Figure 8 compares the thermal performance of the untreated finned heat sink, the 22 mm-high in-house-made aluminum foam heat sink (reticulated22) and the 22 mm-high solid aluminum block to that of their painted/anodized counterparts. By painting or anodizing, the thermal performance improved for each of the tested heat sinks. The aluminum foam heat sink, the finned heat sink and the aluminum block, respectively, dissipated on average 17%, 31% and 44% more heat after they were painted or anodized.

A comparison between the 40 mm-high in-house-made aluminum foam heat sink (reticulated40), the Alveotec® foam heat sink (spheres40) and the 40mm-high solid aluminum block and their painted counterparts is shown in Figure 10. This figure also indicates that the thermal performance of the tested heat sinks in buoyancy-induced convection increases by painting them. The painted version of the 40 mm-high aluminum foam heat sink, the Alveotec® foam heat sink and the 40 mm-high solid aluminum block, respectively, dissipate on average 11%, 21% and 51% more heat than their unpainted versions.

The relative contribution of the thermal radiation decreases with increasing temperature difference, as shown in bothFigure 8 and Figure 10. This trend was also found by Sparrow and Vemuri [34] for pin-fin arrays.

Further, the relative contribution of thermal radiation to the total heat transfer decreases with increasing height of the in-house-manufactured aluminum foam heat sink. In contrast, the relative contribution of thermal radiation to the total heat transfer increases with increasing height for the solid aluminum block. According to De Schampheleire *et al.* [22], the highest aluminum foam has the largest influence by radiation. This conclusion is the opposite of the one found here. The major difference between the experiments of De Schampheleire *et al.* and the ones of this study are the dimensions of the base plate of the heat sink. The base plate in the study of De Schampheleire *et al.* [22] is 254 mm by 25.4 mm, whereas in this study, it is 150 mm by 101 mm. The different ratio between the surface area with a large view factor towards the environment and the total surface area could cause the difference in conclusions.

The improvement by painting/anodizing of the finned heat sink is larger compared to the one of the 22 mm-high in-house-manufactured aluminum foam heat sink (Figure 8). To explain this difference in improvement, the Stefan–Boltzmann law is used, which is defined as:

$$\dot{Q}_{\mathrm{rad}} = A\ F\ \epsilon\ \sigma_b\ \left(T_{hs}^4 - T_{amb}^4\right) \tag{3}$$

Equation (3) calculates the radiative heat transfer towards the environment (Q˙rad) given the total outer surface area of the subject ($A$), the view factor ($F$), the emissivity ($\epsilon$) and the Stefan–Boltzmann constant (σb).

The larger improvement by the finned heat sink can partly be explained by its larger increase in emissivity by painting. The increase in emissivity for the finned heat sink is almost twice as large as the one of the aluminum foam. According to Equation (3), the increase of thermal radiation should be higher for the finned heat sink.

Moreover, in the case of the aluminum foam heat sink, the absolute increase in dissipated heat by painting flattens with increasing temperature difference. This is not expected according to the Stefan–Boltzmann law (Equation (3)). This flattening is caused by the lower thermal conductivity of the metal foam itself. The average object's wall temperature, which is an important variable in Equation (3), strongly depends on the thermal conductivity of the heat sink itself. The total heat dissipated by the heat sink increases with increasing temperature difference. This larger heat flux through the heat sink combined with a low thermal conductivity results in a large temperature gradient over the heat sink. This temperature gradient will lower the object's wall temperature and, therefore, the radiative heat transfer. Because of the fourth power in Equation (3), the effect of the wall temperature is more pronounced than for convective heat transfer, reducing the relative contribution of radiative heat transfer to the total heat transfer in buoyancy-induced convection.

### 3.4. Orientation

Figure 11 represents the thermal performance of a flat plate (bare base plate), a finned heat sink (Aavid® S120) and the in-house-made aluminum foam heat sink, all having the same outer dimensions and all being painted/anodized, in two different orientations: horizontal and vertical. In the vertical orientation, the four-inch side is aligned with gravity.

The aluminum foam heat sink dissipates almost twice as much heat compared to the flat plate at a temperature difference of 70 ∘C. The heat dissipated by the finned heat sink is about three times more than by the flat plate at a temperature difference of 70 ∘C. For both the flat plate and the finned heat sink, no significant difference in thermal performance can be found between the horizontal and vertical orientation. Nonetheless, the thermal performance of the in-house-made aluminum foam heat sink in the horizontal orientation is up to 18% higher than in the vertical orientation for a temperature difference larger than 30 ∘C.

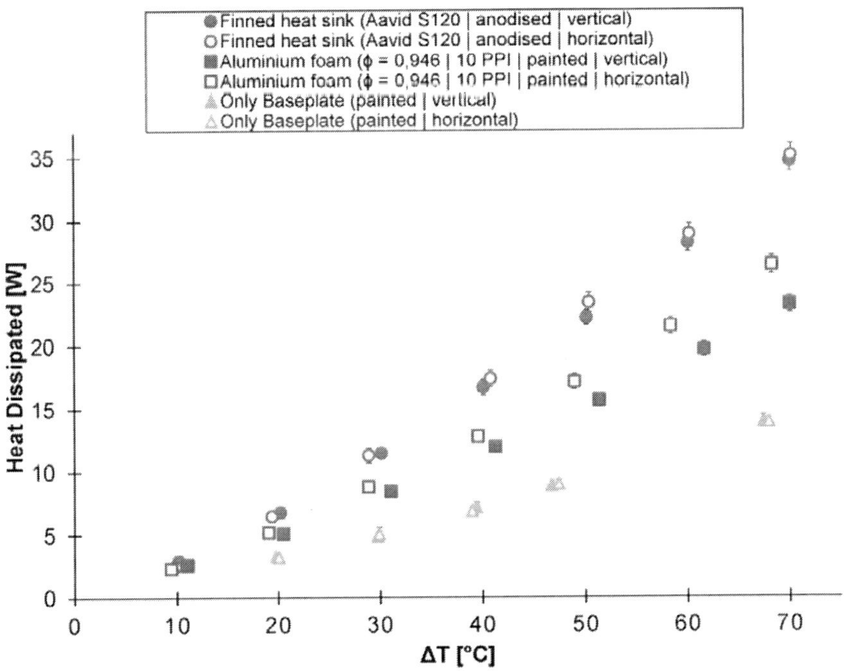


**Figure 11.** The heat dissipated by the heat sink at different orientations as a function of the temperature difference between the ambient air and the heat sink's base plate (heat sink's height = 22 mm; the 4″-side is parallel with the gravity in the vertical orientation).

The results obtained in this study are different from those found in the literature [19,20]. Both Qu *et al.* [20] and Bhattacharya *et al.* [19] did not find any significant difference between the horizontal and vertical orientation. When this study is compared to the ones of Qu *et al.* [20] and Bhattacharya *et al.* [19], it is noted that the temperature test range, foam porosity and pore size are very similar. Yet, the foam used by Qu *et al.* [20] is made of copper instead of aluminum. Further, the experimental setups and the dimensions of the metal foam heat sinks are different for all three studies. Assuming that the experiments found in the literature are conducted correctly, the only difference that can explain the difference in results is the dimensions of the metal foam heat sink. In contrast to the heat sinks tested in this study, which have a rectangular base plate, the heat sinks tested by Qu *et al.* [20] have a square base plate of 100 mm by 100 mm. Furthermore, the height of the heat sinks tested by Qu *et al.* [20] is 55 mm, which is more than double the heat sink's height in this study. No dimensions are specified in the paper of Bhattacharya *et al.* [19].

Figure 9 and Figure 12 shows that in both horizontal and vertical orientations, the metal foam is penetrated very shallowly by the air flow. This means that in the horizontal orientation, only the regions near the four sides of the foam have a large contribution to the convective heat transfer. Hence, having a rectangular base plate is probably beneficial due to the larger perimeter of the base plate for a given foam volume.

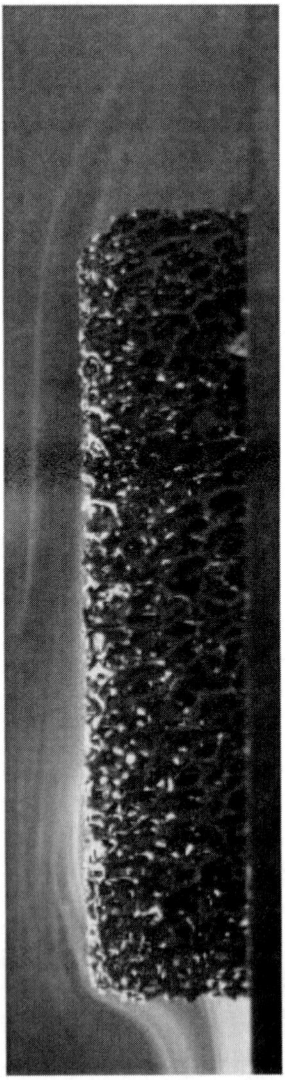

**Figure 12.** Visualization of the air flow through the in-house-made aluminum foam heat sink (reticulated22) using smoke ($\Delta T = 70$ ∘C, vertical orientation).

## 4. CONCLUSIONS

In this study, the thermal performance of seven different aluminum heat sinks in buoyancy-induced convection was tested. All heat sinks had the same base plate dimensions (6″× 4″). During the experiments, the temperature difference between the ambient air and the heat sink's base plate was varied from 10 ∘C to 70 ∘C in steps of 10 ∘C. At each temperature difference, the power dissipated by the heat sink was measured. Based on these experiments, the following conclusions can be made:

- At temperature differences higher than 30 ∘C, significant differences in thermal performance exist between the finned and metal foam heat sinks at the tested conditions. The finned heat sink dissipates on average 17% more heat than the 22 mm-high in-house-manufactured aluminum foam heat sink. The Alveotec® foam heat sink dissipates 12% more heat than the 40 mm-high in-house-made foam heat sink.
- By increasing the emissivity of a heat sink, the power dissipated by a heat sink in buoyancy-induced convection can be significantly increased by up to 50%. The relative attribution of radiative heat transfer to the total heat transfer decreases for all heat sinks with an increasing temperature difference.
- For the tested conditions, metal foam heat sinks dissipate up to 18% more heat in the horizontal orientation.

## ACKNOWLEDGMENTS

Research funded by a Ph.D. grant (No. 141733) of the Agency for Innovation by Science and Technology (IWT).

## AUTHOR CONTRIBUTIONS

Marijn Billiet, Sven De Schampheleire and Henk Huisseune conceived and designed the experiments. Marijn Billiet performed the experiments and analyzed the data. Marijn Billiet, Sven De Schampheleire and Henk Huisseune wrote the paper. Michel De Paepe supervised the study.

# REFERENCES


1. Burris, M. Component Failures: How They Fail and How to Identify Failed Components. Available online: http://components.about.com/od/Components/a/Component-Failures.htm (accessed on 30 July 2015).

2. U.S. Department of Energy. Building Technologies Program: Solid-State Lighting Technology Fact Sheet. Available online: http://www.webcitation.org/6aKbWerxo (accessed on 30 July 2015).

3. Meyaard, D.S.; Shan, Q.; Cho, J.; Fred Schubert, E.; Han, S.H.; Kim, M.H.; Sone, C.; Jae Oh, S.; Kyu Kim, J. Temperature dependent efficiency droop in GaInN light-emitting diodes with different current densities. *Appl. Phys. Lett.* **2012**, *100*.

4. CREE®; CREE Inc., Durham, NC, USA. Thermal Management of Cree® XLamp® LEDs. Available online: http://www.webcitation.org/6aKchRqUw (accessed on 30 July 2015).

5. Kolar, J.; Drofenik, U.; Biela, J.; Heldwein, M.; Ertl, H.; Friedli, T.; Round, S. PWM Converter Power Density Barriers. In Proceedings of the Power Conversion Conference (PCC), Nagoya, Japan, 2–5 April 2007; pp. 9–29.

6. Pop, E. Energy dissipation and transport in nanoscale devices. *Nano Res.* **2010**, *3*, 147–169.

7. Calamas, D.; Baker, J. Behavior of Thermally Radiating Tree-like Fins. *Heat Transfer* **2013**, *135*.

8. Dogan, M.; Sivrioglu, M.; Yılmaz, O. Numerical analysis of natural convection and radiation heat transfer from various shaped thin fin-arrays placed on a horizontal plate-a conjugate analysis. *Energy Convers. Manage.* **2014**, *77*, 78–88.

9. Bornoff, R.; Parry, J. An additive design heat sink geometry topology identification and optimisation algorithm. In Proceedings of the 31st Thermal Measurement, Modeling Management Symposium (SEMI-THERM), San Jose, CA, USA, 15–19 March 2015; pp. 303–308.

10. De Jaeger, P.; T'Joen, C.; Huisseune, H.; Ameel, B.; de Paepe, M. An experimentally validated and parameterized periodic unit-cell reconstruction of open-cell foams. *J. Appl. Phys.* **2011**, *109*.

11. Ashby, F. *Metal Foams: A Design Guide*; Chemical Petrochemical & Process; Butterworth-Heinemann: Oxford, UK, 2000.

12. Banhart, J. Manufacture, characterisation and application of cellular metals and metal foams. *Prog. Mater Sci.* **2001**, *46*, 559–632.

13. Dukhan, N. *Metal Foams: Fundamentals and Applications*; Destech Publications: Lancaster, PA, USA, 2013.

14. Volz, R. Reticulated Polyurethane Foams and Process for Their Production. U.S. Patent 3,171,820, 2 March 1965.

15. Jinnapat, A.; Kennedy, A. The Manufacture and Characterisation of Aluminum Foams Made by Investment Casting Using Dissolvable Spherical Sodium Chloride Bead Preforms. *Metals* **2011**, *1*.

16. Zhao, C. Review on thermal transport in high porosity cellular metal foams with open cells. *Int. J. Heat Mass Transfer* **2012**, *55*, 3618–3632.

17. Mancin, S.; Zilio, C.; Diani, A.; Rossetto, L. Air forced convection through metal foams: Experimental results and modeling. *Int. J. Heat Mass Transfer* **2013**, *62*, 112–123.

18. De Schampheleire, S.; de Jaeger, P.; Huisseune, H.; Ameel, B.; T'Joen, C.; de Kerpel, K.; de Paepe, M. Thermal hydraulic performance of 10 PPI aluminum foam as alternative for louvered fins in an HVAC heat exchanger. *Appl. Therm. Eng.* **2013**, *51*, 371–382.

19. Bhattacharya, A.; Mahajan, R.L. Metal Foam and Finned Metal Foam Heat Sinks for Electronics Cooling in Buoyancy-Induced Convection. *J. Electron. Packag.* **2005**, *128*, 259–266.

20. Qu, Z.; Wang, T.; Tao, W.; Lu, T. Experimental study of air natural convection on metallic foam-sintered plate. *Int. J. Heat Fluid Flow* **2012**, *38*, 126–132.

21. De Schampheleire, S.; de Jaeger, P.; Reynders, R.; de Kerpel, K.; Ameel, B.; T'Joen, C.; Huisseune, H.; Lecompte, S.; de Paepe, M. Experimental study of buoyancy-driven flow in open-cell aluminum foam heat sinks. *Appl. Therm. Eng.* **2013**, *59*, 30–40.

22. De Schampheleire, S.; de Kerpel, K.; de Jaeger, P.; Huisseune, H.; Ameel, B.; de Paepe, M. Buoyancy driven convection in open-cell metal foam using the volume averaging theory. *Appl. Therm. Eng.* **2015**, *79*, 225–233.

23. Sikka, K.; Torrance, K.; Scholler, C.; Salanova, P. Heat sinks with fluted and wavy plate fins in natural and low-velocity forced convection. *IEEE Trans. Compon. Packag. Technol.* **2002**, *25*, 283–292.

24. Khor, Y.K.; Hung, Y.M.; Lim, B.K. On the role of radiation view factor in thermal performance of straight-fin heat sinks. *Int. Commun. Heat Mass Transfer* **2010**, *37*, 1087–1095.

25. Alvéotec; Venissieux, France. Innovation Fonderie. Available online: http://www.alveotec.fr/en/our-news/unique-aluminum-foam-from-goodfellow_41.html (accessed on 24 March 2015).

26. Brun, E. *De L'Imagerie 3D des Structures a L'Etude des Mecanismes de Transport en Milieux Cellulaires*; Institut Universitaire des Systèmes Thermiques Industriels, Aix-Marseille Université: Marseille, France, 2009.

27. Aavid Thermalloy (Laconia, NH, USA). Available online: http://www.aavid.com/ (accessed on 24 March 2015).

28. Taylor, J. *An Introduction to Error Analysis: The Study of Uncertainties in Physical Measurements*; University Science Books: Dulles, VI, USA, 1997.

29. Churchill, S.W.; Chu, H.H. Correlating equations for laminar and turbulent free convection from a vertical plate. *Int. J. Heat Mass Transfer* **1975**, *18*, 1323–1329.

30. Nellis, G.; Klein, S. *Heat Transfer*; Cambridge University Press: Cambridge, UK, 2012.

31. Taji, S.; Parishwad, G.; Sane, N. Experimental investigation of heat transfer and flow pattern from heated horizontal rectangular fin array under natural convection. *Heat Mass Transfer* **2014**, *50*, 1005–1015.

32. De Jaeger, P.; T'Joen, C.; Huisseune, H.; Ameel, B.; de Schampheleire, S.; de Paepe, M. Assessing the influence of four bonding methods on the thermal contact resistance of open-cell aluminum foam. *Int. J. Heat Mass Transfer* **2012**, *55*, 6200–6210.

33. Fluke®; Everett, WA, USA. Emissivity values of common materials. Available online: http://www.webcitation.org/6aKc191IN (accessed on 30 July 2015).

34. Sparrow, E.; Vemuri, S. Orientation effects on natural convection/radiation heat transfer from pin-fin arrays. *Int. J. Heat Mass Transfer* **1986**, *29*, 359–368.

# CHAPTER 14

# Heat Flux Sensors for Infrared Thermography in Convective Heat Transfer

Giovanni Maria Carlomagno [†,*], Luigi de Luca [†], Gennaro Cardone [†] and Tommaso Astarita [†]

*Department of Industrial Engineering, University of Naples Federico II, Piazzale Tecchio 80, Napoli 80125, Italy*

## ABSTRACT

This paper reviews the most dependable heat flux sensors, which can be used with InfraRed (IR) thermography to measure convective heat transfer coefficient distributions, and some of their applications performed by the authors' research group at the University of Naples Federico II. After recalling the basic principles that make IR thermography work, the various heat flux sensors to be used with it are presented and discussed, describing their capability to investigate complex thermo-fluid-dynamic flows. Several applications to streams, which range from natural convection to hypersonic flows, are also described.

## KEYWORDS

infrared thermography; heat flux sensors; convective heat transfer

# 1. INTRODUCTION

The main purpose of this paper is to review on how to take advantage of InfraRed Thermography (IRT) for measuring wall convective heat fluxes, *i.e.*, heat transfer between a body surface and a fluid flowing over it, and to describe some of the IRT applications performed by the authors' research group at the University of Naples Federico II on complex fluid flows.

Measuring heat fluxes is one of the challenging tasks of thermo-fluid-dynamics and requires both a proper *thermal sensor*(which is herein called *heat flux sensor*), with its related thermo-physical model, and, being temperature always involved in heat transfer processes, one or more *temperature transducers*.

In more conventional techniques where temperature is measured with standard transducers (e.g., thermometers, thermocouples, resistance temperature detectors (RTDs), pyrometers, *etc.*), each transducer yields either the temperature at a single point, or its average over a discrete space. Hence, in terms of spatial resolution, the heat flux sensor itself has to be considered as *zero-dimensional*. This constraint makes measurements essentially meaningless whenever the temperature, and/or the heat flux fields exhibit high spatial variations.

Instead, the *infrared* (IR) *camera*, also called *infrared scanner*, constitutes a truly *two-dimensional* temperature transducer since it allows accurate measurements of surface temperature maps even in the presence of relatively high spatial gradients [1]. Accordingly, also the heat flux sensor becomes two-dimensional, as long as the later shown corrections are performed.

Infrared thermography is a methodology which allows for remote detection of thermal energy that is radiated from objects in one of the electromagnetic spectrum infrared bands, conversion of such energy into a video signal, and a two-dimensional representation of the object surface temperature distribution (map). The method can be exploited in many application fields and for many different purposes, as long as surface temperature variations are involved. For example, IRT may be used in various different types of diagnosis (in medicine, architecture, maintenance), or in material characterization and procedures assessment, which can help improve design and manufacturing of products, as well as in non-destructive testing. As the technology evolves, infrared systems offer new prospects for new applications. Almost any process which is temperature-dependent may benefit from the use of an infrared device.

***Infrared Thermography and Thermo-Fluid-Dynamics***

As far as convective heat transfer measurements in thermo-fluid-dynamics are concerned, when compared to standard methods, the use of an infrared camera as a temperature transducer for heat flux sensors appears advantageous from several points of view [1].

Since the IR camera is two-dimensional, with current systems having up to about 1 M pixels per frame, besides producing a whole temperature map, IRT allows an easier evaluation of errors due to radiation and tangential conduction in the sensor. The camera does not disturb the phenomena being measured (*non-intrusive*), and so it does not alter conduction through the test article because of embedded sensors and wiring, it has *high sensitivity* (down to 10 mK) and *low response time*(down to 20 μs). As such, IR thermography can be effectively exploited to measure convective heat fluxes with either steady, or transient, techniques.

The major application of infrared thermography in thermo-fluid-dynamics is the measurement of the *convective heat flux* $q_c$(energy which flows in the heat mode per unit surface and per unit time, W/m$^2$), and/or of the *convective heat transfer coefficient* $h$ (heat flux per unit temperature difference, W/m$^2$K), between a solid surface and a fluid flowing over it.

It has to be explicitly pointed out that the heat flux, as such, is a vectorial quantity. However initially, only its convective component normal to exchanging surface $q_c$ is going to be considered.

The link between the quantities $q_c$ and $h$ is the well-known *Newton's law* [2]:

$$q_c = h(\,T_w - T_r\,) \tag{1}$$

which is herein written in a generalized form and where $T_w$ is the *surface* (wall) *temperature* and $T_r$ is a *reference temperature*that depends on the fluid flow actual conditions.

For example, for simple hyposonic (*i.e.*, small Mach number $M$) external flows, the reference temperature $T_r$ is the static temperature of the undisturbed fluid. Instead, for high Mach number flows, the correct choice for $T_r$ is the so-called *adiabatic wall temperature* [2,3]. It has to be pointed out that the adiabatic wall temperature does not depend only on the Mach number and should be also used in other cases, e.g., for the mixing of two hyposonic streams at different temperatures, such as a warm jet issuing in cold air and impinging on a plate [4]. On the other hand, for hyposonic internal flows, the correct choice for the reference temperature $T_r$ is the local bulk (cup) temperature in the duct cross section.

It is the present authors' opinion that, regardless of the particular flow conditions, $T_r$ should be always referred to with the generalized concept of local adiabatic wall temperature since, as Equation (1) shows, this is the wall temperature $T_w$ that leads to $q_c = 0$.

Within its thermo-fluid-dynamic applications, sometimes, infrared thermography is used also from a more qualitative point of view. This occurs when it is important to characterize the flow field behavior (e.g., transition to turbulence, flow separation and reattachment as well as detection of instability phenomena) but this is of limited interest in the present quantitative frame of reference.

Being $h$ assumed always positive, consequently in Equation (1) the heat flux is considered to be positive if the energy goes from the solid surface to the fluid, *i.e.*, when the fluid is being heated by the wall along which it flows.

Heat transfer theoretical and experimental data is usually reported in terms of dimensionless quantities; in particular, both the *Nusselt number Nu* and the *Stanton number St* are widely used [5]. These two numbers are conventionally defined respectively as:

$$Nu = \frac{hd}{k_f} \tag{2}$$

$$St = \frac{h}{\rho_f c_p V} \tag{3}$$

where: $d$ is a length (m) which, e.g., can be the hydraulic, or equivalent, diameter for internal flows or, in general, a flow characteristic length (such as the gap of an annulus, the distance from a leading edge, the boundary layer thickness, *etc.*); $k_f$(W/m·K), $c_p$ (J/kg·K) and $\rho_f$ (kg/m$^3$) are respectively the fluid thermal conductivity coefficient, specific heat at constant pressure and mass density, all evaluated at film temperature [5]; $V$ (m/s) is a reference velocity, which can be the free stream velocity in external flows, the bulk velocity in a pipe flow and so on.

The Nusselt number is generally employed for internal flows while the Stanton number mostly in the case of external flows.

To measure either $q_c$, or $h$, a thermal sensor, commonly called *heat flux sensor,* is necessary and this justifies the matter treated in this paper for its IRT use.

As it happens when using standard transducers, also for infrared thermography applications, the heat flux sensor generally consists of a slab body with a well-known thermal behavior, whose surface temperature has to be measured by the IR camera. By properly applying to this body a suitable thermo-physical model and the energy conservation equation, it is generally possible to find a relationship between the measured temperature and the convective heat flux, and/or the heat transfer coefficient, between the sensor and the moving fluid.

As far as sensor semantics is concerned, the slab surface the flow is going over, *i.e.*, interacting with the fluid, is herein called *front surface* (its temperature being always indicated with the symbol $T_w$), while the opposite one *back surface* (which may have a temperature $T_l$ different from the front surface).

When the thermo-physical properties (e.g., thermal conductivity coefficient, specific heat, mass density) of the slab can be considered as independent of its thermodynamic state, the sensor is considered to be *ideal*. Frequently, these properties vary only slightly with temperature so that it is often possible to assume the heat flux sensor as ideal. This hypothesis, together with a constant (in time) reference temperature, is mostly adopted from now on.

## 2. BASICS OF INFRARED THERMOGRAPHY

Infrared thermography is based on radiation heat transfer which is an energy transport mechanism that occurs under the form of electromagnetic waves. By way of this heat transfer mode, energy can travel also in vacuum and may partially be absorbed and reflected by a body, or even pass through it [5]. If the intensity of radiation is put equal to unity, and by denoting with $\alpha_r$ the fraction being absorbed by the body, with $\rho_r$ the fraction being reflected by it and with $\tau_r$ the fraction being transmitted (which passes through the body), energy conservation requires:

$$\alpha_r + \rho_r + \tau_r = 1 \tag{4}$$

where: $\alpha_r$, $\rho_r$ and $\tau_r$ are respectively called *absorptivity*, *reflectivity* and *transmissivity coefficients* of the body under consideration. These dimensionless coefficients may depend on both radiation wavelength (*spectral*) and wave propagation direction (*directional*).

Radiation is emitted by all bodies at an absolute temperature $T > 0$ and, for *opaque* (non-transparent) bodies ($\tau_r = 0$), it originates only from their surface.

The body which emits the greatest amount of energy at a given temperature is called *black body.*

The law [5] setting the energy flux (energy rate per unit area) per wavelength (called *spectral hemispherical emissive power*) emitted by a black body $I_b(\lambda)$ [W/m³] is the *Planck's law of radiation*:

$$I_b(\lambda) = \frac{C_1}{\lambda^5 \left(e^{C_2/\lambda T} - 1\right)} \tag{5}$$

where $\lambda$ is the considered radiation wavelength (m), $T$ the absolute black body temperature (K) and $C_1$ and $C_2$ *the first and the second universal radiation constants,* equal, respectively, to $3.7418 \times 10^{-16}$ Wm² and $1.4388 \times 10^{-2}$ Km . Equation (5) shows that $I_b$ goes to zero for both $\lambda \to 0$ and $\lambda \to \infty$.

Generally, the electromagnetic spectrum is roughly divided into a number of wavelength *bands.* The infrared spectral band, of interest within the present framework, is generally sub-divided into four lesser bands with subjectively chosen boundaries: *near infrared* (0.75 ÷ 3 μm), *middle infrared* (3 ÷ 6 μm), *long* (or *far*) *infrared* (6 ÷ 15 μm) and *extreme infrared* (15 ÷ 1000 μm). Most of currently used IR cameras are sensitive in the middle (MWIR, 3 ÷ 5 μm) and the long (LWIR, 8 ÷ 12 μm) spectral bands.

By deriving and integrating Equation (5) with respect to $\lambda$, the following two laws respectively originate:

- *Wien displacement law*: The wavelength $\lambda^*$ at which the black body emits its maximum spectral emissive power is a function of the absolute black body temperature $T$ according to:

$$\lambda^* T = 2897.8\ \mu\text{m K} \tag{6}$$

  *i.e.*, the maximum value of $I_b$ moves toward shorter wavelengths as black body temperature increases.

- *Stefan-Boltzmann law*: The *total* (over all wavelengths) *hemispherical emissive power* $E_b$ (W/m²) also depends on the absolute black body temperature alone:

$$E_b = \sigma T^4 \tag{7}$$

where $\sigma$ is the *Stefan-Boltzmann constant*, which is equal to $5.6704 \times 10^{-8}$ W/m$^2$K$^4$.

Since IR camera detectors capture only a relatively narrow band of the whole electromagnetic spectrum, Planck's law (5), rather than Stefan-Boltzmann law (7), should be applied to the scanner.

Real objects almost never comply with the above described laws even if they may approach blackbody behavior in certain spectral bands and conditions. A real object generally emits only a fraction $I(\lambda)$ of the radiation emitted by the black body$I_b(\lambda)$, at the same temperature and wavelength.

By introducing the *spectral emissivity coefficient*, defined as:

$$\varepsilon(\lambda) = I(\lambda)/I_b(\lambda) \tag{8}$$

Equation (5) can be rewritten for real bodies by simply multiplying its second term by $\varepsilon(\lambda)$:

$$I(\lambda) = \varepsilon(\lambda)\frac{C_1}{\lambda^5\left(e^{C_2/\lambda T} - 1\right)} \tag{9}$$

*Kirchhoff law* states that the spectral emissivity coefficient $\varepsilon(\lambda)$ is equal to the *spectral absorptivity coefficient* $\alpha_r(\lambda)$, which is the absorbed fraction of the incident radiation of wavelength $\lambda$. So, for opaque bodies, such as those mainly used in infrared thermography, Equation (4) becomes:

$$\varepsilon(\lambda) + \rho_r(\lambda) = 1 \tag{10}$$

Therefore, materials with low emissivity $\varepsilon$ (such as polished metallic materials) not only emit less energy but also reflect a large amount of the radiation, coming from the ambient and impinging on them. Whenever possible, they should not be employed in IR thermography or, should they be used, they have to be sandblasted or, if transient heat transfer is not involved, covered with a thin layer of thermally black paint (such as white dull enamel). Bodies with $\varepsilon$ independent of $\lambda$ are called *grey*.

Besides, real objects almost never emit in an isotropic (independent of direction, *diffuse*) way, the emissivity coefficient $\lambda$being dependent also on the angle $\theta$ between the direction of emission and the normal to the emitting surface (*viewing angle*) [1]. Non-metallic materials, which are mostly used in IRT, emit more nigh to the normal direction.

Within the present context, measurement of convective heat fluxes can be performed with infrared thermography by means of a thermal sensor, in which appropriate surface temperatures have to be measured. As it will be shown in the following section, by correctly choosing a suitable heat flux sensor, IR thermography can be successfully exploited to resolve convective heat transfer distributions over a body surface with either steady, or transient, techniques.

# 3. HEAT FLUX SENSORS

In the following, to define the various heat flux sensors, the concept of *thermal thickness* (thin thermal thickness involves$T_w$ practically equal to $T_l$, while, for the thick one, $T_w \neq T_l$), which is more accurately explained in Section 4.2, is used.

With infrared thermography, five different types of heat flux sensors (all under the form of slabs), three steady state and two unsteady, can be essentially employed [1,6,7]. They are:

- *Heated thin foil sensor.* The slab usually consists of a thermally thin metallic sheet (foil), or a printed circuit board, steadily and uniformly (in space) heated by Joule effect [1,6]. Strictly speaking, the foil may be heated also in a different way (e.g., by a radiation heat flux impinging on the foil) but then, the heat flux distribution should be precisely known. The convective heat transfer coefficient can be computed by measuring the heat input, as well as the foil surface temperature with the IR scanner, and by performing a complete energy balance. Due to the foil thermal thinness, the temperature can be measured on either one of the slab surfaces but it is possible to apply this steady-state sensor also to not thermally thin foils (see Section 4.2).

- *Gradient sensor.* In this steady-state sensor, the slab is thermally thick and the temperature difference across the slab thickness is measured. Then, by knowing this thickness and the thermal conductivity coefficient of the slab, the heat flux across it can be computed [6]. In conventional methods, the temperature difference is usually measured by a thermopile of very-thin-ribbon thermocouples or by two thin-film resistance thermometers. To make easier the use of this sensor with infrared thermography, its back surface could be maintained at a given constant temperature (e.g., in contact with a heat sink) so only the front surface (which must be viewed by the scanner) temperature has to be measured.

- *Laplacian sensor.* An unusual steady-state method for measuring convective heat transfer coefficients has been recently described and tested by Carlomagno *et al.* [7]. The proposed technique can be applied to thermally thin sensors, such as slabs made of relatively high thermal conductivity material. Unlike the heated thin foil method demanding a given uniform heating of the slab, this steady sensor is externally (out of the measuring zone) heated, and the heat input is not even to be known. In the energy balance, the tangential conduction (parallel to slab surface) and the convective heat fluxes result to be the predominant contributions. Since currently available IRT allows measuring the two-dimensional temperature distribution with relatively high spatial resolution, this occurrence makes it possible to evaluate the tangential conduction by numerical computation of the temperature distribution Laplacian value. Spatial filtering with a Gaussian window and computation of the numerical derivatives with a relatively large step are needed to deal with the unavoidable noise presence in the acquired data.

- *Thin film sensor.* A thermally thick slab is used as a sensor and the convective heat transfer coefficient is inferred from the theory of unsteady heat conduction in a semi-infinite solid, having a thermal input at its surface. The name of the sensor classically derives from the thin resistance thermometer (typically a platinum very thin film), which is bonded to the low conductivity slab surface. Clearly, the thin film must have negligible heat capacity and thermal resistance as compared to the slab layer affected by the exchanged heat flux. When this unsteady sensor is used in combination with an IR camera, the thin resistance thermometer does not exist but the slab surface in contact with the exchanging fluid must be necessarily viewed by the IR scanner.

- *Thin skin,* or *wall calorimeter sensor.* The sensor is made of a thermally thin slab (skin) and is used as a perfect calorimeter. Being the slab thermally thin, the temperature can be assumed to be constant across its thickness and the convective heat flux is evaluated from the time rate of the slab temperature change. The use of this sensor with IR thermography is straightforward and either one of the slab surfaces can be viewed by the IR scanner. Furthermore, as for the heated thin foil, it is relatively easy to make the slab quite thin because it is not required to include, in this unsteady sensor, also a temperature transducer, such as a thermocouple or else.

To summarize, the above described five types of one-dimensional sensors involve the measurement of the quantities reported in Table 1, where those in the last two lines are functions of time $t$ and those in the third column refer to thermally thin sensors.

**Table 1.** Quantities to be measured in the various heat flux sensors.

| Sensor | Measured Quantity | |
|---|---|---|
| Heated thin foil | $T_w$ | $T_I$ |
| Gradient sensor | $T_w - T_I$ | N. A. |
| Laplacian sensor | $\nabla^2 T_w$ | $\nabla^2 T_1$ |
| Thin film | $T_w(t)$ | N. A. |
| Wall calorimeter | $T_w(t)$ | $T_I(t)$ |

A widely-used classical heat flux sensor is the Gardon gauge (also Schmidt-Boelter) which is typically a zero-dimensional sensor, so of no interest herein. Quite recently, a different type of heat flux sensor, based on a numerical solution of Fourier's law (typically described by an inverse heat transfer model) and surface temperature measurements, has been developed (e.g., Roger [8]). The advantage of using such an approach is that it is possible to take into account the temperature dependence of the material thermo-physical properties and/or to have slabs with high curvatures. However, for the sake of simplicity, this heat flux sensor will not be herein described.

With conventional transducers, it is possible to measure the wall temperature only in a relatively small number of discrete points. Thus, to perform reliable measurements, the hypothesis that, at each point, the heat flux across the slab has to be considered as one-dimensional is normally needed. This requirement obviously implies that the heat flux vector to be measured must be normal to the sensing element surface, *i.e.*, that the components of the temperature gradient, which are parallel to the slab surface (tangential), can be neglected and so the sensor is practically zero-dimensional.

The one-dimensional hypothesis can be dropped when the surface temperature is measured with an infrared scanner because of the high data number and spatial resolution of the measurement. This subject is later addressed and discussed with more details, including the needed corrections.

Very often, the heat flux sensor is flat; however in practice, the slab surfaces can also be curved, but their curvatures can be ignored as long as the thickness of the layer affected by the input heat flux is relatively small as compared with the local radius of curvature of the sensor front surface.

Even if the heat flux, or the convective heat transfer coefficient, may be considered as constant over time, both the thin film and the wall calorimeter intrinsically involve an unsteady measurement procedure of temperature. They definitely operate with a so-called *passive* heating which is due to some already existing temperature difference between the sensor front surface and the flowing fluid. Instead, the heated thin foil, the gradient and the Laplacian sensors are, generally, connected to a steady state procedure that requires some *active* heating.

As a general comment, it has to be pointed out that, while examining gas flows, with the thermally thin heated thin foil, Laplacian or thin skin sensors, one may generally measure the temperature map on either the front, or the back, surface of the sensor (in case, performing the required corrections).

Instead, if liquids are involved, the back surface must be generally viewed by the scanner because liquids are not usually transparent to infrared radiation. This means that thin film and gradient sensors cannot be used in liquid flows because front surface temperature cannot be detected by the camera.

In the following, first the steady sensors and then the two unsteady ones are discussed, all for steady convective heat flux and/or heat transfer coefficient (Sections 4–8); afterwards, the application of some of them to a few thermo-fluid-dynamic problems are presented and discussed in Section 9.

# 4. HEATED THIN FOIL SENSOR

## 4.1. Basics

With IR thermography, the simplest steady state sensor that allows us to measure convective heat transfer coefficients, is the heated thin foil sensor which mostly constitutes also the slab front surface.

Referring to the sketch of Figure 1a, where the thickness is represented much larger, in its classical and simplest accomplishment, the sensor consists of a thin metallic foil (frequently a stainless steel, or constantan, foil, typically tens of microns thick [1,9], e.g., see Figure 2 where thickness is represented to scale), steadily and uniformly (in space) heated by Joule effect. The foil is often thermally insulated at its back surface (right vertical surface in Figure 1a and inner surface ofFigure 2) while its front surface is exposed to the fluid stream.

The constant electric potential difference to the foil can be practically achieved by using a stabilized DC power supply and two couples of bus bars, usually made out of copper, which, as shown in Figure 2, are clamped at two opposite foil edges. The very large equivalent cross section (weighted with small electrical resistivity) of the bus bars, with respect to that of the heated foil, should ensure that the voltage drop along them is very small. In Figure 2, it can be seen that foil wrapping and thermal expansion are taken care of by the lateral stretching screws and the existing springs pushing two rods.

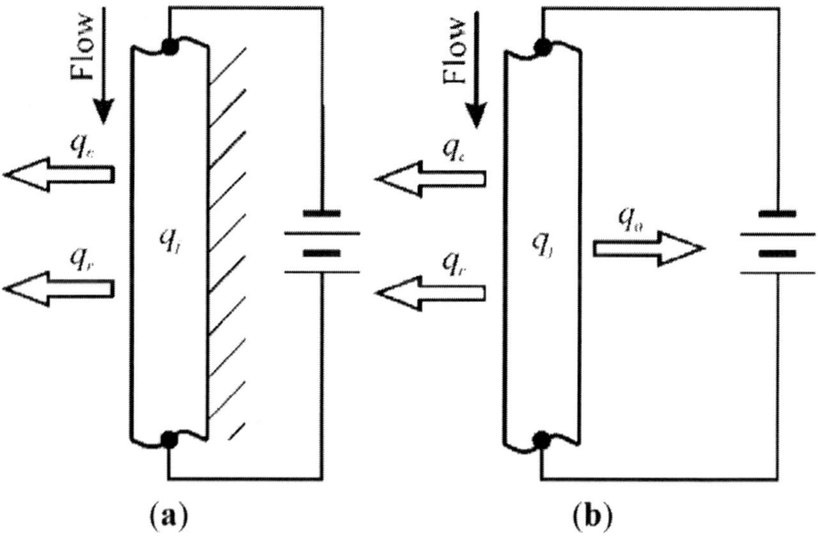


**Figure 1.** Sketch of the heated thin foil sensor: (**a**) adiabatic back surface; (**b**) diabatic back surface.

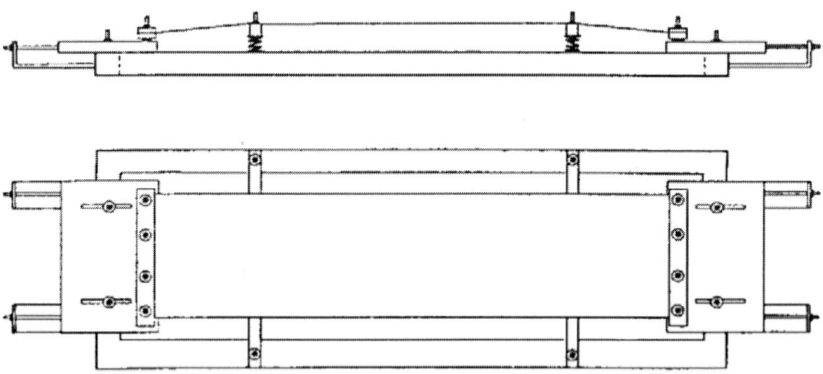

**Figure 2.** Sketch of a foil heater for natural convection studies. From Cardone and Carlomagno [9].

Since the geometries achievable with a thin metallic foil (which has to be uniformly heated) are quite limited (practically, only the rectangular ones), an easily attainable extension of this sensor is to use a printed circuit board as a heating element. Often in this case, the closely spaced copper tracks are 5 to 35 μm thick and arranged in a Greek fret mode over a fiberglass support (see Figure 3a) [10,11].

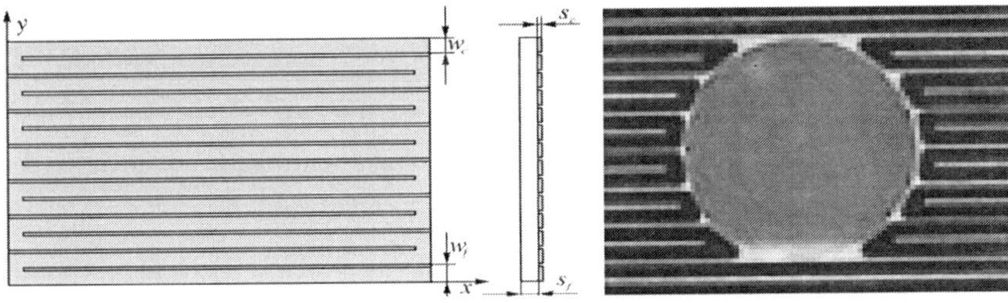

(a) (b)

**Figure 3.** (**a**) Heated thin foil sensor realized with a printed circuit board; (**b**) Close-up of a printed circuit board with a 24 mm hole to study a jet in cross flow ($w_f$ = 2 mm, $w_c$= 1.8 mm) [12].

In both implementations, it is extremely easy to obtain also cylindrical geometries (e.g., see Section 9.5).

A geometry conical and/or with the presence of holes (e.g., see Figure 3b) can be attained in the case of printed circuit boards, while more complicated shapes may be achieved by patching together different *clothes,* as long as the sensor surface has not a high double curvature.

In principle, the foil could be heated by any mean (e.g., also by radiation) and even in a non-uniform way, but in such a case, the input heat flux should be precisely known in every point of the sensor. In many instances, this measurement could be accomplished with the evaluation of the heating distribution by using the foil also as a thin skin sensor (see Section 8).

For the sake of simplicity in the following, it is always supposed that the heating is uniform in space and constant in time.

Apart from the later discussed influence of the electrical resistivity temperature coefficient, from the heat transfer point of view, the Joule heated thin foil experimentally provides an almost constant convective heat flux boundary condition.

By assuming that the back surface of the sensor is adiabatic as in Figure 1a, it is easy to perform a simple steady state one-dimensional energy balance per unit area of the sensor and per unit time:

$$q_j = q_c + q_r \tag{11}$$

where: $q_j$ is the imposed input Joule heat flux (see Equation (17)), $q_c$ is the convective heat flux to the flowing fluid and $q_r$ is the radiative heat flux to the ambient environment on the fluid side. This latter contribution unavoidably exists (because the sensor front surface must be seen) and, from a practical point of view, it has to be considered as a heat loss.

The presence of $q_r$ applies to all sensor types where temperature is measured by infrared thermography over one of the sensor surfaces.

By assuming the fluid ambient environment as a black body at a constant temperature $T_a$ and that the sensor surface is a grey one (assumptions which can be made in most of the cases for the involved wavelengths), the radiative heat flux can be computed by using Equations (7) and (8):

$$q_r = \varepsilon_t \sigma \left(T_w^4 - T_a^4\right) \tag{12}$$

where $\sigma$ is again the Stefan-Boltzmann constant and $\varepsilon_t$ is the front surface total hemispherical emissivity coefficient. Of course, the radiative heat flux has to be computed point by point of the sensor surface by means of the measured local $T_w$.

When more conventional techniques are used to measure the wall temperature, it is possible to have a very low wall emissivity coefficient in the involved wavebands (e.g., by gold plating the exchanging surface) so as to neglect the radiative heat flux to ambient environment. Obviously, this is not the case when measuring temperatures of any surface by means of infrared thermography (the surface emissivity has to be high) and, besides, to view this surface, the adjacent fluid has to be (at least, partially) transparent in the used IR detector band.

From the knowledge of $q_j$, $\varepsilon_t$, $T_a$ and $T_r$ and by measuring $T_w$, through the two previous Equations (11) and (12) and making use of the Newton's law (1), it is possible to directly evaluate the convective heat transfer coefficient [1]:

$$h = \frac{q_j - \varepsilon_t \sigma \left(T_w^4 - T_a^4\right)}{T_w - T_r} \tag{13}$$

By looking at Equation (13), it can be affirmed that, if the radiation contribution can be neglected in the energy balance, the adiabatic wall temperature $T_r$ may be measured for $q_j = 0$, *i.e.*, by switching off the Joule power feeding the foil.

As later shown in Section 4.2 for this sensor, under the assumption that the total *Biot number*, $Bi = h_t s/k_f$ (where $s$ and $k_f$ are respectively the thickness and the thermal conductivity coefficient of the sensor and $h_t$ includes both convection and radiation) is quite small as compared to unity, temperature can be considered practically constant across the foil thickness.

In this instance, it is also possible to measure the temperature of the back surface of the sensor, which, being thereupon viewed obviously becomes diabatic (see Figure 1b).

This occurrence is very convenient when using IR thermography in liquid flows since most liquids are opaque in the used IR bands or whenever the sensor front surface is not accessible to the IR scanner for any other reason.

When also the back surface is diabatic (see Figure 1b), Equation (13) has to be extended by subtracting from $q_j$ also the heat loss $q_a$ from this surface to the external ambient environment, so obtaining [1]:

$$h = \frac{q_j - q_a - \varepsilon_t \sigma \left(T_w^4 - T_a^4\right)}{T_w - T_r} \tag{14}$$

The heat flux towards the external ambient environment via the back surface is usually the sum of radiative and natural convection heat fluxes. Radiative heat flux can again be appraised by means of Equation (12) with the proper external ambient temperature, while convective heat flux to external ambient may be evaluated according to the existing situation by using standard correlations tables [13–15].

However to more carefully evaluate $q_a$, it is much better to perform some *ad hoc* tests with the same IR scanner by thermally insulating the front surface of the sensor [16]. The main advantage of performing such kind of tests is that they include the radiative contribution towards the external ambient environment as well.

It has to be explicitly pointed out that both the heat losses $q_a$ and $q_r$ are to be considered as correction terms and, in order to obtain accurate data, they should be a small fraction of the total Joule heating; otherwise, an error in their evaluation could produce a significant error in the measured $h$. This may be particularly true when performing natural convection studies with gases, where $q_r$ can be of the same order of magnitude of $q_c$.

## 4.2. Limits of the Isothermal Assumption

Often, when using the heated thin foil heat flux sensor, it is possible to detect only the back surface temperature of the sensor, *i.e.*, that opposite to the one the fluid is going over. Since in Equation (1) the front surface temperature $T_w$ must be used, it is necessary to examine what are the limits of the sensor isothermal (across its thickness) assumption.

From a practical point of view, two possible foil heating conditions are examined in the following [1]. The first one regards the case where the imposed heat flux derives from bulk Joule heating of the foil. The second one is encountered when this flux occurs because of an external radiative heating or, in any case, the heat release arises at the sensor back surface. For both conditions, it is assumed that the sensor is ideal and that the measured surface is always the back one.

In the first condition, the heat flux $q_j$ is due to the uniformly generated (except than for small variations due to the temperature coefficient of electrical resistivity) energy rate per unit volume $G$ (heat generation, W/m$^3$) inside the sensor wall because of the electric current passing through it.

In terms of limits of isothermal condition and unless there is a heat input at the back surface, the worst case is when the back surface is adiabatic while the front one (*i.e.*, that exchanging energy with the fluid) is diabatic. For such boundary conditions, the solution of the steady state Fourier equation leads to the following parabolic temperature distribution inside the sensor:

$$T - T_w = \frac{G}{k_s}\left(sx - \frac{x^2}{2}\right) \tag{15}$$

where $k$ is the thermal conductivity coefficient of the foil material, $s$ the foil thickness and the coordinate $x$ starts from the sensor front surface being directed towards the back one.

Therefore, the maximum temperature difference inside the slab is that which occurs between the front surface $T_w$ and the back one $T_l$ and is equal to:

$$T_1 - T_w - \frac{Gs^2}{2k_s} \quad (16)$$

In the meantime, the heat flux exchanged with the fluid is equal to:

$$q_j = Gs = \frac{2k_s}{s}(T_1 - T_w) \quad (17)$$

By neglecting the radiative contribution $q_r$, the above quantity must be equal to the convective heat flux at the fluid side given by Equation (1) and, by recalling the definition of the Biot number, it is obtained:

$$Bi = \frac{hs}{k_s} = 2\frac{T_1 - T_w}{T_w - T_r} \quad (18)$$

Therefore, the Biot number can be regarded as a measure of the relative importance of the temperature difference in the sensor with respect to that between the sensor itself and the fluid.

In a sense, the Biot number measures the above mentioned thermal thickness of the foil (slab). A relatively low Biot number implies an essentially constant temperature across the foil and, therefore, its low thermal thickness. This is true also in the case of no Joule heating as for the thin skin and the Laplacian sensors.

With regard to the second situation, when the Biot number can get to significant values (*i.e.*, the foil is not thermally thin) it is still possible to detect the temperature from the back surface of the sensor but, in this case, an extension of Equation (14)is needed [17].

E.g., this extension is necessary if the sensor is made of a printed circuit board, or the metallic foil is bonded to a support, and the heating element is placed at the sensor back surface. These two configurations are usual dealing with corrosive (or electrically conducting) fluids and, for this reason, the metallic heating element must be protected. It is easy to imagine that a similar situation may also develop while heating a thermally thick foil by radiation with an external lamp or else.

The related geometry is sketched in Figure 4. In this case, it is reasonable to assume that at least the Biot number of the metallic heating component can be considered very small, which is easy to achieve both on account of its thinness and because of its relatively high thermal conductivity coefficient. Of course, in radiation heating the metallic heating element does not exist.

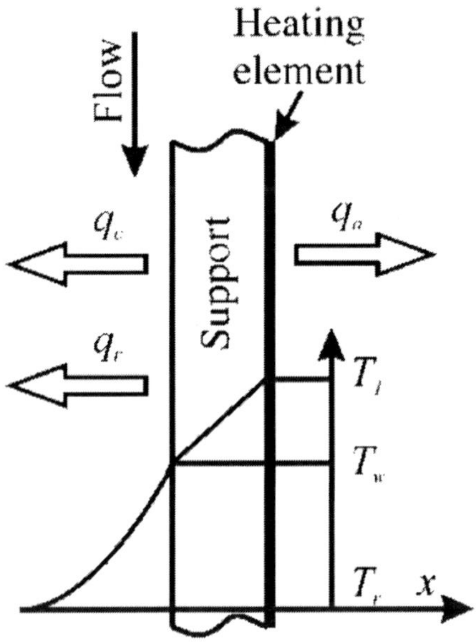


**Figure 4.** Thin foil heated from the sensor back surface for a relatively high *Bi*. Adapted from [1].

By assuming the sensor ideal, the steady state energy balances applied to both its front and back surfaces become:

$$q_c + q_r = k_s \frac{dT}{dx} = q_j - q_a \tag{19}$$

where $x$ is the spatial coordinate, normal to the sensor surface that is directed away from the fluid, $k_s$ is the support thermal conductivity coefficient, which is assumed to be precisely known, and the derivative is clearly the same on both support sides.

Under the hypothesis of one-dimensional steady heat flux, the derivative of Equation (19) can be easily calculated by noting that the temperature profile within the support thickness $s$ may be considered as a linear one:

$$h\left(T_w - T_r\right) + q_r = \frac{k_s}{s}\left(T_1 - T_w\right) = q_j - q_a \qquad (20)$$

Since the IR scanner detects the heating element temperature $T_l$ (at the back surface), the previous relation consists of a system of two equations (generally, non-linear) in the two unknown quantities $h$ and $T_w$ that can be easily solved by using standard methods.

### 4.3. Tangential Conduction within the Sensor

The assumption of a one-dimensional heat flux sensor is strictly satisfied only if the temperature within the slab constituting the sensor has negligible gradient components along the sensor surface (*i.e.*, in the tangential to surface directions). However, the infrared camera allows researchers to better explore 2D effects, especially in studying complex thermo-fluid-dynamic phenomena, where the temperature of the sensor generally varies over its measured surface. For a heated thin foil sensor made of isotropic material (such as a thin metal sheet), by retaining the assumption that the slab is thermally thin (*i.e.*, isothermal across its thickness) and ideal, it is possible to evaluate the tangential conduction heat flux (*i.e.*, referred to the unit sensor area, $W/m^2$) by means of the Fourier's law [1]:

$$q_k = -sk_s \nabla^2 T_w \qquad (21)$$

where: $\nabla^2$ is the two-dimensional Laplacian operator evaluated in the heat flux sensor plane, $s$ and $k_s$ are the thickness and the thermal conductivity coefficient of the foil, respectively.

When using an infrared scanner, a two-dimensional detailed distribution of the surface temperature is directly measured and, in principle, it should be straightforward to evaluate the conductive heat flux by numerically approximating the two-dimensional Laplacian of Equation (21) with the classical 5 points formula. However, the infrared scanner unavoidably embodies in its signal high frequency random noise that is obviously amplified by the numerical derivatives. Therefore, it is indispensable to calculate the Laplacian only after an adequate filtering of the temperature map. One procedure for this signal smoothing can be accomplished by a simple convolution with a Gaussian-filtering window.

In any case, when performing a steady state measurement, to reduce the random noise, it is always very helpful to make an average of several temperature maps acquired in a sequence.

Another kind of smoothing can be accomplished before derivation in some peculiar situations. E.g., when experimental conditions involve temperature variations in only one direction along the sensor surface (such as testing a two-dimensional airfoil in a wind tunnel, or the example of Figure 2), a span wise average can be performed, even for unsteady conditions.

Instead, when the investigated geometry exhibits an axial symmetry (e.g., a jet normally impinging on a flat surface or a disk rotating in still air [1]), it is advantageous to perform an azimuthal average of the temperature map so as to calculate the convective heat transfer coefficient only along the radial coordinate $r$ (radial profile). In particular, in this case the Laplacian of Equation (21) reduces to the much simpler formula:

$$\nabla^2 T_w = \frac{\partial^2 T}{\partial r^2} + \frac{1}{r}\frac{\partial T}{\partial r} \tag{22}$$

which, however, has a singularity at $r = 0$ that has to be carefully handled.

Once the tangential conduction heat flux is evaluated with Equation (21), it is easy to extend the approach presented in the previous chapter to the multi-dimensional case. In fact, by including the tangential conduction additional term in the energy balance (in particular, in Equation (14), where the tangential conduction does not exist), the convective heat transfer coefficient can be evaluated by means of [1]:

$$h = \frac{q_j - q_a - \varepsilon_t \sigma \left(T_w^4 - T_a^4\right) + s k_s \nabla^2 T_w}{T_w - T_r} \tag{23}$$

It will be later shown that Equation (23) allows us to generalize the heated thin foil sensor to a more comprehensive and less complex heating condition, *i.e.*, the Laplacian sensor, by putting $q_j = 0$.

As previously mentioned, in many practical realizations of the heated thin foil sensor, a spatially quite constant Joule heating can be achieved by using a copper-clad laminate where a printed circuit is carved, *i.e.*, a printed circuit board. The printed circuit board is generally made of a not electrically conductive fiberglass support (typically 0.2 ÷ 0.5 mm thick) to which extremely thin (generally from 5 to 35 μm thick) conductive pure copper tracks are bonded. However, notwithstanding the copper layer thinness, because of the extremely high thermal conductivity coefficient of the copper ($k \cong 390$ W/mK), the board exhibits an anisotropic thermal conduction behavior (along, or across, the tracks). Therefore, in order to effectively evaluate the tangential conduction term, it is necessary to generalize Equation (21) as extensively reported in [1,16].

### 4.4. Data Analysis and Experimental Procedure

For what already said in Section 4.2, when using the electrically heated thin foil and if the Biot number is not very small, to evaluate $T_w$ to be used in Equation (1) it is necessary to subtract the quantity $Gs^2/2k$ from the measured $T_1$. If it is necessary to consider also the heat losses to ambient environment $q_a$, computation gets more involved. Luckily enough, when using the electrically heated metallic thin foil, the Biot number turns out to be often very small (generally, less than 1%) so that such a correction is not usually required. Instead, for the second situation described in Section 4.2, a correction according to what suggested there has to be performed.

The other operations to correct acquired data, such as taking into account the radiation terms or the heat flux to ambient environment, are already comprehensively addressed in Section 4.1.

As far as the experimental procedures are concerned, the simple ones described in the following may improve the process of obtaining more reliable data.

As an absolute and often overlooked general comment, it must be pointed out that, as previously affirmed, being the heated thin foil, a steady technique, most of the times it is strongly recommended to first acquire with the IR scanner a relatively large number of temperature distributions (*thermograms*) and then to average them in order to decrease the existing temporal random noise. Of course, in doing so the flow conditions must be kept rigorously constant.

Furthermore, it is suggested that, in hyposonic external flows (which is an often encountered situation), it is much better to previously measure the adiabatic wall temperature $T_r$ of Equation (1), without heating the foil but with on-going normal testing flow conditions, before measuring temperatures while heating the foil. This procedure can compensate the IR scanner limited thermal accuracy (which is worse than its sensitivity), so eliminating the unavoidable bias.

The heat transfer boundary condition that the heated thin foil practically enforces is the constant heat flux one. However, since the convective heat transfer coefficient is not generally uniform over the whole foil, temperature differences may arise (see Equation (1)). They induce local changes of the electrical resistivity of the foil material that, in turn, may produce local variations of the power dissipated by Joule effect. Clearly, this effect is strictly linked to the magnitude of the temperature variations over the foil and to the temperature coefficient of the electrical resistivity but it can be normally neglected by using a proper material (e.g., constantan or manganin alloys).

However also for different types of alloys, it has to be evidenced that, when the temperature differences over the sensor are relatively small, this effect is often negligible. E.g., for a standard nichrome alloy, the temperature coefficient is of the order of half a thousandth per Kelvin and, if the maximum temperature difference over the foil turns out to be not larger than 20 K, the error in the heat flux evaluation is less than 1%.

Instead, for printed circuit boards, it has to be stressed that copper metal has a relatively high temperature coefficient of electrical resistivity (about 0.004 $K^{-1}$), about ten times that of nichrome, *i.e.*, a temperature difference of 20 °C would cause an uncertainty in the heat flux evaluation of about ±4%. Therefore while conducting experiments with printed circuit boards, large temperature differences should be avoided or, should they exist, an adequate correction must be performed on the basis of the measured temperature distribution of the circuit.

A very simple and direct way to accomplish this objective is to multiply, segment by segment of the circuit, the initially estimated Joule average heat flux by the sum of one plus the product of the copper electrical resistivity temperature coefficient times the difference between the locally measured temperature and the mean (over the whole heated board surface) one.

# 5. GRADIENT SENSOR

## 5.1. Basics

As already stated, for this steady state sensor, the temperature difference across the sensor thickness is measured and, by knowing the thickness *s* and the thermal conductivity coefficient $k_s$ of the slab, the heat flux across it can be computed [6,18].

Obviously, the assumption that the slab thermal conductivity coefficient does not depend on temperature makes linear the temperature profile across the slab thickness.

Therefore, the energy balance equation to be used for the gradient sensor is:

$$q_c + q_r = k_s \frac{dT}{dx} = \frac{k_s}{s}(T_1 - T_w) \tag{24}$$

so that, by recalling Equation (1), the convective heat transfer coefficient may be inferred from the following relationship:

$$h = \frac{k_s}{s}\left(\frac{T_1 - T_w}{T_w - T_r}\right) - \frac{q_r}{T_w - T_r} \tag{25}$$

One of the major problems arising when the IR camera is used as a temperature transducer for the gradient sensor lies in the incapability of the camera to look at two different temperature distributions (maps), the first over the front surface and the second over the back one. This difficulty can be bypassed either by scanning the wall surfaces one at a time or by viewing both of them by means of mirrors; the first technique is more advantageous because it allows a higher spatial resolution but requires a definite steady state, the second one involves the mirror calibration.

An alternative method to make use of this sensor with IRT consists of viewing with the IR camera only the front sensor surface while keeping the back one at a given constant temperature $T_l$ (e.g., in contact with a heat sink). If $q_c$ is not relatively large, this could be achieved by exposing the back surface to a boiling liquid or a condensing vapor. The high heat transfer coefficient obtained in these conditions allows $T_l$ to coincide with the phase-change temperature of the fluid. Then in order to compensate the IR camera limited thermal accuracy (which is worse than its sensitivity), therefore eliminating the unavoidable bias, a small part of the slab could be removed to directly observe the heat sink, which emissivity however should be exactly equal to that of the sensor front surface.

### 5.2. Data Analysis, Tangential Conduction and Experimental Procedure

The gradient sensor method can also be used to a certain extent in the heat transfer unsteady regime. In particular, if the functions $T_w(t)$ and $T_l(t)$, which describe the time-temperature histories, do not contain harmonic components with frequencies greater than 0.1 $\alpha/s^2$ ($\alpha$ being the slab material thermal diffusivity coefficient), the heat flux can be inferred from the formula [19]:

$$q_c + q_r = k_s \frac{T_1\left(t + s^2/3\alpha\right) - T_w\left(t - s^2/6\alpha\right)}{s} \tag{26}$$

which, in particular, for $s^2/\alpha \to 0$, reduces to Equation (24).

Even at steady state, the tangential conduction of this sensor cannot be longer appraised with Equation (21) that assumes a constant temperature across the slab, while in the gradient sensor, for multi-dimensional operation, a quasi-linear temperature profile exists. For a constant $T_l$ and a harmonic variation of the front surface temperature, the tangential conduction depends on the ratio R between the wavelength of the variation and the sensor thickness $s$ according to the approximate formula:

$$q_k = -s k_s \nabla^2 T_w / \chi \tag{27}$$

where the dimensionless parameter $\chi$ is a function of R.

So, since it occurs that $\chi$ is always greater than 1, only a fraction of the Laplacian at the front side $\nabla^2 T_w$ contributes to the tangential conduction along the slab.

Values of the coefficient $\chi$, computed to be used in Equation (27), are reported in Figure 5, as a function of the dimensionless wavelength R, which shows that, at high wavelengths, only one third of the Laplacian results to be active for tangential conduction. Obviously, for more complex conditions the inverse heat transfer problem should be solved.

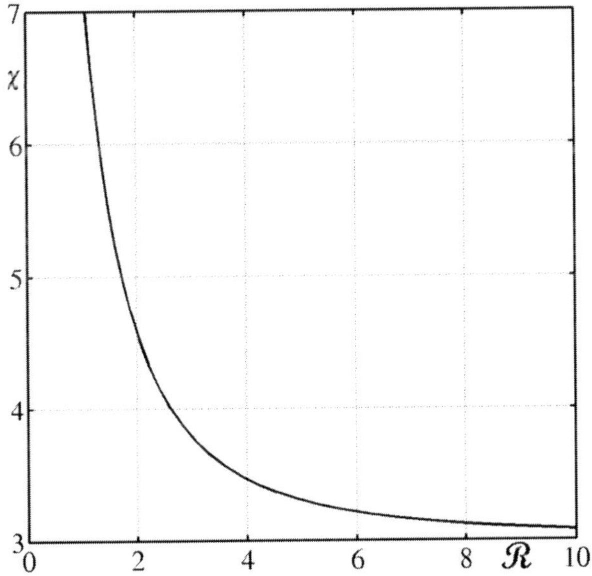


**Figure 5.** Value of χ for Equation (27) as a function of the dimensionless wavelength R.

As far as the experimental procedures are concerned, if they apply also to the gradient sensor, the same considerations already made for the heated thin foil may be employed; e.g., when performing a steady state temperature acquisition, it is always very helpful to make an average of several temperature maps picked up in a sequence so as to reduce the random temporal noise.

# 6. LAPLACIAN SENSOR

## 6.1. Basics

Noticeably enough, the Laplacian sensor is governed by the same energy balance of the heated thin foil sensor but with a different relevance of the involved terms and the absence of Joule heating.

As already mentioned in Section 4.3, the capability of an IR scanner, to evaluate the two-dimensional distribution of the conductive heat flux, allows us to generalize the heated thin foil sensor to a more comprehensive and less complex heating condition. In particular, it is possible to heat the slab not in the measurement zone but near it and this external heating can be even non-uniform and unknown. In this way, another peculiar heat flux sensor is practically generated: the Laplacian sensor (see Carlomagno *et al.* [7]).

This (still steady state) sensor is based on Equation (23) where the term $q_j$ does not appear in the equation itself since the slab is externally heated and no Joule heating term exists in the energy balance equation within the measurement zone. In fact, the flow over the sensor surface induces the convective heat transfer that, besides the heat losses, mainly derives from the derivative of the tangential conduction heat flux within the slab, which can be appreciated by computing the two-dimensional Laplacian of the temperature distribution.

A peculiar aspect of such a sensor is therefore associated to the fact that, while for the heated thin foil sensor, $q_j$ is the main term in the energy balance equation, for this sensor, the Joule heating in the measured zone is not needed so that $q_j$ is identically zero; hence, the most important term in Equation (23) becomes indeed the derivative of the tangential conduction heat flux $s$ $k_s \nabla^2 T_w$. So it is obtained:

$$h = \frac{s k_s \nabla^2 T_w - q_a - \varepsilon_t \sigma \left( T_w^4 - T_a^4 \right)}{T_w - T_r} \tag{28}$$

Consequently, the critical feature of this sensor is the accurate evaluation of the temperature two-dimensional Laplacian because its erroneous evaluation has a direct consequence on the computation of the convective heat transfer coefficient $h$. Being dependent on the Laplacian evaluation, this heat flux sensor is also herein called *Laplacian sensor* as proposed in [7].

### 6.2. Data Analysis and Experimental Procedure

The main hindrance of the Laplacian sensor is therefore related to the importance of a correct estimation of the derivative of the tangential conduction heat flux which, with the convective one $q_c$, are the most relevant contributions in Equation (28). As a matter of fact, being the tangential conduction usually a correction term, the heated thin foil sensor is less affected by this issue, the most important contributions arising from the convective term $q_c$ and from the imposed Joule heating $q_j$. Therefore, because of the aforementioned reasons, for the Laplacian sensor the numerical computation of the Laplacian ofEquation (28) involves some critical aspects to be considered.

An adaptive filtering technique, consisting of consecutive applications of the Wiener filter to the temperature distributions, is proposed by Rainieri *et al.* [20]. However, in the work of Carlomagno *et al.* [7], where a quite complex flow field is analyzed, a different approach is adopted, involving spatial filtering with a

Gaussian window and calculating the numerical derivatives with a relatively large spatial distance.

This method implies a certain reduction of the spatial resolution of the measurement so that an operating compromise has to be necessarily found. Practically, the authors propose to correlate the dimensions of the spatial filtering window and of the numerical derivatives computational step to the physical parameters of the investigated phenomenology.

Besides, the use of an IR camera with Focal Plane Array introduces another error critical source, *i.e.*, the spatial measurement noise generated by the different t response of the several pixels of the IR detector. Therefore, an accurate and reliable process of Non-Uniformity Correction (NUC) is of fundamental importance. As a matter of fact, both spatial filtering and computation of the derivatives, with a relative large step are not often sufficient to compensate the decay of the signal-to-noise ratio due to the non-uniformities between the various pixels response.

Of course, the issue of tangential conduction correction is not herein treated because it constitutes the basic principle on which the Laplacian sensor operates.

In the application Section 9.3, the use of this sensor and its sensitivity to the heating conditions are analyzed in much more detail.

# 7. THIN FILM SENSOR

## 7.1. Basics

Since it measures a map of surface temperatures, the IR scanner output can be thought as originating from a two-dimensional array of an extremely large number of very small thin resistance thermometers with the advantage that the thickness of the equivalent thin films is exactly zero.

A very schematic sketch of the thin film sensor is depicted in Figure 6, where the temperature to be measured as a function of time $t$ is $T(x = 0,t) = T_w(t)$, so that of the sensor front surface. With this sensor, since necessarily the viewed surface is that along which the fluid is flowing over, it is compulsory that the heat exchanging fluid has to be (at least partially) transparent in the IR band detected by the scanner.

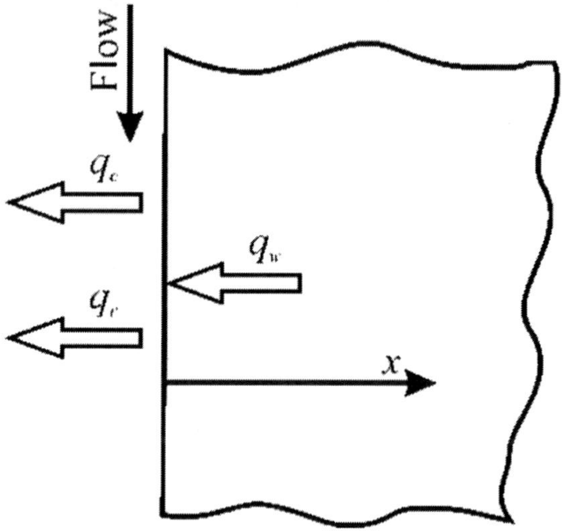


**Figure 6.** Sketch of the thin film sensor. Adapted from [1].

It is obvious that the wall conductive heat flux within the solid $q_w$ at the fluid/solid interface, must balance the sum of the radiative $q_r$ and convective $q_c$ heat fluxes towards the fluid:

$$q_w = q_c + q_r \tag{29}$$

As for the thin skin, the major part of the works that use the thin film, to measure convective heat transfer coefficients, are associated with hypersonic flows (high values of $T_r$), so that, in Equation (1)$q_w$ becomes a negative quantity. Therefore, in the following for both this two sensors, the role of $T_r$ and $T_w$ in Equation (1) are reversed, resulting in an increase of the slab surface temperature with time.

From the classical heat conduction books (e.g., Carslaw and Jaeger [21]), it is possible to recover the one-dimensional solution for a semi-infinite wall having a constant initial temperature $T_{wi}$ $(x, t = 0)$ and, at following times $(t > 0)$, subjected on its surface to a uniform (in space) convective heat flux governed by Equation (1) with constant $h$ and $T_r$ (which is the most commonly encountered case):

$$\theta = \frac{T - T_{wi}}{T_r - T_{wi}} = erfc(\xi) - \exp\left(Bi_x + \beta^2\right) erfc(\xi + \beta) \tag{30}$$

$T$ being the generic slab temperature at a certain depth $x$; the dimensionless quantities $\xi$, $\beta$ and $Bi_x$ are defined as:

$$\xi = x/2\sqrt{\alpha t}\,,\ \beta = h\sqrt{t/\rho_s c k_s}\ \text{and}\ Bi_x = hx/k_s.$$

The latter quantity represents the local Biot number, *i.e.*, the Biot number based on $x$.

In the previous notations: $\alpha$, $\rho_s$, $k_s$ and $c$ indicate respectively the thermal diffusivity coefficient, the mass density, the thermal conductivity coefficient and the specific heat of the sensor material; $x$ and $t$ are respectively the spatial coordinate (starting at the interface and directed as in Figure 6) and the time variable. Often, the product $\rho_s c k_s$ is called *thermal inertia*(or *thermal effusivity*); others use these names for the square root of the same product.

The dimensionless parameter $\xi$ is proportional to the square root of the reciprocal of the local *Fourier number* ($Fo_x = \alpha t/x^2$), also based on $x$. The Fourier number can be thought as measuring the relative importance of the conductive heat flux with respect to the rate of thermal energy storage but it may be certainly regarded also as a dimensionless time.

Alternatively, the dimensionless parameter $\beta$ can be also expressed in terms of the local Biot and Fourier numbers, or $Bi_x$and $\xi$, since:

$$\beta = Bi_x\sqrt{Fo_x} = \frac{Bi_x}{2\xi} \qquad (31)$$

The dimensionless temperature profiles into a semi-infinite wall at a constant initial temperature are plotted in Figure 7, as a function of $\xi$, for several different values of $\beta$.

At a given time $t$, the increase of the front surface temperature ($\xi = 0$) with $\beta$ can be associated to either an increase of $h$ or a decrease of the sensor thermal inertia $\rho_s c k_s$. This makes the sensor, at initial times, more sensitive if it is made of low thermal conductivity $k_s$, and/or low thermal capacitance per unit volume $\rho_s c$, material. On the other hand, for constant $h$ and$\rho_s c k_s$, the increase of $\theta$ is connected to a time increase and, thus, to the total heat input.

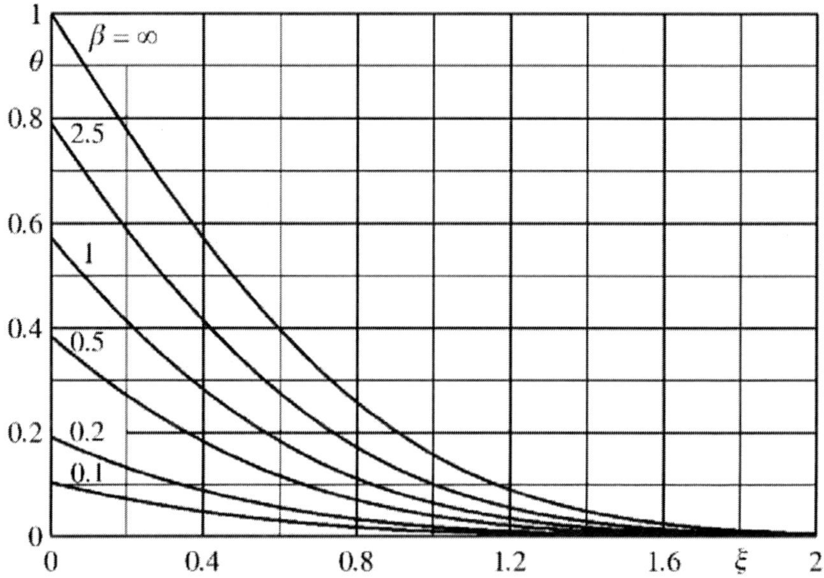


**Figure 7.** Dimensionless temperature profiles, for several $\beta$ values, in a semi-infinite wall at a constant initial temperature and subjected to a constant $h$ and $T_r$. Adapted from [1].

The top curve of Figure 7 corresponds to $\beta \to \infty$ (e.g., $h \to \infty$) and, in this case, the boundary condition at the wall surface for $t > 0$ reduces to a constant fluid/solid interface temperature $T_w = T_r$, so that Equation (30) simplifies to:

$$\theta = erfc\left(\xi\right) \tag{32}$$

However, the case of $\beta \to \infty$ is of no interest to measure convective heat transfer coefficients since it gives a constant temperature $T_w$ with time. This condition may be rather exploited to measure the adiabatic wall temperature $T_r$ of Equation (1), as long as the radiative contribution $q_r$ of Equation (29) can be assumed negligible.

### 7.2. Effects of the Finite Thickness of the Sensor

In practice, the thin film sensor cannot be semi-infinite but it is always made of a slab of finite thickness $s$ (so also a sensor back surface does exist) and, as it can be easily understood, the equivalence to the semi-infinite wall model is valid only during a relatively small measurement time interval $t_m$, before the thermal wave reaches the back surface. This being the case, the boundary condition on the existing back surface (e.g., $q = 0$ or $T_l = \text{const}$) becomes practically irrelevant.

As a matter of fact, for very large values of $\xi$ (e.g., large values of $x$ and/or small time values), the temperature of the slab coincides with the initial one (*i.e.*, $\theta = 0$), while, for $\xi$ decreasing, $\theta$ increases.

By first supposing $\beta \to \infty$, Equation (32) can be used to find a very conservative time limit for the correct application of the semi-infinite wall model to a finite thickness sensor.

Really, by fixing a maximum value of $\theta$ (for instance, 1%) which can be accepted at the sensor back surface without substantially altering the measurement, it is possible to find (for a semi-infinite sensor model) the corresponding value of $\xi$(indicated with $\bar{\xi}$) and, therefore, the maximum measurement time that can be practically used with a finite slab [1]:

$$t_m = \frac{s^2}{4\bar{\xi}^2 \alpha} = \frac{s^2}{\alpha p} \tag{33}$$

where $p=4\,\bar{\xi}^2$ represents a constant to be computed, which is later evaluated also in less conservative limits. On a quantitative basis for $\beta \to \infty$, by putting in Equation (32) $\bar{\xi} = 1.82$, since $erfc(1.82) \approx 0.01$, it is readily found $p \approx 13.3$.

This time limit is clearly independent of the Biot number and is a very conservative one because, besides the assumption $\beta \to \infty$ (which, as stated before, is useless to measure convective heat transfer coefficients), it considers the departure from the semi-infinite model at the sensor back surface. As a matter of fact, in the thin film sensor, the monitored temperature, which is useful to compute the heat flux, is always the front surface one.

The found time limit is in marked contrast with what affirmed by Carlomagno and de Luca [6] who suggest a $p$ value equal to 2, even without specifying a back surface temperature increase as low as 1%. However, it has to be stressed that this point is often presented in the literature in a controversial way (some researchers fix $p = 16$, e.g., see Gülhan *et al.* [22]) so, in the following, the problem is examined on the condition of a $\theta$ accuracy at the front surface of 1%, which can represent a rather satisfactory value.

Being the slab back surface ($x = s$) adiabatic (which is an often encountered experimental condition) and the front one ($x = 0$) almost always subjected to a convective heat flux with constant $h$ and $T_r$, it is possible to retrieve the following exact solution [21] for the slab temperature distribution:

$$\theta = 1 - \sum_{n=1}^{\infty} \frac{2Bi}{\cos(\gamma_n)\left(Bi\left(Bi+1\right)+\gamma_n^2\right)} \cos\left(\gamma_n \frac{1-x}{s}\right) e^{-\gamma_n^2 \frac{\alpha t}{s^2}} \quad (34)$$

where now $Bi$ (based on the sensor thickness) is defined as in Section 4.2 by the formula (18) and $\gamma_n$ are the positive roots of the equation $\gamma \tan \gamma = Bi$. Values of the first six roots of this equation can be found in [21].

Therefore, with regard to Equation (33), a less conservative measurement time limit can be found by introducing the ratio:

$$\Xi = \frac{\theta_s - \theta_\infty}{\theta_\infty(0)} \quad (35)$$

which represents the percentage departure between the two solutions (for the finite slab $\theta_s$—Equation (34)—and for the semi-infinite one $\theta_\infty$—Equation (30)), where the subscript $s$ and $\infty$ refer to the finite slab and to the semi-infinite wall, respectively. The quantity $\Xi$ can be either referred to the front sensor surface or to the back one but, as already said, only the one referring to the front surface is of interest for the thin film heat flux sensor.

The time at which $\Xi$ is equal to a prefixed threshold (namely 0.01) can be used again as an estimate for the evaluation of the maximum measurement time. It turns out [1] that $p$ is practically independent on $Bi$ (the maximum value for low $Bi$being $p \approx 3.1$). Therefore, a dependable value for $p$ in Equation (29) can be definitely established at $p \approx 3$, in most of the encountered experimental conditions.

### 7.3. Data Analysis and Experimental Procedure

In the previous sub-section it has been just assessed that, if the measurement time is sufficiently small ($t < s^2/3\alpha$), a slab, which is used as a thin film heat flux sensor, is well approximated by the semi-infinite wall model. Then, by supposing that the slab at the initial time ($t = 0$) is isothermal at a temperature $T_{wi}$, from the front surface temperature evolution $T_w(t)$, measured with the infrared camera, it is possible to use the classical formula of Cook and Felderman [23] to evaluate the wall heat flux $q_w$, as a function of time:

$$q_w = \sqrt{\frac{\rho_s c k_s}{\pi}} \left( \frac{\varphi(t)}{\sqrt{t}} + \int_0^t \frac{\varphi(t) - \varphi(\tau)}{(t-\tau)^{\frac{3}{2}}} d\tau \right) \tag{36}$$

where $\varphi(t) = T_w(t) - T_{wi}$ and, obviously, $q_w$ is the total conductive wall heat flux ($q_w = q_c + q_r$) at the fluid/solid interface.

The previous equation is valid for a wall heat flux generally varying with time but a much simpler formula can be found when a constant wall heat flux is imposed:

$$\varphi(t) = \frac{q_w}{2} \sqrt{\frac{t}{\pi \rho_s c k_s}} \tag{37}$$

where the front temperature rise is proportional to the square root of time.

Notwithstanding its simplicity, Equation (37) is of seldom met relevance in convective heat transfer measurements because, with the thin film sensor, the most commonly encountered boundary condition is of constant convective heat transfer coefficient and reference temperature in Equation (1), and not of constant wall heat flux. Anyway, a best fit, of the type later given by Equation (39), as suggested later, could be simply applied also in this case.

The solution given by Equation (37) could be considered acceptable only on condition that the quantity $\varphi$ remains very small as compared to $T_r - T_{wi}$, during all the experimental test, so that the exchanged heat flux may be consequently considered almost constant.

Instead, by numerically evaluating the integral of Equation (36), it is possible to determine the wall conductive heat flux but, when using standard methods, the singularity that is present for $\tau = t$, may reduce the results accuracy. Cook and Felderman [23] assumed that the temperature could be approximated by a piecewise linear function and found the following stable method:

$$q_w = \sqrt{\frac{\rho_s c k_s}{\pi}} \, 2 \sum_{i=1}^{n} \frac{\varphi(t_i) - \varphi(t_{i-1})}{\sqrt{t_n - t_i} + \sqrt{t_n - t_{i-1}}} \tag{38}$$

with $t_i = i\Delta t$, where $\Delta t$ is the time interval between the acquired data and $n$ represents the total number of measured data.

Since the wall temperature history is measured with the infrared scanner, the calculation of the radiative heat flux can be made with the help of Equation (12), then, by using Equations (1) and (29), it is straightforward to evaluate also the convective heat transfer coefficient.

However, the evaluation of the conductive wall heat flux with Equation (38) is particularly suited only when a large number of measurement points in time are available and both the initial time and model temperature are well known. This is not always the case when performing tests in relatively fast transient conditions (e.g., in short duration hypersonic blow down tunnels or, in the worst instance, in shock tubes or tunnels).

In fact, in many actual unsteady measurements, it is not possible to either insert immediately the model in the main stream or to accelerate instantaneously the fluid to its final velocity value, so the initial time may be itself partially unknown.

In these circumstances, the data acquisition frequency, even of a modern IR scanner which is often of the order of 100 *Hz*, may be not large enough to accurately evaluate the wall heat flux with Equation (38).

Then, for a constant $h$ and reference temperature $T_r$, a different approach, based on a non-linear least square fit, can be more advantageous as suggested by de Luca *et al.* [24].

By writing Equation (30) at the fluid/solid interface, it is found:

$$\theta_w = \frac{T_w - T_{wi}}{T_r - T_{wi}} = 1 - e^{\beta^2} erfc(\beta) \tag{39}$$

From this equation, it may be assumed that, apart from some known slab thermo-physical properties and time intervals, the convective heat transfer coefficient as well as the initial time and sensor temperature are unknown. The best fit of the measured wall temperature $T_{wm}$ to the exact solution described by Equation (39)$T_w$ can be found by varying $h$ (in $\beta$) and $T_{wi}$in order to minimize the following functional:

$$\sum_{i=1}^{n} \left(T_w - T_{wm}\right)^2 \tag{40}$$

A second problem is the radiative heat flux that, when using IR thermography, is not generally possible to neglect. Under the assumption that the convective and radiative contributions are uncoupled, de Luca *et al.* [24] propose to modifyEquation (39) with the following one:

$$T_w - T_{wi} = (T_r - T_{wi})\left[1 - e^{\beta^2} erfc(\beta)\right] - \frac{q_r}{h} \tag{41}$$

Then, by using this equation, it is possible to take easily into account also the radiative heat flux in the minimization of the Functional (36).

The previous approach is, in every respect, an inverse heat transfer problem and can be extended, by a numerical solution of the Fourier heat equation. This may be necessary in order to include the temperature dependence of the slab thermo-physical properties and/or to have, in case, slabs with high curvature (e.g., Mulcahy *et al.* [25]).

### 7.4. Tangential Conduction within the Sensor

In the previous sub-sections, also for thin film sensor, the temperature distribution within the slab is supposed to be just one-dimensional (along coordinate $x$). However, as said before, this is not the most usual case, since variations often occur over the sensor surface due to complex flow fields.

The analysis that follows is developed under the assumption that the sensor material is isotropic or, by choosing a Cartesian coordinate system with its axes directed as the two principal axes of the thermal conductivity tensor, it is always possible to split the conduction effects along the two tangential directions.

Therefore, since the extension to any arbitrary convective heat flux distribution is straightforward, for the sake of simplicity, in the following it is assumed that the convective heat flux harmonically varies only along a direction $y$ parallel to the front surface of the sensor, that is:

$$q_c = A_q \cos(\varpi y) \tag{42}$$

where $A_q$ is the heat flux amplitude and $\varpi_x$ is the wave number.

In Equation (42) it is not included any possible unsteady but uniform in space part of the convective heat flux, which is already looked upon in detail in the previous sub-sections. In fact, by considering that the involved phenomenology is linear (for an ideal sensor), the two effects can be treated separately and successively summed up.

The solution with the boundary condition (42) at the front surface and initial sensor temperature spatially constant, is given by de Felice *et al.* [26] in terms of the difference $\varphi$ between the front surface temperature $T_w$ at time $t$ and the initial ($t = 0$) sensor temperature $T_{wi}$ and can be put in the form:

$$\varphi(y,t) = T_w - T_{wi} = A\, f(Fo_{\varpi})\cos(\varpi y) \tag{43}$$

where $Fo_{\varpi} = \varpi_x^2\, \alpha\, t$ is a modified Fourier number, $A$ is a constant *reference temperature amplitude* and $f$ is a function of $Fo_{\varpi}$.$A$ and $f$ are equal to:

$$A = \frac{A_q}{k_s \varpi} \quad ; \quad f = erf\sqrt{Fo_{\varpi}} \tag{44}$$

So, there is no phase difference (in space) between the harmonic heat flux and the front surface temperature response and $f$ turns out to be an increasing function of $Fo_{\varpi}$ that varies between 0 and 1. Therefore, in Equation (43), $A$ indicates the maximum amplitude (attained for $Fo_{\varpi} \to \infty$) of the cosine wave while, for smaller values of $Fo_{\varpi}$, the amplitudes are reduced by the *attenuation factor f*.

It has to be noticed that, despite the fact that the attenuation factor is an increasing functions of the Fourier number, the effective temperature amplitudes $Af$ increase for decreasing spatial frequencies and the opposite is true for $\varpi_x \to 0$. In particular, in this latter case, the effective amplitude limit is:

$$\lim_{\varpi \to 0} \frac{Af}{A_q} = \frac{2}{\sqrt{\pi}} \frac{\sqrt{\alpha t}}{k_s} = \frac{2}{\sqrt{\pi}} \sqrt{\frac{t}{\rho_s c k_s}} \tag{45}$$

As expected, the limits is unbounded for increasing $t$ since the problem is reduced to a constant (in time and space) boundary condition.

To correct the measured temperatures so as to take into account tangential conduction effects, it is convenient to evaluate the ratio between the effective temperature amplitude $A\ f(Fo_{\varpi})$ (as computed from Equation (44)) and that corresponding to the same value of $A_q$ but in absence of tangential conduction (*i.e.*, that given by the one-dimensional solution or, analogously, by the limit of Equation (45).

By defining this ratio as the *temperature modulation transfer function F*, it results:

$$F = \frac{\sqrt{\pi}}{2} \frac{erf\left(\sqrt{Fo_{\varpi}}\right)}{\sqrt{Fo_{\varpi}}} \tag{46}$$

Once the function $F$ represented in Figure 8 is known, the amplitude of each harmonic component of the measured temperature may be corrected and the restored temperature maps can be used to compute the effective heat flux by using the formulae already presented in Section 7.3.

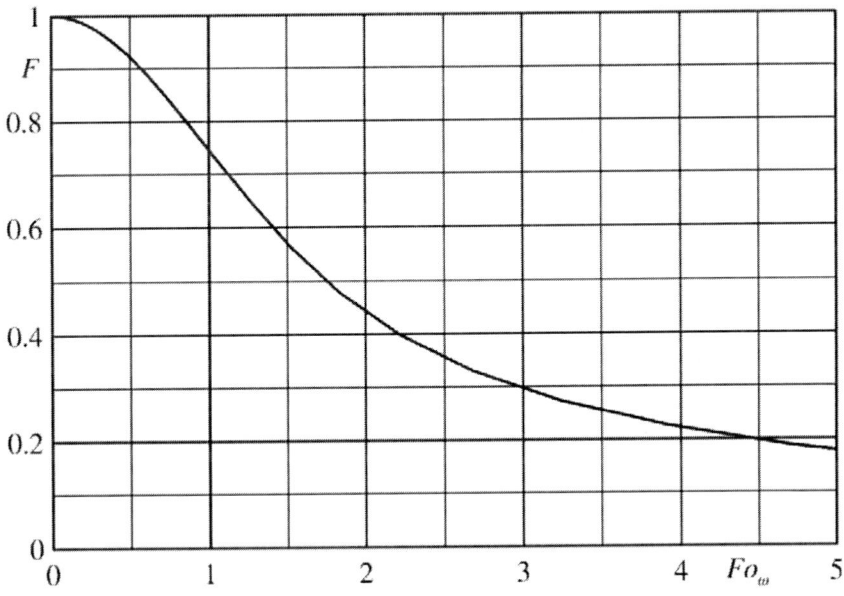


**Figure 8.** Temperature modulation transfer function $F$ for the thin film sensor. Adapted from [1].

As a matter of fact, it turns out that, for the tangential conduction in a thin film sensor, there is a kind of modulation function in time ($Fo_{\varpi}$) and not one in space as it happens for the modulation transfer function that occurs for the scanner spatial resolution [1].

An example of image restoration for the thin film sensor is reported in Section 9.2.

# 8. THIN SKIN SENSOR

## 8.1. Basics

In the case of the thin skin (or wall calorimeter), the sensor, practically a thin slab (see the sketch of Figure 9) of thicknesss, is usually modelled as a perfect calorimeter (isothermal across its thickness, so thermally thin) which is heated at the front surface and thermally insulated at the back one.

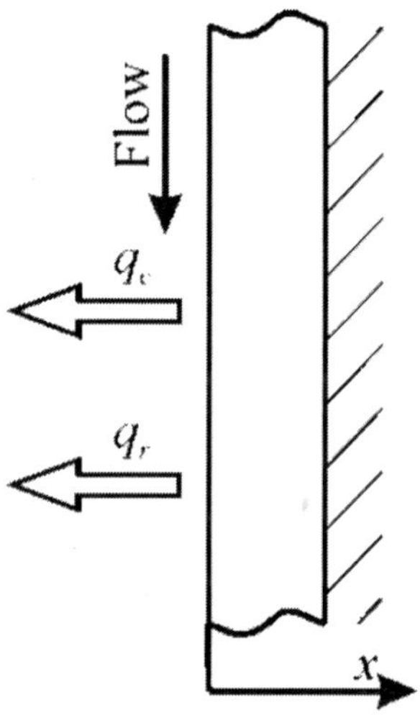


**Figure 9.** Sketch of the thin skin sensor.

As it will be shown later, the isothermal condition involves that the slab has to be not only thin but also with a high thermal conductivity coefficient as it happens, e.g., for metals.

Even if the isothermal condition is not fulfilled, the unsteady one-dimensional energy balance applied to the slab is:

$$q_w = -\rho_s cs \frac{dT_m}{dt} \tag{47}$$

where $T_m$ is the mean temperature across the slab thickness and $q_w$ is the external heating which must always fulfil the condition $q_w = q_c + q_r$. In the following subsections, *Bi* and *Fo* are always based on the slab thickness *s*.

Having assumed the slab isothermal (which as it will be seen is true as long as $Bi \ll 1$ and for relatively large *Fo* values), the wall heat flux $q_w$ and consequently the convective heat flux can be quite easily computed by numerically evaluating the time derivative of the temperature measured on either side of the sensor.

If the back surface of the sensor is not adiabatic for measurement needs, Equation (47) can be extended, as already discussed for the heated thin foil sensor, by including the total heat flux $q_a$ to the ambient environment.

When the imposed wall heat flux is constant with time and the sensor is ideal, Equation (47) can be easily integrated and the result is practically a linear increase of the slab mean temperature with time. In particular, in this case, as in the following one, it is assumed that the temperature rise is the same across the slab thickness.

As shown later and apart from the influence of the radiative heat flux, an essentially exponential rise of the temperature with time can be easily found when a convective heat flux, with constant *h* and reference temperature $T_r$, is enforced at the wall (see Equation (51) in the following).

However, in both the above described cases, the simple steady state solution is preceded by a transient (for low *Fo*values), where the temperature at every slab point rises from its initial value to an asymptotic dependence. As it will be seen in the next section, during this transient, the rate of variation of the wall temperature (on both sides of the sensor but especially at the front one) can be significantly different from the mean one.

Furthermore, it has to be stressed that it is not generally convenient to directly evaluate either heat fluxes or convective heat transfer coefficients with derivatives of acquired temperatures because the measured experimental signal is unavoidably affected by noise.

Some solutions to overcome this problem are to either previously filtering the signal or adopting an integral approach or else minimize a functional of the type given by Equation (40).

## 8.2. Isothermal Assumption

There are many experimental circumstances where the testing time is relatively short, such as short running time blow down wind tunnels or shock tunnels, and the limits of the thin skin sensor isothermal conditions have to be correctly assessed.

The isothermal condition of the thin skin can be expressed by introducing the dimensionless parameter:

$$\Theta = \frac{\theta_w - \theta_1}{1 - \theta_w} \tag{48}$$

Where $\theta$ has the same definition given in Section 7.1 and the subscripts $w$ and 1 again respectively indicate the front and back surfaces of the sensor. As it will be seen later, the condition $|\Theta| \ll 1$ is realized, also at the asymptotic state (*i.e.*, relatively large *Fo* values), as long as the Biot number is sufficiently small.

By supposing again the slab adiabatic at its back surface and subjected to a convective heat flux, with constant $h$ and $T_r$, and by using Equation (34), in Figure 10 the parameter $\Theta$ is plotted as a function of *Fo* for different values of the Biot number. Initially, the temperature is constant inside the slab so the parameter $\Theta$ is small regardless of the Biot number. Afterwards, $\Theta$ increases and reaches an asymptotic value, which is practically attained for $Fo > 0.5$ and appears to be an almost linear function of *Bi*, as it should be expected.

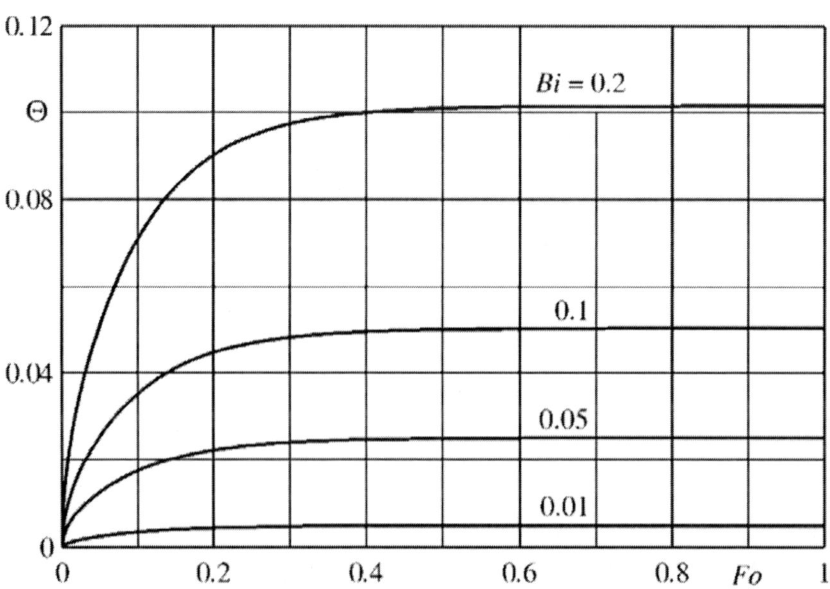


**Figure 10.** $\Theta$ as a function of *Fo* for several *Bi* values. Adapted from [1].

The isothermal condition, previously enforced, does not imply that, as shown in Figure 11, the temperature across the slab, even for a quite small Biot number (e.g., $Bi = 0.05$), is practically constant. Furthermore, for very small $Fo$ values, the thermal wave is not able to yet reach the opposite wall (sensor back surface).

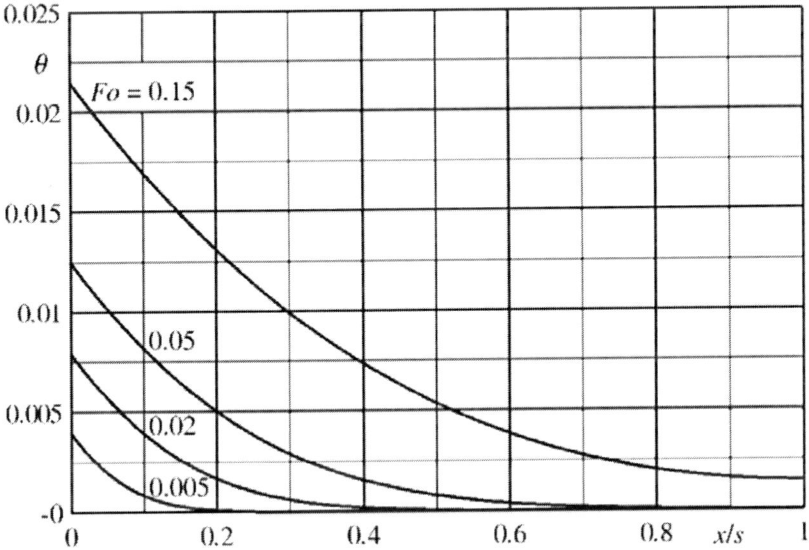


**Figure 11.** Temperature profiles in the slab ($Bi = 0.05$) at several $Fo$ values for adiabatic back surface.

Clearly, if the IR scanner measures the back temperature $T_l$, the time derivative of the measured temperature is, initially, equal to zero, thus it is necessary to wait some time before acquiring the measurement.

On the contrary, the sensor front surface temperature initially increases more rapidly than the average one and, for small Fourier numbers, also this temperature evolution does not enable to correctly evaluate the time derivative of the mean slab temperature (which is the one that monitors the heat flux, see Equation (47)).

The error made in the evaluation of the mean temperature time derivative can be expressed in a quantitative way with the dimensionless ratio:

$$\Pi = \left( \frac{dT}{dt} - \frac{dT_m}{dt} \right) \Big/ \frac{dT_m}{dt} \tag{49}$$

which is a function of the position in the slab and of the Fourier and Biot numbers.

In Figure 12 for both the front and back surfaces, the ratio Π is plotted as a function of the Fourier number for two *Bi* values. For very small times (*i.e.*, *Fo*), at the back surface the condition dT/dx=0 occurs and this event explains the initial 100% constant value of the negative error. On the contrary, at the front surface the temperature variation is much larger than the mean one and this is the cause for the extremely large positive error at very low *Fo* values.

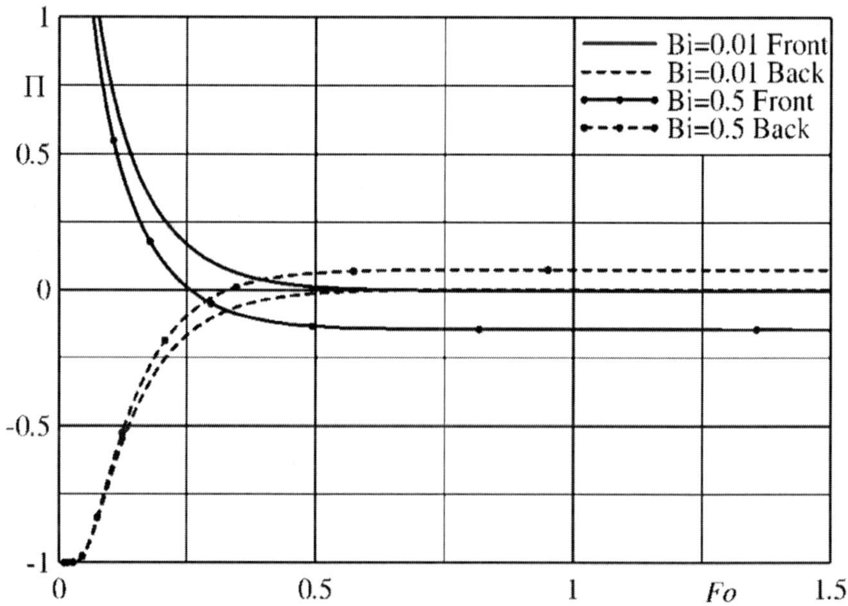


**Figure 12.** Error in the evaluation of the derivative of the mean slab temperature. Adapted from [1].

From the graph, it is evident that both absolute errors decrease for increasing *Fo*; they become smaller than 1%, for $Bi = 0.01$ and Fourier numbers larger than 0.5. The error curves remain practically unchanged as long as $Bi < 0.05$ while, as it can be noticed from the figure, for the larger Biot number the asymptotic error increases significantly but the time needed to reach its final value remains practically the same.

It is interesting to notice that, for the high Biot number, the asymptotic error for the back surface is smaller (about 0.075) than that for the front one (about

−0.15) and this because its temperature is always closer to the mean one (e.g., see Figure 11).

At the asymptotic state and when the Biot number is not appropriately small, it is still possible to use the thin skin sensor but it is necessary to adopt a more accurate approximation of the mean slab temperature or, as it is shown in the following sub-section, of the time at which evaluating it from the measured data.

### 8.3. Data Analysis and Experimental Procedure

The computation of the convective heat transfer coefficient can be performed by rearranging Equations (1), (25) and (47)so as to obtain [1]:

$$h = \frac{-\rho_s c s \dfrac{dT_m}{dt} - q_r}{T_w - T_r} \tag{50}$$

For quite small Biot numbers and $Fo > 0.5$, the slab can be considered as isothermal and it is possible to evaluate $T_m$ with either the front $T_w$ or the back $T_l$ surface temperature. The solution of Equation (50) for a constant convective heat transfer coefficient and for $q_r = 0$ (but a similar result is also obtained, whenever possible, by linearizing $q_r$ and adding to $h$ the radiative heat transfer coefficient $h_r$) is given by:

$$\frac{T_w - T_r}{T_{wi} - T_r} = \exp\left(-\frac{ht}{\rho_s c s}\right) \tag{51}$$

This relation can be used to implement a regression process to measure $h$ as in the case of the thin film previously described.

A different approach is based on the numerical solution of Equation (50); a simple central difference formula may be normally sufficient:

$$\left(\frac{dT}{dt}\right)_i = \frac{T(t_{i+1}) - T(t_{i-1})}{2\Delta t} \tag{52}$$

where the time interval is defined as $2\Delta t = t_{i+1} - t_{i-1}$.

However, as said above it is advisable, before using Equation (52), to filter (in time) the temperature signal to avoid errors due to the captured noise.

As previously mentioned, when the slab cannot be considered as isothermal ($Bi > 0.05$), or the measurement is unsteady (as it may happen because of the continuous change of $T_w$ in Equation (1) and the non-linear change of $q_r$), approximating the mean temperature with that of one of the surfaces may lead to significant errors in the evaluation of the wall heat flux.

In the following, some formulae, which are useful to reduce unsteady convective heat transfer measurements with the thin skin sensor (such as the case of constant $h$ and $T_r$), are reported.

By making use of the Laplace transform and considering a first order expansion, it is possible to find the following approximate relations (Douglas [19]) between the front surface temperature $T_w$, the mean $T_m$ and the back surface one $T_l$(the right insulated surface of Figure 9):

$$T_w\left(t-\frac{s^2}{3\alpha}\right)\cong T_m\cong T_1\left(t+\frac{s^2}{6\alpha}\right) \tag{53}$$

where $s$ is the sensor thickness and $\alpha$ its thermal diffusivity coefficient.

By numerically differentiating Equation (53), a better estimate of the mean temperature derivative can be easily accomplished for unsteady measurements. In particular, if the IR scanner measures the back surface temperature, the use of Equation (53) enables to link easily $T_l$ to $T_m$ and the front surface temperature $T_w$ needed in Equation (50) to evaluate the convective heat transfer coefficient.

In the conventional heat flux measurement techniques, the temperature is normally measured with a thermocouple bonded to the rear of the sensor (the back surface) and higher accuracy reduction methods are reported. The main idea is to use a Taylor expansion (truncated at the order $2n$) of the temperature distribution inside the slab and to exploit the adiabatic boundary condition and the Fourier equation so as to obtain:

$$T(x)\cong T_1+\sum_{i=1}^{n}\frac{d^{2i}T_1}{dx^{2i}}\frac{(x-s)^{2i}}{(2i)!}=T_1+\sum_{i=1}^{n}\frac{d^{i}T_1}{dt^{i}}\frac{(x-s)^{2i}}{\alpha^{i}(2i)!} \tag{54}$$

where the coordinate $x$ is oriented as in Figure 9 and the second equality derives from the Fourier's equation.

Equation (54), where the dependence from $t$ has been dropped for sake of ease, can be used (by simply imposing $x = 0$) to evaluate the temperature $T_w$ on the opposite side of the slab.

Jepps [27] proposes the following second order discretisation:

$$T_w(t) \cong 3T_1\left(t + \frac{s^2}{6\alpha}\right) - 2T_1(t) \tag{55}$$

By integrating Equation (54) in space, the result can be used to compute also the mean temperature:

$$T_m \cong T_1 + \sum_{i=1}^{n} \frac{d^i T_1}{dt^i} \frac{s^{2i}}{\alpha^i (2i+1)!} \tag{56}$$

By differentiating in time the previous equation, one obtains the following relation that, for unsteady measurements, extends to higher frequencies the field of validity of the thin skin sensor:

$$\frac{dT_m}{dt} \cong \sum_{i=0}^{n} \frac{d^{i+1} T_1}{dt^{i+1}} \frac{s^{2i}}{\alpha^i (2i+1)!} \tag{57}$$

In the first order approximation ($n = 0$), the previous formula obviously reduces to the case of an isothermal slab, while for higher orders, the time derivatives can be easily numerically evaluated.

For $n = 1$ Jepps [27] proposes the following discretisation (that is equivalent to a first order numerical approximation of Equation (53)):

$$\frac{dT_m}{dt} \cong \frac{3\alpha}{s^2}\left(T_1\left(t + \frac{s^2}{3\alpha}\right) - T_1(t)\right) \tag{58}$$

Generally, higher order approximations are not required but, should it be necessary, they can be easily implemented from (57).

As already said, since with conventional techniques the temperature that is measured in wall calorimeters is generally the back surface one (e.g., with a leaf-type thermocouple), in the classical literature there is less interest in formulae based on the front surface temperature. Instead, the latter is helpful with IR thermography where $T_w$ can be easily measured when dealing with gas flows.

Nevertheless, a formula equivalent to Equation (54) can be found by again imposing the adiabatic condition on the back surface:

$$T(x) \cong T_w + \sum_{i=1}^{n} b_i \frac{d^i T_w}{dt^i} \frac{x^{2i}}{\alpha^i} \tag{59}$$

where the first five coefficients $b_i$ are equal to 1, −1/3, 2/15, −17/315, 62/2835. Again, Equation (59) can be used directly (by simply imposing $x = s$) to evaluate the temperature on the back sensor surface or, after integrating in space, to compute the mean temperature and its derivative with numerical formulae.

### 8.4. Tangential Conduction

A solution in the form of (43) has been also developed by de Felice *et al.* [26] (see [11]). It has, however, to be explicitly noted that, when the temporal variations are not too fast, the tangential conduction correction for the thin skin sensor can be more directly appreciated as for the case of the heated thin foil with Equation (21) but, in this case, as a function of time.

## 9. APPLICATIONS

The applications of IR thermography to thermo-fluid-dynamics encompass a very much diversified phenomenology which spans from turbine cooling including film cooling, to transition separation and reattachment, natural and forced convection, enhanced heat transfer, micro systems, rotating bodies, impinging jets, flow instabilities, two-phase and hypersonic flows, *etc.*; a review is presented in [1].

The authors' research group has been involved in IR thermography since about 30 years. In the following, a few significant papers of this group, essentially regarding complex fluid flows and specifically investigated with IR thermography, are presented and reviewed, also with the aim of pointing out in many of them a few relevant aspects of the used sensors.

## 9.1. Natural Convection

When studying with IR thermography natural convection in fluids transparent to infrared radiation (such as in gas flows), since the radiative heat flux may be often of the same order of magnitude of the convective one, this occurrence has to be carefully considered. For example, when determining the convective heat transfer coefficient between the sensor and a gas with the heated thin foil sensor, the substantial radiative contribution must be accurately subtracted to the Joule heat input (see Section 7.1).

The experimental apparatus, used by Cardone and Carlomagno [9], to analyze transient and steady one-dimensional natural convection on a vertical plate with the heated thin foil, is shown in Figure 2 and also reported in [28].

The sensor is made of a vertical stainless steel foil (245 mm high, 960 mm wide and 40 μm thick), coated on one side with a thin layer of high emissivity ($\varepsilon_t$ = 0.95) paint. The foil is Joule heated with a DC stabilized power supply and a step initial condition can be imposed by activating a relay with a relatively small (with respect to the foil characteristic time) activation time. The reached heat flux is varied in the range 80 ÷ 270 W/m$^2$. By measuring apparent temperatures at both foil sides, the uncoated side emissivity is evaluated to correct for radiation losses also from this side.

For a step Joule heat flux input 0 → 130.5 W/m$^2$, the evolution with time of the temperature vertical profile along the foil central segment is shown in the pseudo-thermogram of Figure 13 where: abscissa is time (starting from the relay activation and for a total of 80 s), ordinate is position along the foil central vertical segment (0 ÷ 245 mm) and colors indicate local temperatures (21 ÷ 42 °C).

The thermogram left region, characterized by a sequence of colored vertical bands, points out an initial uniform time increasing of foil temperature which indicates that the foil is acting essentially as a thin skin sensor with respect to the Joule input, Equation (47) with $q_w = q_j - q_r$. Besides, this means also that a prevalent radiative and conductive heat transfer regime is initially established between the foil and the ambient air that produces a constant temperature along the vertical direction; so, it appears that the convective flow is not started yet. This initial foil temperature evolution well agrees with theoretical calculations which include conduction and radiation from the foil to ambient.

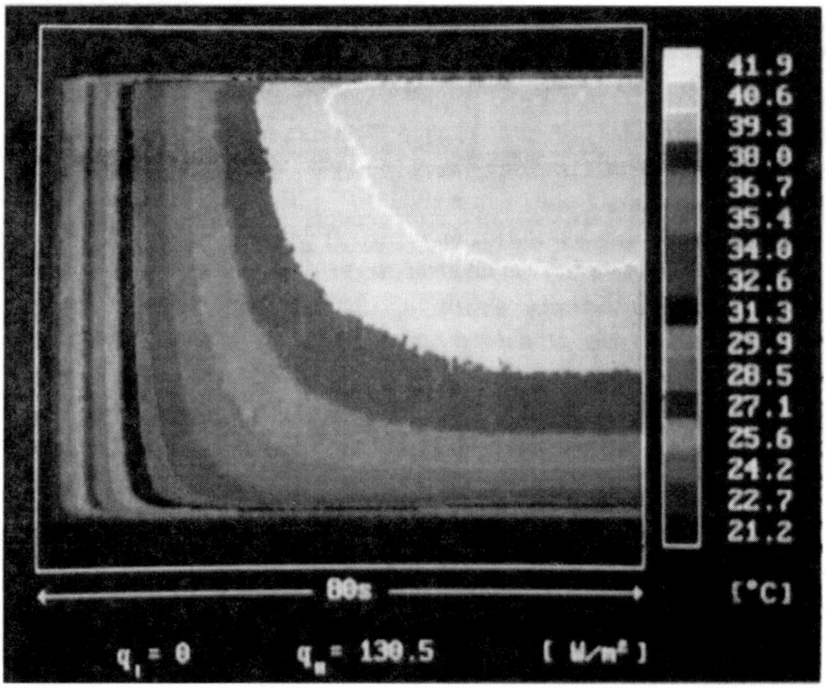


**Figure 13.** Time evolution of the temperature vertical profile on the foil under transient natural convection. From Carlomagno and de Luca [28].

The temperature variation which later occurs along the vertical direction shows the progressive onset of the natural convection boundary layer. When steady state is reached, Equation (14) applies and, as represented in Figure 14, where local *Nu* data are plotted as a function of the local modified Rayleigh number *Ra* defined in the standard way, the measured data well agrees (within ±5%) with the theoretical prediction of Sparrow and Gregg [29] for laminar flow.

Obviously, all data is corrected for the radiative contribution, that is subtracted to the Joule input as a function of position, but not for tangential conduction, which mainly affects the low *Ra* data.

### 9.2. Hypersonic Flow over a Ramp

A relatively old example of the restoration of a degraded image, which considers both the IR system modulation transfer function (MTF) and the tangential conduction in the thin film, is reported by de Luca *et al.* [30]. They perform heat transfer tests on a ramp which follows a delta wing in a hypersonic wind tunnel. Since heat transfer measurements are performed with the thin film sensor, after starting of the tunnel, the model (which is initially in a remote position at room temperature) is suddenly injected into the high stagnation temperature (800 K) hypersonic stream and its temperature monitored.

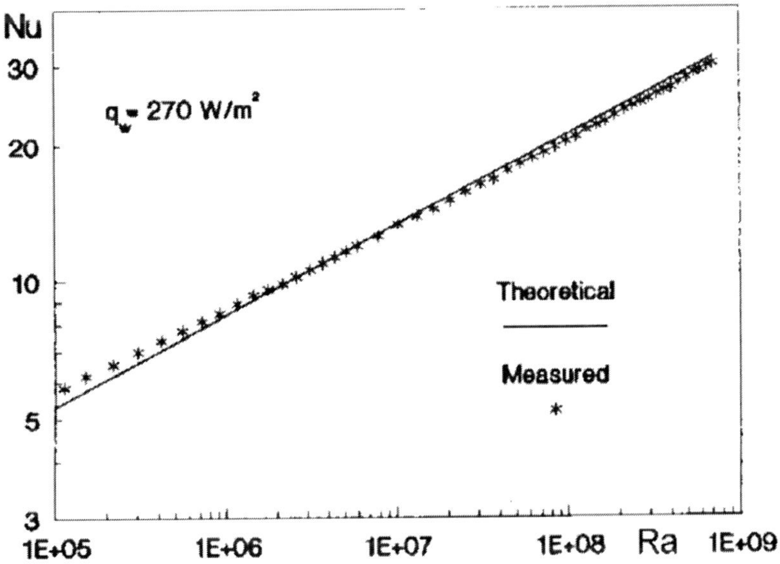


**Figure 14.** Nusselt number on a vertical plate cooled by natural convection as a function of the Rayleigh number. From Carlomagno and de Luca [28].

The gas flows over a ramp, placed downstream of a 70° delta wing at zero angle of attack, at Mach number $M$ = 8.15.Figure 15a shows a detected thermogram with the coarse footprints ($T$ in °C) of the Görtler vortices on the solid 15° RTV elastomer ramp as recorded by the camera. The Görtler vortices develop on the ramp reattachment zone due to the streamlines curvature induced by the ramp presence (flow going from the thermogram bottom, hinge line, upwards). In the raw thermogram of Figure 15a, the vortices are barely evident in the span wise quasi-periodic horizontal variation of wall temperature.

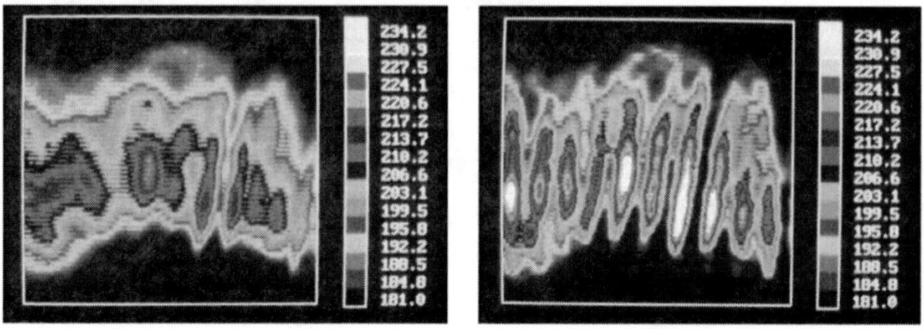


(a) (b)

**Figure 15.** Images of Görtler vortices on a 15° ramp placed downstream of a 70° delta wing in a hypersonic flow at $M = 8.15$: (**a**) coarse; (**b**) restored. Flow from bottom to top. From de Luca *et al.* [30].

The amplitude of each harmonic component of measured temperatures has been corrected for the MTF and by means of the function $F$ (see Equation (46). The restored temperature map, that shows the effective temperatures, is represented inFigure 15b. It has to be observed that, even if the used interlaced 880LW camera (AGEMA, Sweden) has a limited spatial resolution (140 × 140 pixels), being operated with (36 + 20) mm extension rings at the minimum focus distance (with a 21 mm × 21 mm field of view), the image is well sampled so degradation is due to camera and sensor. Anyhow, the restored image allows us to correctly identify the rather regular Görtler vortices structure which exhibits a mean pitch of approximately 2 mm.

The horizontal striping, which is evident in the vortices prints for both the coarse and restored thermograms, is exclusively due to the camera interlacing and to the transient heating of the ramp that is connected to the unsteady thin film sensor functioning mode. In their work, de Luca *et al.* [30] address also the problem of the signal digital sampling, which is not herein analyzed. A similar investigation has been performed by de la Chevalerie *et al.* [31].

De Luca *et al.* [32] report flow visualizations and convective heat transfer measurements on a double ellipsoid model in a hypersonic stream at $M = 8.15$ and several angles of attack $\alpha$. To obtain heat transfer data, the models are made of NORCOAT® 4000 (Norco Industries, Compton, CA, USA) which is a silicon elastomer filled with hollow silica microspheres that has a relatively low thermal conductivity coefficient ($k = 0.129$ W/m K) and a high emissivity coefficient ($\varepsilon_t$ = 0.93). Besides performing standard oil film visualizations and in order to exploit the capability of the infrared camera to detect an induced transition, several trip cylindrical wires (0.5 mm high, 0.22 mm in diameter and placed at 5 mm steps) are implanted 30 mm away from the ellipsoid nose, on the windward surface, with their axes normal to this surface.

The thermogram of Figure 16, where temperatures are in Celsius, shows the effect produced by the trip wires on the temperature distribution over the model windward surface, 0.48 s after model injection. To better distinguish the flow behavior downstream of the trip wires, the temperature range of the used 880LW camera (AGEMA, Sweden) is set so as to have the ellipsoid forepart in saturated conditions.

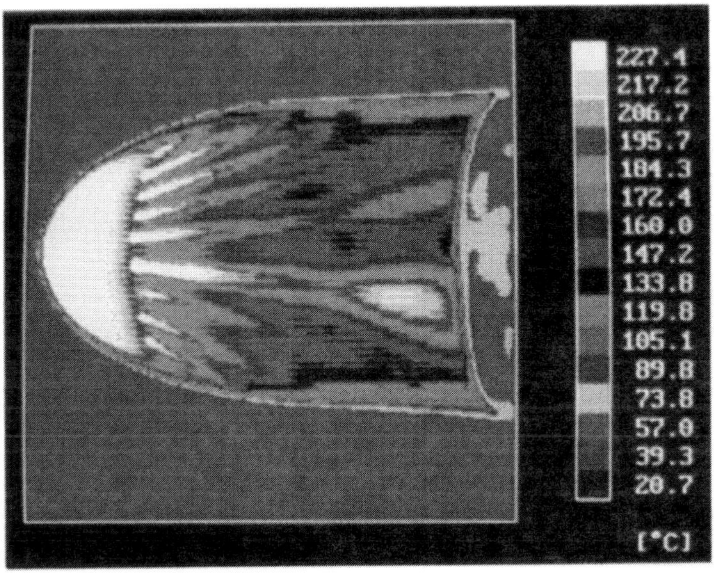


**Figure 16.** Temperature (in Celsius) map of the ellipsoid side (windward) of the double ellipsoid model at M = 8.15, $\alpha = 30°$, 0.48 s after model injection. Flow from left to right. From de Luca *et al.* [32].

The wakes of the wires are made evident by the various stripes, which indicate the surface flow streamlines direction over the model surface. They start downstream of the transversal violet line of the thermogram, where the transition trips are approximately located. In the model lateral zones, where the fluid particles following curved divergent path lines accelerate, the turbulisation induced to the flow by the trips seems to damp out, denoting a flow re-laminarization. On the contrary, near the symmetry axis, the flow after re-laminarization becomes downstream unstable and the occurrence of turbulence seems evident in the two hot red and pink spots located close to the model trailing edge.

Unlike standard oil film visualizations, this procedure does not need to have the model surface sullied and renovated after each test; e.g., if the angle of attack has to be changed, it is only necessary to retrieve the model back in the remote position and let it recover a uniform temperature.

### 9.3. Jet Impinging on a Plate

The group of Carlomagno has extensively studied jets impinging on a plate with IR Thermography [4,6,7,28,33–35]; herein first, the measurement of the convective heat transfer coefficient by means of the Laplacian sensor [7] is presented, then the adiabatic wall temperature detection [4] is treated.

The first tackled application task is the measurement of the distribution of the convective heat transfer coefficient over a thermally thin plate subjected to a jet normally impinging on it, at a short nozzle-to-plate distance, which originates a quite known but rather complex thermal field.

The experimental procedure and apparatus are the same used by Meola *et al.* [4]. The main difference is that the impinging sensor plate is no longer a heated thin foil as in [4] but rather an externally heated 1.1 mm thick metal sheet horizontally placed, with the jet normally impinging underneath, in order to minimize the effects of hot air recirculation in the measurement zone.

An aluminum alloy (Al–3105, $k = 185$ W/mK) hexagonal plate, constitutes the impinging target which has a free circumscribed circle about 300 mm in diameter. To correctly evaluate tangential conduction, both thickness and thermal conductivity of the sensor plate are precisely known.

The heat input is provided by means of electric cartridge heaters, placed inside copper bars, which are positioned at the plate sides. In order to investigate the sensitivity of the sensor to the heating geometry, three different peripheral positions of the external heaters are tested (see Figure 17): *hexagonal* (six heathers on all hexagon sides, Figure 17a), *triangular* (three heathers on each other side, Figure 17b) and *one-side* (three heaters placed on three adjacent sides, Figure 17c). The black-painted hexagonal zones of Figure 17 coincide with the measurement region. In order to reduce the total radiative heat flux towards the ambient, only these zones observed by the IR camera are covered with black paint.

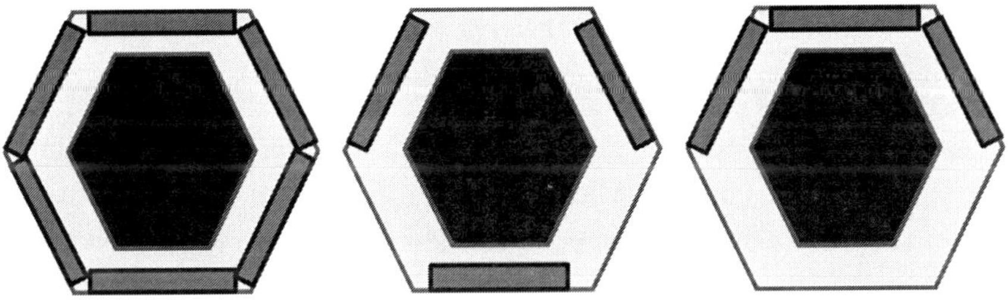


**Figure 17.** Schematic configurations of the peripheral heaters in the plan view of the plate: (**a**) Hexagonal; (**b**) Triangular; (**c**) One-side. From Carlomagno *et al.* [7].

Even if compared with the relatively high local convective heat transfer coefficient, the rather small thermal thickness of the aluminum plate determines a very small Biot number. Consequently, the temperature can be considered as practically constant across the plate, so allowing for measuring the surface temperature distribution with the IR camera by observing the sensor back surface, *i.e.*, that opposite to the jet impinging one. An infrared camera QWIP FLIR SC6000 LW (640 × 512 pixels resolution) operating in the LWIR band is employed to measure the surface temperature map.

The jet issues from a slightly convergent axisymmetric nozzle with an exit diameter $D = 18.7$ mm. The Reynolds number$Re = 30{,}000$ is based on the nozzle diameter $D$ and jet initial velocity. The convective heat transfer coefficients are presented in terms of Nusselt number (2), also based on $D$.

Tests are carried out by varying the dimensionless nozzle-to-plate distance $z/D$ but herein only results for $z/D = 2$ are reported, being them the most intriguing ones in terms of $Nu$ radial profile. For each test, the heat input is chosen by trying to maintain on the slab a minimum temperature difference of at least 15 K with respect to the jet temperature. Obviously, as shown by Equation (28), the higher the minimum temperature difference, the more accurate the results while evaluating $h$.

Anyhow, a limit is imposed on the maximum operative temperature of the plate; e.g., restriction are imposed by the camera calibration range, the maximum heat input of the heating elements and, overall, the melting point of the black paint. Since in the performed experiments the calibration of the IR camera system is made between 15 and 80 °C, the heaters are shielded with insulated material in order to avoid saturated pixel values.

The physical and the geometrical parameters of the experimental apparatus are carefully chosen. E.g., the distance between the heaters, the slab thermal thickness and the outflow conditions directly influence the distribution of the convective heat transfer and, consequently, a suitable temperature distribution for the Laplacian numerical computation. For these reasons, it is suggested that an accurate preliminary study of the phenomenology has to forego the application of this sensor.

First, the effect of different sources of measurement noise, affecting the Laplacian computation, is investigated. Results show that the temporal noise can be consistently reduced if each temperature map is obtained by averaging a relatively high number of images. In the present investigation, 100 instantaneous images are averaged to obtain each steady-state temperature map.

Then as already stated, the use of an IR camera with Focal Plane Array introduces another error critical source, *i.e.*, the spatial measurement noise generated by differences in the response of each detector pixel. Therefore, the application of an accurate and reliable process of Non-Uniformity Correction (NUC) is of fundamental importance. In fact, tests demonstrate that both spatial filtering and computation of the derivatives with a relative large step are not sufficient to compensate the decay of the signal-to-noise ratio due to the non-uniformities between the various pixels response. Carlomagno *et al.* [7] show that even a small local error (of the order of 0.1 K) can completely jeopardize the accuracy of the computed Nusselt number distribution in a region whose dimension is connected to both the spatial filtering window and the step for the derivatives numerical computation.

In Figure 18, the steady state temperature maps for the three configurations without the impinging jet are reported. As previously stated, the slab sides are shielded in order to avoid saturated pixel values in the viewed scene. The temperature distributions are convolved with a 30 × 30 pixels wide (about $0.5D \times 0.5D$) Gaussian filtering window, with a standard deviation equal to 6. From the temperature maps, it is evident that a more regular distribution occurs for the hexagonal heating configuration that allows us to better evaluate the convective heat transfer coefficient from the temperature Laplacian.

Further increase of the signal-to-noise ratio influence is recovered by computing the numerical derivatives with a spatial step of one nozzle diameter.

In Figure 19, the computed Nusselt number distributions for the three tested configurations are shown. The distributions seem to be quite independent of the heaters arrangement (see the figures relative to the hexagonal and triangular configurations), unless the strongly asymmetric heating (one-side) is adopted. In this latter case, it is clear that, since the condition of a maximum temperature is imposed, the minimum temperature is lower than in the case of more symmetric configurations (see the bottom zone of the right image of Figure 18). As can be seen, this decreases the tangential heat transfer and accordingly increases the uncertainty due to the lower difference between wall and jet temperatures.

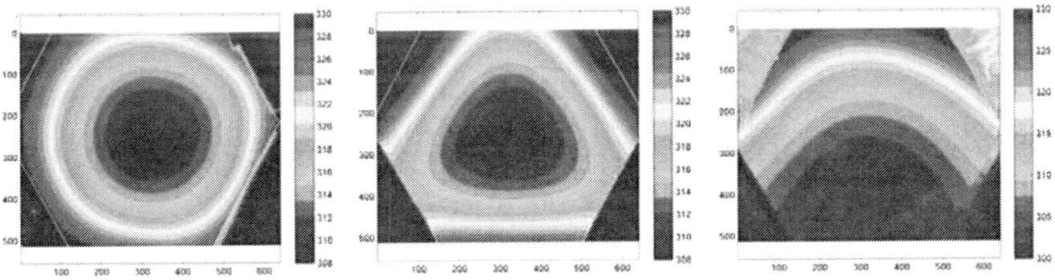

**Figure 18.** Steady state temperature (K) of the plate without the jet. From left to right: hexagonal heating configuration; triangular; one-side. From Carlomagno *et al.* [7].

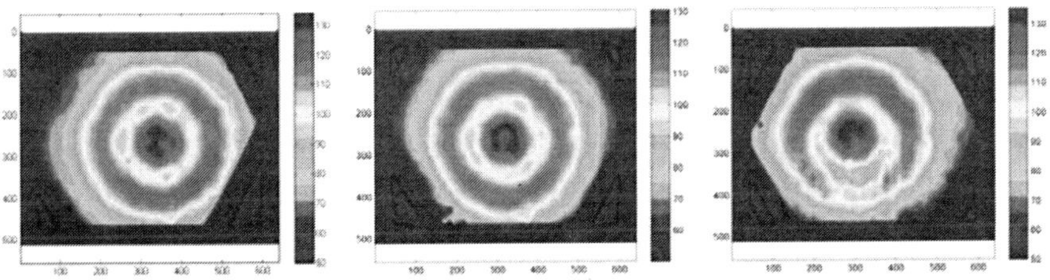

**Figure 19.** Nusselt number distribution for $Re = 30{,}000$ and $z/D = 2$. From left to right: hexagonal heating configuration; triangular; one-side. From Carlomagno *et al.* [7].

In Figure 20, the azimuthal averages of the Nusselt number as a function of the dimensionless radial coordinate $r$ (from the jet axis) are reported for the three heating configurations and $z/D = 2$. The obtained averaged profiles are in quite good agreement with each other and with that of the literature for the same $z/D$ value and, approximately, the same Reynolds number [28].

In spite of the reduction of the spatial resolution, connected to the use of a smoothing Gaussian filter and the calculation of the numerical derivatives with a relatively large step, the Laplacian sensor shows to be able to detect all the local variations of $Nu$, providing quite reliable results.

Anyhow, averages in the region close to the jet stagnation point are affected by a higher uncertainty since the number of available samples for the azimuthal average is not sufficient to smooth the noise of the computed Nusselt number distribution, there.

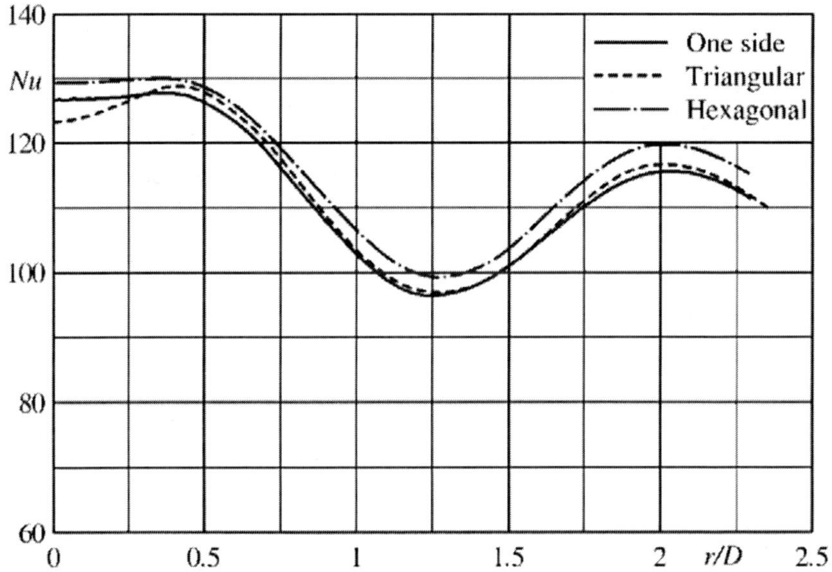


**Figure 20.** Azimuthally averaged profiles of the Nusselt number radial distributions ofFigure 19 for the three heating configurations; $Re = 30{,}000$, $z/D = 2$. Data from Carlomagno *et al.* [7].

This partially justifies also the not exactly horizontal tangent and the larger differences between the Nusselt number profiles, at $r/D = 0$, in one of the curves reported in Figure 20. Nevertheless, on the average, the differences are contained within a few per cent.

Meola *et al.* [4], by measuring the adiabatic wall temperature $T_{aw}$ on a plate with impinging jets, observe with IRT the instabilities developing at relatively high Mach numbers $M$. The phenomenology is strongly dependent on the impingement distance $z$; in particular, for $z/D < 6$, as $M$ increases, the vortex ring, which is located in the shear layer at about $1.2D$ from the jet axis, strengthens up to its highest magnitude ($M \approx 0.7$). Then for even larger Mach numbers, the vortex ring breaks up (Widnall instability, Widnall *et al.* [36]) with the formation of secondary flow structures.

At low Mach number the $T_{aw}$ map is practically axisymmetric. But, as $M$ increases, the vortex ring reinforces and, for $M \approx 0.7$, breaks up when impacting onto the plate, entraining warmer ambient air and giving rise to secondary minima, with maxima in between them. This is shown in the adiabatic wall temperature map of Figure 21, which refers to $D = 5$ mm, $z/D = 4$, $M = 0.78$, $Re = 86{,}400$. For higher $M$, the structures first strengthen up and reach their highest magnitude and, with a further Mach number increase, they break up into numerous smaller structures, which tend to coalesce giving rise to a transient alternate circumferential movement.

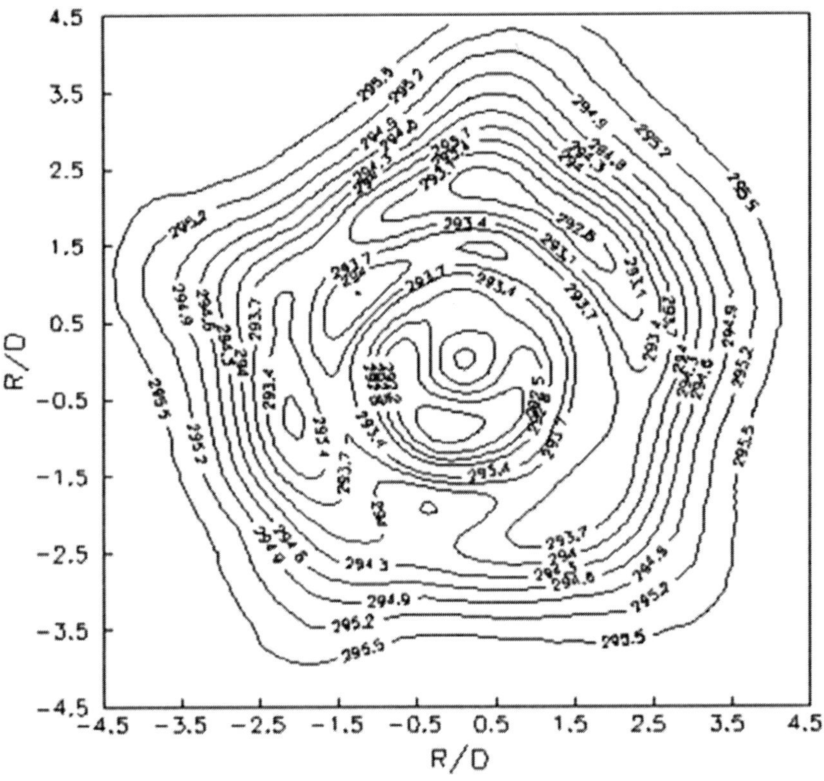


**Figure 21.** Adiabatic wall temperature on a plate with a normally impinging jet. $D = 5$ mm; $z/D = 4$; $M = 0.78$; $Re = 86{,}400$. From Meola *et al.* [4].

Since the temperature fluctuations at $M > 0.7$ for $T_{aw}$ and $T_w$ are of the same order of magnitude, the azimuthally averaged radial distribution of the Nusselt number does not show additional peaks. Nevertheless, some azimuthal structures are visible in the two-dimensional map of the Nusselt number. In fact, local $Nu$ contours exhibit loss of symmetry about the stagnation point, this effect being due to the transient alternate circumferential movement of the instability structures [4].

It is interesting to point out that, to perform the qualitative studies on the adiabatic wall temperature distribution such as the map presented above, the sensor to be used (on which the jet impinges) could be simply constituted by a very low thermal conductivity flat surface.

## 9.4. Rotating Channels

As it is well known, to cool gas turbine vanes and blade, air from the high pressure compressor stage goes through the hub section into the component interior and, after flowing through serpentine passages, is discharged into the main flow to provide film cooling as well. These passages are mostly made of several adjacent straight ducts, spanwise aligned along the blade and connected by 180° turns (also called *U turns*). Turns cause flow separation/reattachment and induce secondary flows so the convective heat transfer coefficient exhibits high variations with consequent wall thermal stresses. For a rotating channel, it is common usage to call *leading* the wall that goes ahead during rotation and *trailing* the one that follows.

One of the first attempts to measure convective heat transfer coefficients in a rotating air channel with IR thermography is reported by Cardone *et al.* [37]. The apparatus concept is a direct consequence of the used heated thin foil sensor. Since the foil back surface (to be viewed by the IR camera) cannot be thermally insulated, the only way to prevent high thermal losses by forced convection over this surface (see Equation (14), Section 4.1) was to have the channel rotating in a vacuum chamber. In the pioneering work by Cardone *et al.*, the spatial resolution of the results was relatively poor on account of the strong influence of the tangential conduction heat fluxes which were due to the relatively small dimension of the channel (22 mm × 22 mm). Therefore, the presented results were not the detailed ones that can be obtained with IR thermography but this work is reported because it offers a useful option to study flows into moving bodies.

A different approach, in order to reduce the relative importance of the external convection with respect to the internal one is chosen by Gallo *et al.* [17] to obtain detailed Nusselt number maps near a 180° sharp turn of a rotating U channel with the heated thin foil sensor. The authors decide to significantly increase $h$ at the sensor front surface by using water as a working fluid as well as to use a much larger channel in order to reduce its rotational speed and, therefore, $q_a$ at the back surface. In this way, they are able to obtain a quite good spatial resolution and low tangential conduction in the acquired measurements.

The experimental apparatus, represented in Figure 22, consists of a Plexiglas® two-pass water channel with a sharp 180° turn, mounted on a revolving platform whose rotational speed can be continuously varied and precisely monitored in the range 0 ÷ 60 rpm. The channel has a square cross section with a side $d = 60$ mm, its length of 1200 mm ahead of the 180° turn ensuring an almost dynamically fully developed flow before the turn. The central partition wall dividing two adjacent ducts is 12 mm thick. Water from a tank is pumped through an orifice meter, a rotating hydraulic coupling and, after flowing in the test channel, is discharged back into the tank. Mass flow rate can be varied with a by-pass circuit and the inlet to channel water temperature is kept constant with a heat exchanger. A magnetic pick-up allows the synchronization of the IR image acquisition.

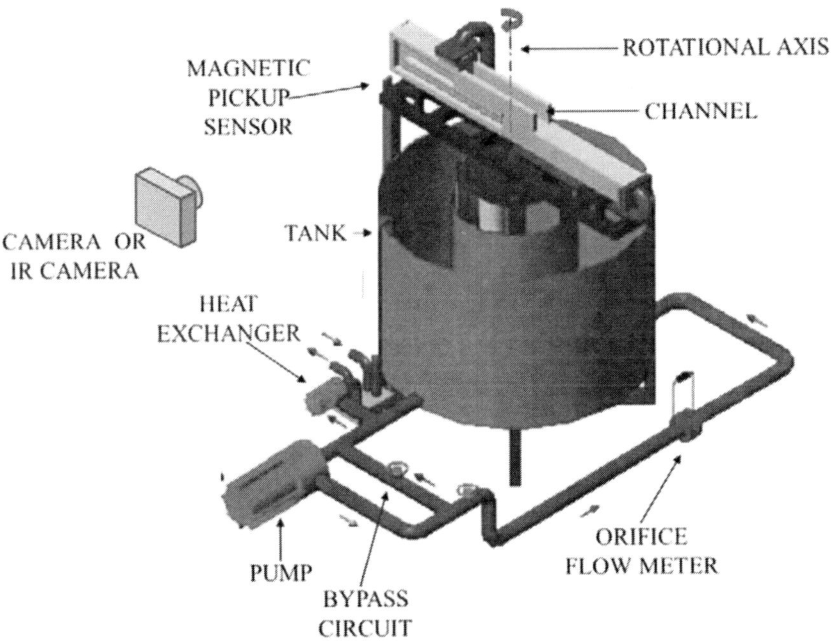


**Figure 22.** Experimental apparatus for a rotating channel of Gallo *et al.* [17].

The apparatus is capable of simulating both Reynolds number *Re* and *Rotation number* $Ro = \omega\, d/V$ (where $\omega$ is the angular speed of the channel) values typical of turbine blades. The *d* increase and the *V* decrease (to maintain a given *Re*) allow a drastic reduction in $\omega$ by keeping a constant *Ro*. Results relative to the static channel (no rotation) are in good agreement with the measurements performed with air by Astarita and Cardone [10].

Heat transfer coefficients are obtained for two heating conditions (from one or both channel sides) and presented in terms of the normalised local Nusselt number $Nu/Nu^*$ (which can be considered also as $h/h^*$). $Nu^*$ and $h^*$ are respectively the Nusselt number and the convective heat transfer coefficient values predicted by the Dittus and Bölter correlation for fully developed channel flows as interpreted by Kakac *et al.* [15].

In Figure 23a, the normalised Nusselt number $Nu/Nu^*$ (as earlier defined) distribution over the leading wall, for $Re =$ 20,000 and $Ro$ equal to 0.3, is represented. In the inlet duct, the flow appears to be fully developed, also from the thermal point of view, since the normalised Nusselt number is practically constant. The $Nu/Nu^*$ values are lower than those relative to the static case and, following Gallo *et al.* [17], decrease with the increasing of the rotation number *Ro*.

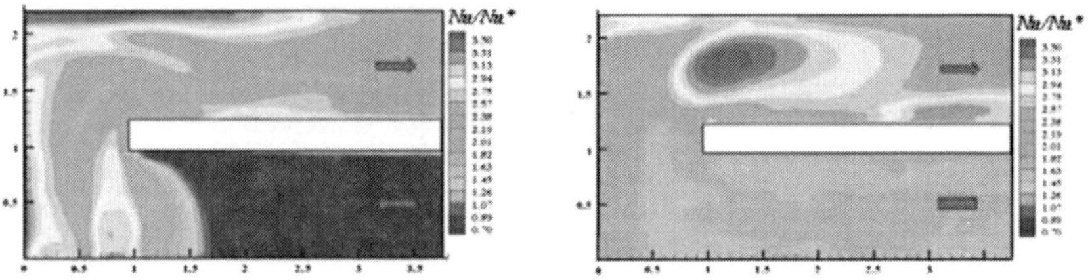


(a) (b)

**Figure 23.** Normalised Nusselt number Nu/Nu* maps for a two-side heated rotating channel; $Re =$ 20,000; $Ro =$ 0.3: (**a**) Leading wall; (**b**) Trailing wall. Flow enters from bottom and exit from top. Adapted from Gallo *et al.* [17].

In the first half of the first corner, it is possible to notice a high heat transfer zone that is caused by the inversion of the Coriolis force in the turn zone. Really in the turn zone, the radial velocity component suddenly decreases and changes sign with a consequent decrease and inversion of the Coriolis force. This inversion makes the flow separate by the trailing side, and to abruptly reattach toward the leading one, with a strong increase of the normalised Nusselt number at the reattachment point. Other low and high heat transfer zones are clearly visible in the normalised Nusselt map and their cause is explained in detail by flow field measurements made with particle image velocimetry in the work of Gallo *et al.* [38].

As it is possible to see from Figure 23b, the normalised Nusselt number distribution over the trailing wall appears to be completely different from that on the leading wall. In the inlet channel, the $Nu/Nu^*$ values are again uniform, but much higher than those for the static case and for the leading wall; besides, as reported by the authors [17], they tend to increase for increasing rotation number *Ro*.

In the turn region, it is possible to note that the iso-Nusselt zone tend to advance into the first corner of the turn and to insinuate in the first half of the second corner. On the second outer angle, it is possible to note a high heat transfer zone that, near the frontal wall, results as being adjacent to a relatively lower heat transfer zone. In the outlet duct, the Nusselt number distribution relative to the trailing wall exhibits two high heat transfer coefficient zones located respectively on the center and downstream near the partition wall. Again, the reason for such behavior can be found in the paper of Gallo *et al.* [38].

It should be noted that the measurements by Gallo *et al.*, are performed at relatively high Biot number because of the presence of water in the channel and since the heated tracks of the printed circuit are placed at the sensor back surface for electrical insulation reasons. Therefore, data are reduced according to the specific procedure described in Section 4.2.

## 9.5. Spray Tubes

As far as aircraft wings are concerned, de Luca and co-workers have studied both the boundary layer transition from laminar to turbulent flow on an airfoil for varying angle of attack [39] as well as the phenomena of flow separation and reattachment over a delta wings [40], both with the heated thin foil sensor.

With the same sensor, Imbriale *et al.* [41] went inside the leading edge of an airfoil to analyze wing de-icing aspects. This investigation represents a cylindrical geometry heated thin foil sensor realized with a metallic foil inserted in an *ad-hoc*fixture.

In fact, spray tubes (often called *piccolo tubes*) are amongst the most widely used anti-icing devices for wings and engine nacelles of commercial aircrafts [42]. In such devices, hot air is extracted from the compressor and blown on the inside surface of the wing leading edge, through small holes drilled in a pipe. The aim is to supply enough energy to keep the wing surface above the freezing point of water and to liquefy impinging ice crystals.

Sometimes, also to cool leading edges of turbine blades, a spray tube inside and parallel to them, with an array of aligned holes, generates a jet row which blows cold air to maintain the blade surface temperature below critical values [43]. In these devices, the impingement distance is relatively short, this problem being addressed for single jets by Carlomagno and Ianiro [35].

A piccolo tube is experimentally analyzed by Imbriale *et al.* [41]. The test article, using the fixture of Meola *et al.* [44], includes the leading edge of a NACA 0012 wing profile with a 1.50 m chord, with a spray tube located inside at 4% of the chord profile. The profile is 0.20 m span-wise long and it is stopped at about 1/10 of the chord with an open side to facilitate discharge of injected gas so to avoid recirculation effects. To allow measurements with the heated thin foil sensor, the leading edge section consists of a thin stainless steel sheet (40 μm thick) lodged inside an *ad-hoc* fixture that provides bus bars exactly shaped to the wing profile. A small indium wire is inserted between bus bars and metal sheet to assure a good electrical contact.

In particular, different spray tubes are used with a different number of holes from 3 to 5, orifice diameter $D$ from 2 mm to 4 mm, and orifice pitch-to-diameter ratio $p/D$ between 5 and 15. In addition, the jet inclination $\varphi$ is varied from 0° to 50° by rotating the spray tube around its axis. The jets exit Mach number $M$ is varied from 0.6 to 1.0. Herein, only results relative to 3 orifices having $D = 4$ mm, $\varphi = 30°$, $p/D = 15$ and M = 1 are reported.

In order to avoid measurement errors due to the surface directional emissivity decay, the camera viewing angle is always kept lower than 55°. Raw data is corrected for radiation, tangential conduction within the sensor and natural convection at the foil viewed side.

Due to the curvature of the viewed surface, to obtain the temperature distribution over the entire leading edge surface, at least two images must be acquired; temperature maps are reconstructed on the object mesh grid according to Cardone *et al.*[45]. This involves not only a different data reduction, but mainly a geometrical calibration of the IR camera. Then, before performing calculation with Equation (23), the temperature maps of $T_w$ and $T_{aw}$ have to be reconstructed on a 3D mesh grid from the 2D acquired IR images.

An example of such a map reconstruction is given in Figure 24 for the adiabatic wall temperature $T_{aw}$ (Figure 24a, foil not heated) and for the wall temperature $T_w$ (Figure 24b, heated foil), taken for three aligned holes, $D = 4$ mm, inclination angle$\varphi = 30°$, $M = 1.0$ and orifice pitch-to-diameter ratio $p/D =$ 15. Maps clearly show the jet impingement zones which are not centered with respect to chord due to the 30° jet inclination and, again, the influence of the relatively high Mach number on the adiabatic wall temperature distribution is recognized.

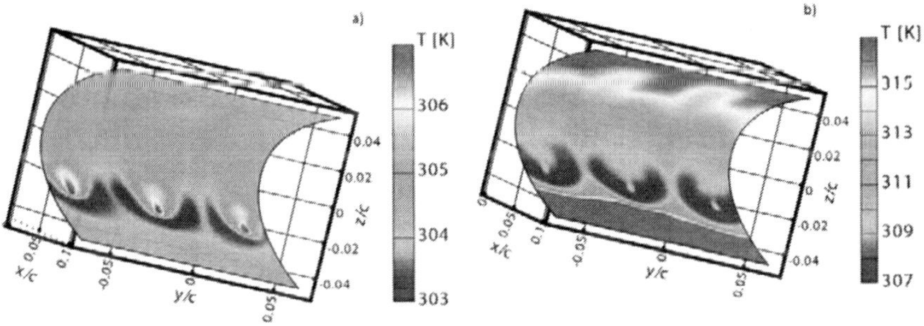


**Figure 24.** 3D temperature reconstruction for $D = 4$ mm, $\varphi = 30°$, $M = 1.0$ and $p/D = 15$: (**a**) adiabatic wall temperature $T_{aw}$ (foil not heated); (**b**) wall temperature $T_w$ (heated foil), Adapted from Imbriale *et al.* [41].

A typical $Nu$ map, for the same testing conditions of Figure 24, is shown in Figure 25. The impinging jets entail very high Nusselt number values with local peaks in a small region matching with the jets centers; so, these peaks clearly locate the area of jet impingement on the front-side surface. Even though the holes are perfectly circular, the high $Nu$ region somehow stretches in chord-wise direction. This behavior is due to the jets inclination with respect to the foil surface; in fact, only jets with inclination $\varphi$ either equal to 0° or 65° are perpendicular to the foil; for other $\varphi$ values, the impingement is affected by inclination effects.

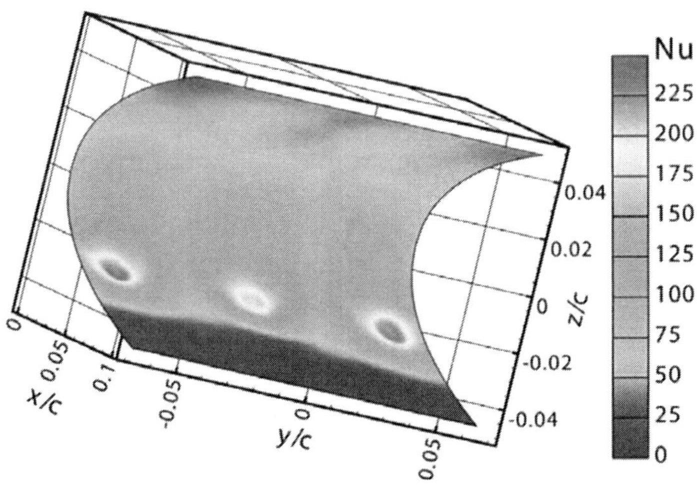


**Figure 25.** Nusselt number reconstruction for $D = 4$ mm, $\varphi = 30°$, $M = 1.0$ and $p/D = 15$. Adapted from Imbriale *et al.* [41].

In addition, variations in the $Nu$ distribution are present on the backside region ($z/c > 0.2$). In particular, it is possible to recognize local $Nu$ increase in the spanwise direction between two contiguous jets, at the same locations where the two green streaks are visible in the wall temperature maps of Figure 24. This behavior resembles the fountain effect, already described in literature [46]. However, some fundamental differences regarding position and appearance of such local maxima exist. In fact, the fountain effect, described in literature, is exactly localized between contiguous jets; while, in the present case, local heat transfer maxima seem to originate between jets but they extend and strengthen on the backside region, far from the impingement.

## 9.6. Chevron Nozzles

The great implications that vortical features can have on surface heat transfer rate and distribution motivated the flourishing of countless investigations devoted to passive and active strategies to enhance the heat flux between a jet and an impinged surface.

Passive strategies are mainly based on the shape of the nozzle. In particular, Gao *et al.* [47] show that, for a nozzle-to-plate distance of 4 diameters, triangular tabs placed around a circular orifice (*chevron nozzle*) lead to an enhancement higher than 25% with respect to the round configuration for the same nozzle-to-plate distance.

Besides in recent years, a special attention is devoted to a particular, but very relevant, technique in the field of modern experimental thermo-fluid-dynamics, namely Tomographic Particle Image Velocimetry (Tomo-PIV). The outcomes of its visualization and flow inspection capabilities can be compared with those of recent advanced computational methods. In particular, while infrared thermography gives the outcome of the flow field on the skimmed over surface, Tomo-PIV contributes to understand the reasons for such an outcome.

Tomographic PIV is based on the double-pulsed laser illumination of a fluid volume seeded with particles; the particles distribution is reconstructed from multiple camera views and then the volumetric flow field is obtained through volume cross-correlation. Tomographic PIV is proved to be capable of high resolution volumetric flow measurements which allow for the computation also of derivative quantities giving a detailed description of vortex and turbulence dynamics.

The chevron arrangement is also exploited in [48] where it is shown that, compared to the circular nozzles, heat transfer at impingement is enhanced by using the chevron ones; in the center of the impinged area, the chevron jet exhibits heat transfer values up to 44% higher than those provided by the circular jet. This is shown by the bottom images of Figure 26where the maps of the average Nusselt number on the impinging plate for $Re = 5000$ and $z/D = 4$ (symbols as in Section 9.3) are presented. This event is addressed to the development in the free jet region of streamwise vortices which, compared with the toroidal vortices of the circular nozzles, are associated with a deeper penetration of turbulent induced mixing and to a higher arrival speed (see top images of Figure 26).

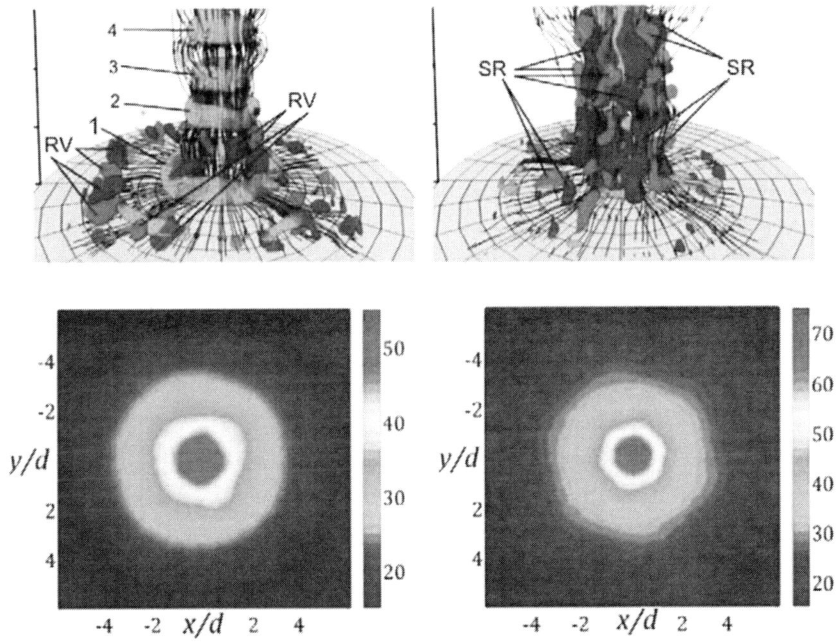


**Figure 26.** Instantaneous flow patterns (**top**) and Nusselt number distributions on the impinging plate (**bottom**) for a cylindrical (**left**) and a chevron (**right**) nozzle; $Re =$ 5000, $z/D =$ 4; RV radial vortices, SR streamwise structures. Adapted from Violato *et al.*[48].

The coupling of the inspection capabilities of the two experimental techniques, such as IR thermography and Tomographic PIV, allows for a deeper understanding of the involved thermo-fluid-dynamic phenomena in this very complex flow field.

# 10. CONCLUSIONS

This paper has the aim of reviewing and assessing the most common heat flux sensors, which can be used with infrared thermography to measure convective heat transfer coefficient distributions between a body and a fluid flowing over it. After recalling the basic principles that make IR thermography work, the various heat flux sensors to be used with it are presented and discussed, namely: heated thin foil, gradient sensor, Laplacian sensor, thin film and thin skin. For each sensor, the basic model, its limits, the data analysis and the experimental procedure are delineated.

Applications to complex thermo-fluid-dynamic flows that range from natural convection to hypersonic flows are finally outlined by describing some of the heat flux sensors exploitations performed by the authors' research group at the University of Naples Federico II.

For all the presented applications, the infrared adopted technique proves its capability to accurately measure the convective heat transfer coefficient distributions generated by the fluid flows in the examined complex geometries and to be a very effective investigation tool for thermo-fluid-dynamic experimental research.

When compared to standard techniques, the use of an infrared camera as a temperature transducer in convective heat transfer measurements appears advantageous from several points of view. In fact, since the IR camera is two-dimensional, with current systems having up to about 1 M pixels per frame, besides producing a whole temperature map, IRT allows an easier evaluation of errors due to radiation and tangential conduction. The camera does not disturb the phenomena being measured and does not alter conduction through test article embedded sensors and wiring, it has high sensitivity (down to 10 mK) and low response time (down to 20 ms). As such, IR thermography can be effectively exploited to measure convective heat fluxes (with either steady, or transient, techniques) even in circumstances where they undergo drastic variations.

# AUTHOR CONTRIBUTIONS

All the authors equally conceived, designed and performed the experiments, and analyzed the data; T. Astarita and G. M. Carlomagno wrote the paper.

# REFERENCES

1. Astarita, T.; Carlomagno, G.M. *Infrared Thermography for Thermo-Fluid-Dynamics*; Springer: Berlin/Heidelberg, Germany, 2013.
2. Schlichting, H. *Boundary-Layer Theory*; McGraw-Hill: New York, NY, USA, 1979.
3. Shapiro, A.H. *The Dynamics and Thermodynamics of Compressible Fluid Flow*; Ronald Press: New York, NY, USA, 1954; Volumes I and II.
4. Meola, C.; de Luca, L.; Carlomagno, G.M. Influence of shear layer dynamics on impingement heat transfer. *Exp. Therm. Fluid Sci.* **1996**, *13*, 29–37.
5. Rohsenow, W.M.; Hartnett, J.P. *Handbook of Heat Transfer*; McGraw-Hill: New York, NY, USA, 1973.
6. Carlomagno, G.M.; de Luca, L. Infrared Thermography in Heat Transfer. In *Handbook of Flow Visualization*; Yang, W.J., Ed.; Hemisphere: London, GB, UK, 1989; pp. 531–553.
7. Carlomagno, G.M.; Discetti, S.; Astarita, T. Experimental assessment of a new heat flux sensor for measuring convective heat transfer coefficients. *Quant. Infr. Therm. J.* **2011**, *8*, 37–49.
8. Roger, M. A periodic-transient method for high-resolution heat transfer measurement on two dimensional curved surfaces. *J. Heat Transf. Trans. ASME* **2007**, *129*, 1638–1654.
9. Cardone, G.; Carlomagno, G.M. Convezione naturale su lastra piana nei regimi stazionario ed instazionario. Proceedings of the IX Congresso Nazionale UIT, Pisa, Italy, 13–14 June 1991; pp. 397–410, In Italian.
10. Astarita, T.; Cardone, G.; Carlomagno, G.M. Convective heat transfer in ribbed channels with a 180 degrees turn. *Exp. Fluids* **2002**, *33*, 90–100.
11. Carlomagno, G.M.; Cardone, G. Infrared thermography for convective heat transfer measurements. *Exp. Fluids* **2010**,*49*, 1187–1218.
12. Carlomagno, G.M.; Nese, F.G.; Cardone, G.; Astarita, T. Thermo-fluid-dynamics of a complex fluid flow. *Infrared Phys. Technol.* **2004**, *46*, 31–39.
13. Perry, J.H. *Chemical Engineers' Handbook*; McGraw-Hill: New York, NY, USA, 1963.
14. Kays, W.M.; Crawford, M.E. *Convective Heat and Mass Transfer*; McGraw-Hill: New York, NY, USA, 1993.

15. Kakac, S.; Shah, R.K.; Aung, W. *Handbook of Single Phase Flow Convective Heat Transfer*; Wiley: New York, NY, USA, 1987.

16. Astarita, T.; Cardone, G. Thermofluidynamic analysis of the flow in a sharp 180 degrees turn channel. *Exp. Therm. Fluid Sci.* **2000**, *20*, 188–200.

17. Gallo, M.; Astarita, T.; Carlomagno, G.M. Heat transfer measurements in a rotating two-pass square channel. *Quant. Infr. Therm. J.* **2007**, *4*, 41–62.

18. Kaiser, R.; Du Puits, R. Local wall heat flux in confined thermal convection. *Int. J. Heat Mass Transf.* **2014**, *73*, 752–760.

19. Douglas, B. Selecting Unsteady Heat Flux Sensors. *Instr. Contr. Syst.* **1972**, *45*, 80–83.

20. Rainieri, S.; Bozzoli, F.; Pagliarini, G. Wiener filtering technique applied to thermographic data reduction intended for the estimation of plate fins performance. *Exp. Therm. Fluid Sci.* **2004**, *28*, 179–183.

21. Carslaw, H.S.; Jaeger, J.C. *Conduction of Heat in Solids*; Oxford University Press: New York, NY, USA, 1959.

22. Gulhan, A.; Schutte, G.; Stahl, B. Experimental Study on Aerothermal Heating Caused by Jet-Hypersonic Crossflow Interaction. *J. Spacecr. Rocket.* **2008**, *45*, 891–899.

23. Cook, W.J.; Felderman, E.J. Reduction of data from thin-film heat-transfer gages: A concise numerical technique.*AIAA J.* **1966**, *4*, 561–562.

24. De Luca, L.; Cardone, G.; de la Chevalerie, D.A.; Fonteneau, A. Viscous interaction phenomena in hypersonic wedge flow. *AIAA J.* **1995**, *33*, 2293–2298.

25. Mulcahy, J.M.; Browne, D.J.; Stanton, K.T.; Diaz, F.R.C.; Cassady, L.D.; Berisford, D.F.; Bengtson, R.D. Heat flux estimation of a plasma rocket helicon source by solution of the inverse heat conduction problem. *Int. J. Heat Mass Transf.* **2009**, *52*, 2343–2357.

26. De Felice, G.; de Luca, L.; Carlomagno, G.M. La misura dei flussi convettivi nel caso di distribuzioni non uniformi. Proceedings of the VII Congresso Nazionale UIT Sulla Trasmissione del Calore, Firenze, Italy, 15–17 June 1989; pp. 591–599, In Italian.

27. Jepps, G. Heat Conduction in Single-layer and Double-layer Walls, with Boundary Conditions. In *Australian Aeronautical Research Committee Report ACA-66*; Department of Supply: Melbourne, Victoria, Australia, 1965.

28. Carlomagno, G.M.; de Luca, L. Infrared thermography for flow visualization and heat transfer measurements. Proceedings of the Workshop Stato Dell'arte del Rilevamento con Camere Termiche Nella Banda 8–15 Micron, Firenze, Italy, 4 March 1991; pp. 191–219.

29. Sparrow, E.M.; Gregg, J.I. Laminar free convection from a vertical plate with uniform surface heat flux. *Trans. ASME***1956**, *78*, 435–440.

30. De Luca, L.; Cardone, G.; Carlomagno, G.M. Image Restoration in Thermo-Fluid-Dynamic Applications of IR Digital Imagery. Proceedings of the Infrared Technology and Applications SPIE, London, UK, 26 June 1990; Volume 1320, pp. 448–457.

31. De la Chevalerie, D.A.; Fonteneau, A.; de Luca, L.; Cardone, G. Görtler-type vortices in hypersonic flows: The ramp problem. *Exp. Therm. Fluid Sci.* **1997**, *15*, 69–81.

32. De Luca, L.; Cardone, G.; Carlomagno, G.M.; de la Chevalerie, D.A.; de Roquefort, T.A. Flow visualization and heat transfer measurement in a hypersonic wind tunnel. *Exp. Heat Transf.* **1992**, *5*, 65–78.

33. Meola, C.; de Luca, L.; Carlomagno, G.M. Azimuthal instability in an impinging jet-adiabatic wall temperature distribution. *Exp. Fluids* **1995**, *18*, 303–310.

34. Carlomagno, G.M.; de Luca, L. Heat Transfer Measurements by Means of Infrared Thermography. In *Flow Visualization IV*; Veret, C., Ed.; Hemisphere: Washington, DC, USA, 1987; pp. 611–616.

35. Carlomagno, G.M.; Ianiro, A. Thermo-fluid-dynamics of submerged jets impinging at short nozzle-to-plate distance: A review. *Exp. Thermal Fluid Sci.* **2014**, *58*, 15–35.

36. Widnall, S.E.; Bliss, D.B.; Tsai, C. The instability of short waves on a vortex ring. *J. Fluid Mech.* **1974**, *66*, 35–47.

37. Cardone, G.; Astarita, T.; Carlomagno, G.M. Wall heat transfer in static and rotating 180 degrees turn channels by quantitative infrared thermography. *Rev. Gen. Therm.* **1998**, *37*, 644–652.

38. Gallo, M.; Astarita, T.; Carlomagno, G.M. Thermo-fluid-dynamic analysis of the flow in a rotating channel with a sharp "U" turn. *Exp. Fluids* **2012**, *53*, 201–219.

39. De Luca, L.; Carlomagno, G.M.; Buresti, G. Boundary-layer diagnostics by means of an infrared scanning radiometer.*Exp. Fluids* **1990**, *9*, 121–128.

40. De Luca, L.; Guglieri, G.; Cardone, G.; Carlomagno, G.M. Experimental analysis of surface flow on a delta wing by infrared thermography. *AIAA J.* **1995**, *33*, 1510–1512.

41. Imbriale, M.; Ianiro, A.; Meola, C.; Cardone, G. Convective heat transfer by a row of jets impinging on a concave surface. *Int. J. Thermal Sci.* **2014**, *75*, 153–163.

42. Cebeci, T.; Kafyeke, F. Aircraft Icing. *Ann. Rev. Fluid Mech.* **2003**, *35*, 11–21.

43. Iacovides, H.; Launder, B.E. Internal blade cooling: The Cinderella of computational ad experimental fluid dynamics research in gas turbines. *Proc. Inst. Mech. Eng. Part A J. Power Energy* **2007**, *221*, 265–290.

44. Meola, C.; Carlomagno, G.M.; Riegel, E.; Salvato, F. An Experimental Study of an Anti-Icing Hot Air Spray-Tube System. Proceedings of the 19th Congress International Council Aeronautical Sciences, Anaheim, CA, USA, 18–23 September 1994; Volume 3, pp. 2345–2351.

45. Cardone, G.; Ianiro, A.; dello Ioio, G.; Passaro, A. Temperature maps measurements on 3D surfaces with infrared thermography. *Exp. Fluids* **2012**, *52*, 375–385.

46. Iacovides, H.; Kounadis, D.; Launder, B.E.; Li, J.; Xu, Z. Experimental study of the flow and thermal development of a row of cooling jets impinging on a rotating concave surface. *J. Turbomach.* **2005**, *127*, 222–229.

47. Gao, N.; Sun, H.; Ewing, D. Heat transfer to impinging round jets with triangular tabs. *Int. J. Heat Mass Transf.* **2003**,*46*, 2557–2569.

48. Violato, D.; Ianiro, A.; Cardone, G.; Scarano, F. Three-dimensional vortex dynamics and convective heat transfer in circular and chevron impinging jets. *Int. J. Heat Fluid Fl.* **2012**, *37*, 22–36.


## LIST OF SYMBOLS

### Roman letters

| | |
|---|---|
| $A$ | Temperature amplitude Equation (39) |
| $A_q$ | Heat flux amplitude, Equation (38) |
| $c$ | Sensor specific heat, Airfoil chord |

| | |
|---|---|
| $c_p$ | Fluid specific heat at constant pressure |
| $C_1$ | First radiation constant |
| $C_2$ | Second radiation constant |
| $d$ | Channel diameter |
| $D$ | Nozzle diameter |
| $E_b$ | Emissivc power of black body |
| $f$ | Attenuation factor Equation (39) |
| $F$ | Temperature modulation transfer function |
| $g$ | Gravity acceleration |
| $G$ | Heat generation per unit volume |
| $I$ | Spectral emissive power |
| $h$ | Convective heat transfer coefficient |
| $k$ | Thermal conductivity coefficient |
| $p$ | Pitch, Dimensionless quantity, Equation (29) |
| $q$ | Heat flux |
| $r$ | Radial coordinate |
| R | Dimensionless wavelength |
| $s$ | Sensor thickness, Thickness |
| $t$ | Time |

| | |
|---|---|
| $t_m$ | Maximum measurement time (thin film) |
| $T$ | Temperature |
| $V$ | Fluid velocity |
| $w$ | Width |
| $x$ | Coordinate |
| $y$ | Coordinate |
| $z$ | Coordinate Nozzle-to-plate-distance |

## Greek letters

| | |
|---|---|
| $\alpha$ | Sensor thermal diffusivity coefficient, Angle of attack |
| $\alpha_f$ | Fluid thermal diffusivity coefficient |
| $\alpha_r$ | Absorptivity coefficient |
| $\beta$ | β=ht/(ρscks)−−−−−−−−√ |
| $\varepsilon$ | Emissivity coefficient |
| $\lambda$ | Wavelength of electromagnetic wave |
| $\nu$ | Fluid kinematic viscosity coefficient |
| $\xi$ | ξ=x/2αt−−√ |
| $\Xi$ | Dimensionless quantity, Equation (31) |
| $\Pi$ | Dimensionless ratio, Equation (45) |
| $\theta$ | Dimensionless temperature, Equation (26) |

| | |
|---|---|
| $\Theta$ | Dimensionless quantity, Equation (44) |
| $\rho$ | Mass density |
| $\rho_r$ | Reflectivity coefficient |
| $\sigma$ | Stefan-Boltzmann constant |
| $\tau_r$ | Transmissivity coefficient |
| $\varphi$ | Temperature difference, $\varphi(t) = T_w(t) - T_{wi}$ |
| $\varphi$ | Jet inclination to airfoil chord |
| $\omega$ | Angular speed |

## Subscripts

| | |
|---|---|
| $a$ | Ambient |
| $aw$ | Adiabatic wall |
| $b$ | Black body |
| $c$ | Convective, Copper |
| $f$ | Fluid, Fiberglass |
| $j$ | Joule |
| $m$ | Mean |
| $r$ | Radiative, Reference |
| $s$ | Sensor |
| $t$ | Tangential, Total |

| | |
|---|---|
| $x$ | Local |
| $w$ | Wall |
| $\varpi_x$ | Wave number |

**Apices**

| | |
|---|---|
| $m$ | At maximum emissive power |
| * | As per Dittus and Bölter correlation |

**Dimensionless groups**

| | |
|---|---|
| $Bi$ | Biot number, $hs/k_s$ |
| $Bi_x$ | Local Biot number, $hx/k_s$ |
| $Fo$ | Fourier number, $^{\alpha t}/_{s2}$ |
| $Fo_x$ | Local Fourier number, $\alpha t/x^2$ |
| $Fo_\varpi$ | Modified Fourier number, ϖ2xαt |
| $M$ | Mach number |
| $Nu$ | Nusselt number, $hd/k_f$ |
| $Pr$ | Prandtl number, $c_p\mu/k_f$ |
| $Ra$ | Rayleigh number, $g\beta qx^4/(\nu\ \alpha_f k_f)$ |
| $Re$ | Reynolds number, $Vd/\ \nu$ |
| $Ro$ | Rotational number, $\nu\ d/V$ |
| $St$ | Stanton number, $h/(\rho_f c_p V)$ |

# Index

## P



## R



## S



## T



## U



## V



## W